KT-462-318

OMICS IN CLINICAL PRACTICE

Genomics, Pharmacogenomics, Proteomics, and Transcriptomics in Clinical Research

OMICS IN CLINICAL PRACTICE

Genomics, Pharmacogenomics, Proteomics, and Transcriptomics in Clinical Research

Edited by
Yu Liu, PhD

Apple Academic Press

TORONTO   NEW JERSEY

Apple Academic Press Inc.	Apple Academic Press Inc.
3333 Mistwell Crescent	9 Spinnaker Way
Oakville, ON L6L 0A2	Waretown, NJ 08758
Canada	USA

©2014 by Apple Academic Press, Inc.
Exclusive worldwide distribution by CRC Press, a member of Taylor & Francis Group

No claim to original U.S. Government works
Printed in the United States of America on acid-free paper

International Standard Book Number-13: 978-1-77188-060-2 (Hardcover)

Library of Congress Control Number: 2014937166

Library and Archives Canada Cataloguing in Publication

Omics in clinical practice: genomics, pharmacogenomics, proteomics, and transcriptomics in clinical research/edited by Yu Liu, PhD.

Includes bibliographical references and index.
ISBN 978-1-77188-060-2 (bound)
1. Medical genetics. 2. Diseases. 3. Genomics. 4. Pharmacogenomics. 5. Proteomics. 6. Diagnosis. I. Liu, Yu, 1971-, editor

RB155.O45 2014   616'.042   C2014-902295-6

Apple Academic Press also publishes its books in a variety of electronic formats. Some content that appears in print may not be available in electronic format. For information about Apple Academic Press products, visit our website at **www.appleacademicpress.com** and the CRC Press website at **www.crcpress.com**

ABOUT THE EDITOR

YU LIU, PhD

As a bioinformatician, Dr. Yu Liu's research is centered on the development and application of computational tools for the study of complex diseases. He has extensive experience with data generated from microarrays, next generation sequencing and high-resolution mass spectrometry, and developing bioinformatics tools and applying system biology approach to study complex diseases, like sleep apnea, neurodegenerative diseases, and cancers. More recently, he developed a systems biology approach that enables the discovery of high-level disease mechanisms and provides testing hypotheses for further research. Currently, Dr. Liu is a senior research associate at the Center for Proteomics and Bioinformatics at Case Western Reserve University, Cleveland, Ohio. He received a PhD in Bioinformatics from Montreal University, Montreal, Canada, and has postdoc training from the University of Toronto, Canada.

CONTENTS

ACKNOWLEDGMENT AND HOW TO CITE

The editor and publisher thank each of the authors who contributed to this book, whether by granting their permission individually or by releasing their research as open source articles or under a license that permits free use provided that attribution is made. The chapters in this book were previously published in various places in various formats. To cite the work contained in this book and to view the individual permissions, please refer to the "How to Cite" notes at the beginning of each chapter. Each chapter was read individually and carefully selected by the editor. The result is a book provides a comprehensive introduction to genomics, proteomics, and transcriptomics in relation to human health and disease.

LIST OF CONTRIBUTORS

Luisa Barzon
Department of Histology, Microbiology, and Medical Biotechnologies, University of Padova, I-35121 Padova, Italy

Claire Bertelli
Center for Research on Intracellular Bacteria (CRIB), Institute of Microbiology, University Hospital Center, University of Lausanne, Lausanne, Switzerland

Stephanie A. Boot
Molecular PathoBiology, National Microbiology Laboratory, Public Health Agency of Canada, Winnipeg, MB, Canada R3E 3R2 and Department of Medical Microbiology, University of Manitoba, Winnipeg, MB, Canada R3E 0W3

Marian P. Brennan
Molecular Modelling Group, Royal College of Surgeons in Ireland, Dublin, Ireland

Anthony J. Chubb
Molecular Modelling Group, Royal College of Surgeons in Ireland, Dublin, Ireland

Federica Cioffi
Dipartimento di Scienze della Vita, Seconda Università di Napoli, Via Vivaldi 43, 81100 Caserta, Italy

François Collyn
Center for Research on Intracellular Bacteria (CRIB), Institute of Microbiology, University Hospital Center, University of Lausanne, Lausanne, Switzerland

Antony Croxatto
Center for Research on Intracellular Bacteria (CRIB), Institute of Microbiology, University Hospital Center, University of Lausanne, Lausanne, Switzerland

Pieter de Lange
Dipartimento di Scienze della Vita, Seconda Università di Napoli, Via Vivaldi 43, 81100 Caserta, Italy

Elisabeth Drucker
Department of Molecular Biotechnology, University of Applied Science Vienna, Helmut-Qualtinger-Gasse 2, Vienna A-1030, Austria

Birgit Eisenhaber
Bioinformatics Institute (BII), Agency for Science, Technology and Research (A*STAR), 30 Biopolis Street, #07-01, Matrix, 138671, Singapore

Frank Eisenhaber
Bioinformatics Institute (BII), Agency for Science, Technology and Research (A*STAR), 30 Biopolis Street, #07-01, Matrix, 138671, Singapore
School of Computer Engineering (SCE), Nanyang Technological University (NTU), 50 Nanyang Drive, Singapore, 637553, Singapore
Department of Biological Sciences (DBS), National University of Singapore (NUS), 8 Medical Drive, Singapore, 117597, Singapore

Akshada Gajbhiye
National Centre for Cell Science, University of Pune Campus, Ganeshkhind, Pune, Maharashtra 411007, India

Daniela Glinni
Dipartimento di Scienze Biologiche ed Ambientali, Università degli Studi del Sannio, Via Port'Arsa 11, 82100 Benevento, Italy

Fernando Goglia
Dipartimento di Scienze Biologiche ed Ambientali, Università degli Studi del Sannio, Via Port'Arsa 11, 82100 Benevento, Italy

Gilbert Greub
Center for Research on Intracellular Bacteria (CRIB), Institute of Microbiology, University Hospital Center, University of Lausanne, Lausanne, Switzerland

Heinrich C. Hoppe
CSIR Biosciences, Pretoria, South Africa

Yi Huang
Department of Microbiology, The University of Hong Kong, Hong Kong

Rhiannon L. C. H. Huzarewich
Molecular PathoBiology, National Microbiology Laboratory, Public Health Agency of Canada, Winnipeg, MB, Canada R3E 3R2

Yujie Jiang
Department of Hematology, Provincial Hospital affiliated to Shandong University, Jinan, China

Carole Kebbi-Beghdadi
Center for Research on Intracellular Bacteria (CRIB), Institute of Microbiology, University Hospital Center, University of Lausanne, Lausanne, Switzerland

Kurt Krapfenbauer
Department of Otorhinolaryngology, Medical University of Vienna, Head and Neck Surgery, Waehringer Guertel 18-20, Vienna A-1090, Austria

Vladimir Kuznetsov
Bioinformatics Institute (BII), Agency for Science, Technology and Research (A*STAR), 30 Biopolis Street, #07-01, Matrix, 138671, Singapore
School of Computer Engineering (SCE), Nanyang Technological University (NTU), 50 Nanyang Drive, Singapore, 637553, Singapore

Y. W. Francis Lam
Department of Pharmacology, School of Medicine, University of Texas Health Science Center San Antonio, 7703 Floyd Curl Drive, San Antonio, TX 78229-3900, USA

Antonia Lanni
Dipartimento di Scienze della Vita, Seconda Università di Napoli, Via Vivaldi 43, 81100 Caserta, Italy

Susanna K. P. Lau
State Key Laboratory of Emerging Infectious Diseases, The University of Hong Kong, Hong Kong
Research Centre of Infection and Immunology, Carol Yu Centre of Infection, Department of Microbiology, The University of Hong Kong, Hong Kong

Martin Latterich
Proteogenomics Research Institute for Systems Medicine, 11107 Roselle Street, San Diego, CA, 92121-1206, USA

Enrico Lavezzo
Department of Histology, Microbiology, and Medical Biotechnologies, University of Padova, I-35121 Padova, Italy

Hwee Kuan Lee
Bioinformatics Institute (BII), Agency for Science, Technology and Research (A*STAR), 30 Biopolis Street, #07-01, Matrix, 138671, Singapore

Avinash Kumar
National Centre for Cell Science, University of Pune Campus, Ganeshkhind, Pune, Maharashtra 411007, India

Qi Liu
Center for Quantitative Sciences, Department of Biomedical Informatics, Vanderbilt University School of Medicine, 37232, Nashville, TN, USA

Assunta Lombardi
Dipartimento delle Scienze Biologiche, Sezione Fisiologia, Università degli Studi di Napoli "Federico II", Via Mezzocannone 8, 80134 Napoli, Italy

Elena López
Centro de Investigación i+12, Hospital 12 de Octubre, Av. De Córdoba s/n, 28040, Madrid, Spain

Juan López-Pascual
Hospital Universitario 12 de Octubre, Av. De Córdoba s/n, 28040, Madrid, Spain

Luis Madero
Hospital Infantil Universitario Niño Jesús, Av. Menéndez Pelayo 65, 28009, Madrid, Spain

Sebastian Maurer-Stroh
Bioinformatics Institute (BII), Agency for Science, Technology and Research (A*STAR), 30 Biopolis Street, #07-01, Matrix, 138671, Singapore
School of Biological Sciences (SBS), Nanyang Technological University (NTU), 60 Nanyang Drive, Singapore, 637551, Singapore

Valentina Militello
Department of Histology, Microbiology, and Medical Biotechnologies, University of Padova, I-35121 Padova, Italy

Maria Judit Molnár
Institute of Genomic Medicine and Rare Disorders, Tömö Street 25-29, 1083, Budapest, Hungary

Maria Moreno
Dipartimento di Scienze Biologiche ed Ambientali, Università degli Studi del Sannio, Via Port'Arsa 11, 82100 Benevento, Italy

Kevin B. Nolan
Pharmaceutical and Medicinal Chemistry, Royal College of Surgeons in Ireland (RCSI), Dublin, Ireland

Giorgio Palù
Department of Histology, Microbiology, and Medical Biotechnologies, University of Padova, I-35121 Padova, Italy

Debasish Paul
National Centre for Cell Science, University of Pune Campus, Ganeshkhind, Pune, Maharashtra 411007, India

Sandor Pongor
Faculty of Information Technology, Pázmány Péter Catholic University, Budapest, Hungary (PPKE), Práter u. 50/a, 1083, Budapest, Hungary

Didier Raoult
Unité des Rickettsies, Faculté de Médecine, Université de la Méditerranée, Marseille, France

Beat M. Riederer
Department of Cellular Biology and Morphology, University of Lausanne, Lausanne, Switzerland and Proteomics Unit, Department of Psychiatric Neurosciences, Cery, Prilly-Lausanne, Switzerland

Manas K. Santra
National Centre for Cell Science, University of Pune Campus, Ganeshkhind, Pune, Maharashtra 411007, India

Rosalba Senese
Dipartimento di Scienze della Vita, Seconda Università di Napoli, Via Vivaldi 43, 81100 Caserta, Italy

Derek Shyr
College of Arts & Sciences, Washington Universityin St. Louis, 63130, St. Louis, MO, USA

Christine G. Siemens
Molecular PathoBiology, National Microbiology Laboratory, Public Health Agency of Canada, Winnipeg, MB, Canada R3E 3R2

Elena Silvestri
Dipartimento di Scienze Biologiche ed Ambientali, Università degli Studi del Sannio, Via Port'Arsa 11, 82100 Benevento, Italy

Rapole Srikanth
National Centre for Cell Science, University of Pune Campus, Ganeshkhind, Pune, Maharashtra 411007, India

David Toomey
Molecular Modelling Group, Royal College of Surgeons in Ireland, Dublin, Ireland

Stefano Toppo
Department of Biological Chemistry, University of Padova, I-35121 Padova, Italy

Xin Wang
Department of Hematology, Provincial Hospital affiliated to Shandong University, Jinan, China

Daniel J. Wilson
Wellcome Trust Centre for Human Genetics, Nuffield Department of Clinical Medicine, Experimental Medicine Division, University of Oxford, Oxford, United Kingdom

Patrick C. Y. Woo
State Key Laboratory of Emerging Infectious Diseases, Research Centre of Infection and Immunology, Carol YuCentre of Infection, and Department of Microbiology, The University of Hong Kong, Hong Kong

Camille Yersin
Center for Research on Intracellular Bacteria (CRIB), Institute of Microbiology, University Hospital Center, University of Lausanne, Lausanne, Switzerland

Kwok-Yung Yuen
State Key Laboratory of Emerging Infectious Diseases, Research Centre of Infection and Immunology, Carol Yu Centre of Infection, and Department of Microbiology, The University of Hong Kong, Hong Kong

INTRODUCTION

This book serves as an introduction to genomics, proteomics, and transcriptomics, putting these fields in relation to human disease and ailments. The various chapters consider the role of translation and personalized medicine, as well as pathogen detection, evolution, and infection, in relation to genomics, proteomics, and transcriptomics. The topic of companion diagnostics is also covered.

The book is broken into five sections. Part I examines the connection between Omics and Human disease. Part II looks at the applications for the fields of translational and personalized medicine. Part II focuses on molecular and genetic markers. Part IV describes the use of omics while studying pathogens, and Part V examines the applications for companion diagnostics.

Mitochondria are the most complex and the most important organelles of eukaryotic cells, which are involved in many cellular processes, including energy metabolism, apoptosis, and aging. And mitochondria have been identified as the "hot spot" by researchers for exploring relevant associated dysfunctions in many fields. In Chapter 1, the emergence of comparative proteomics enables Jiang and Wang to have a close look at the mitochondrial proteome in a comprehensive and effective manner under various conditions and cellular circumstances. Two-dimensional electrophoresis combined with mass spectrometry is still the most popular techniques to study comparative mitochondrial proteomics. Furthermore, many new techniques, such as ICAT, MudPIT, and SILAC, equip researchers with more flexibilities inselecting proper methods. This article also reviews the recent development of comparative mitochondrial proteomics on diverse human diseases. And the results of mitochondrial proteomics enhance a better understanding of the pathogenesis associated with mitochondria and provide promising therapeutic targets.

Omics approaches to the study of complex biological systems with potential applications to molecular medicine are attracting great interest in clinical as well as in basic biological research. Genomics, transcriptomics and proteomics are characterized by the lack of an a priori definition of scope, and this gives sufficient leeway for investigators (a) to discern all at once a globally altered pattern of gene/protein expression and (b) to examine the complex interactions that regulate entire biological processes. Two popular platforms in "omics" are DNA microarrays, which measure messenger RNA transcript levels, and proteomic analyses, which identify and quantify proteins. Because of their intrinsic strengths and weaknesses, no single approach can fully unravel the complexities of fundamental biological events. However, an appropriate combination of different tools could lead to integrative analyses that would furnish new insights not accessible through one-dimensional datasets. In Chapter 2, Silvestri and colleagues outline some of the challenges associated with integrative analyses relating to the changes in metabolic pathways that occur in complex pathophysiological conditions (viz. ageing and altered thyroid state) in relevant metabolically active tissues. In addition, the authors discuss several new applications of proteomic analysis to the investigation of mitochondrial activity.

The wide application of next-generation sequencing (NGS), mainly through whole genome, exome and transcriptome sequencing, provides a high-resolution and global view of the cancer genome. Coupled with powerful bioinformatics tools, NGS promises to revolutionize cancer research, diagnosis and therapy. In Chapter 3, Shyr and Liu review the recent advances in NGS-based cancer genomic research as well as clinical application, summarize the current integrative oncogenomic projects, resources and computational algorithms, and discuss the challenge and future directions in the research and clinical application of cancer genomic sequencing.

The mapping of the human genome and subsequent advancements in genetic technology had provided clinicians and scientists an understanding of the genetic basis of altered drug pharmacokinetics and pharmacodynamics, as well as some examples of applying genomic data in clinical practice. This has raised the public expectation that predicting patients' responses to drug therapy is now possible in every therapeutic area, and personalized drug therapy would come sooner than later. However, debate continues among most stakeholders involved in drug development and clinical decision-making on whether pharmacogenomic biomarkers should be used in patient assessment, as well as when and in whom to use the biomarker-based diagnostic tests. Currently, most would agree that achieving the goal of personalized therapy remains years, if not decades, away. Realistic application of genomic findings and technologies in clinical practice and drug development require addressing multiple logistics and challenges that go beyond discovery of gene variants and/or completion of prospective controlled clinical trials. In Chapter 4, Lam argues that the goal of personalized medicine can only be achieved when all stakeholders in the field work together, with willingness to accept occasional paradigm change in their current approach.

Since the advent of the new proteomics era more than a decade ago, large-scale studies of protein profiling have been used to identify distinctive molecular signatures in a wide array of biological systems, spanning areas of basic biological research, clinical diagnostics, and biomarker discovery directed toward therapeutic applications. Recent advances in protein separation and identification techniques have significantly improved proteomic approaches, leading to enhancement of the depth and breadth of proteome coverage. Proteomic signatures, specific for multiple diseases, including cancer and pre-invasive lesions, are emerging. Chapter 5, by López and colleagues, combines, in a simple manner, relevant proteomic and OMICS clues used in the discovery and development of diagnostic and prognostic biomarkers that are applicable to all clinical fields, thus helping to improve applications of clinical proteomic strategies for translational medicine research.

Genome sequencing and bioinformatics have provided the full hypothetical proteome of many pathogenic organisms. Advances in microarray and mass spectrometry have also yielded large output datasets of possible target proteins/genes. However, the challenge remains to identify new targets for drug discovery from this wealth of information. Further analysis includes bioinformatics and/or molecular biology tools to validate the findings. This is time consuming and expensive, and could fail to yield novel drugs if protein purification and crystallography is impossible. To pre-empt this, a researcher may want to rapidly filter the output datasets for proteins that show good homology to proteins that have already been structurally characterised or proteins that are already targets for known drugs. Critically, those researchers developing novel antibiotics need to select out the proteins that show close homology to any human proteins, as future inhibitors are likely to cross-react with the host protein, causing off-target toxicity effects later in clinical trials. To solve many of these issues, in Chapter 6, Toomey and colleagues developed a free online resource called Genomes2Drugs which ranks sequences to identify proteins that are (i) homologous to previously crystallized proteins or (ii) targets of known drugs, but are (iii) not homologous to human proteins. When tested using the *Plasmodium falciparum* malarial genome the program correctly enriched the ranked list of proteins with known drug target proteins. Genomes2Drugs rapidly identifies proteins that are likely to succeed in drug discovery pipelines. This free online resource helps in the identification of potential drug targets. Importantly, the program further highlights proteins that are likely to be inhibited by FDA-approved drugs. These drugs can then be rapidly moved into Phase IV clinical studies under 'change-of-application' patents.

Since the emergence of the so-called omics technology, thousands of putative biomarkers have been identified and published, which have dramatically increased the opportunities for developing more effective therapeutics. These opportunities can have profound benefits for patients and for the economics of healthcare. However, the transfer of biomarkers from discovery to clinical practice is still a process filled with lots of pitfalls and limitations, mostly limited by structural and scientific factors. To become a clinically approved test, a potential biomarker should be confirmed and validated using hundreds of specimens and should be reproducible, specific and sensitive. Besides the lack of quality in biomarker validation, a number of other key issues can be identified and should be addressed. Therefore, the aim of Drucker and Krapfenbauer in Chapter 7 is to discuss a series of interpretative and practical issues that need to be understood and resolved before potential biomarkers become a clinically approved test or are already on the diagnostic market. Some of these issues are shortly discussed here.

The currently hyped expectation of personalized medicine is often associated with just achieving the information technology led integration of biomolecular sequencing, expression and histopathological bioimaging data with clinical records at the individual patients' level as if the significant biomedical conclusions would be its more or less mandatory result. It remains a sad fact that many, if not most biomolecular mechanisms that translate the human genomic information into phenotypes are not known and, thus, most of the molecular and cellular data cannot be interpreted in terms of biomedically relevant conclusions. Whereas the historical trend will certainly be into the general direction of personalized diagnostics and cures, the temperate view suggests that biomedical applications that rely either on the comparison of biomolecular sequences and/or on the already known biomolecular mechanisms have much greater chances to enter clinical practice soon. In addition to considering the general trends, in Chapter 8, Kuznetsov et al. review advances in the area of cancer biomarker discovery, in the clinically relevant characterization of patient-specific viral and bacterial pathogens (with emphasis on drug selection for influenza and enterohemorrhagic *E. coli*) as well as progress in the automated assessment of histopathological images. As molecular and cellular data analysis will become instrumental for achieving desirable clinical outcomes, the role of bioinformatics and computational biology approaches will dramatically grow.

The advent of genomics and proteomics has been a catalyst for the discovery of biomarkers able to discriminate biological processes such as the pathogenesis of complex diseases. Prompt detection of prion diseases is particularly desirable given their transmissibility, which is responsible for a number of human health risks stemming from exogenous sources of prion protein. Diagnosis relies on the ability to detect the biomarker PrP^{Sc}, a pathological isoform of the host protein PrP^{c}, which is an essential component of the infectious prion. Immunochemical detection of PrP^{Sc} is specific and sensitive enough for antemortem testing of brain tissue, however, this is not the case in accessible biological fluids or for the detection of recently identified novel prions with unique biochemical properties. A complementary approach to the detection of PrP^{Sc} itself is to identify alternative, "surrogate" gene or protein biomarkers indicative of disease. Biomarkers are also useful to track the progress of disease, especially important in the assessment of therapies, or to identify individuals "at risk". In Chapter 9, Huzarewich and colleagues provide perspective on current progress and pitfalls in the use of "omics" technologies to screen body fluids and tissues for biomarker discovery in prion diseases.

Bacterial pathogens impose a heavy burden of disease on human populations worldwide. The gravest threats are posed by highly virulent respiratory pathogens, enteric pathogens, and HIV-associated infections. Tuberculosis alone is responsible for the deaths of 1.5 million people annually. Treatment options for bacterial pathogens are being steadily eroded by the evolution and spread of drug resistance. However, population-level whole genome sequencing offers new hope in the fight against pathogenic bacteria. By providing insights into bacterial evolution and disease etiology, these approaches pave the way for novel interventions and therapeutic targets. Sequencing populations of bacteria across the whole genome provides unprecedented resolution to investigate (i) within-host evolution, (ii) transmission history, and (iii) population structure. Moreover, advances in rapid benchtop sequencing herald a new era of real-time genomics in which sequencing and analysis can be deployed within hours in response to rapidly changing public health emergencies. The purpose of Chapter 10 by Dr. Wilson, is to highlight the transformative effect of population genomics on bacteriology, and to consider the prospects for answering abiding questions such as why bacteria cause disease.

With the availability of new generation sequencing technologies, bacterial genome projects have undergone a major boost. Still, chromosome completion needs a costly and time-consuming gap closure, especially when containing highly repetitive elements. However, incomplete genome data may be sufficiently informative to derive the pursued information. For emerging pathogens, i.e. newly identified pathogens, lack of release of genome data during gap closure stage is clearly medically counterproductive. Thus, in Chapter 11, Greub and colleagues investigated the feasibility of a dirty genome approach, i.e. the release of unfinished genome sequences to develop serological diagnostic tools. They showed that almost the whole genome sequence of the emerging pathogen *Parachlamydia acanthamoebae* was retrieved even with relatively short reads from Genome Sequencer 20 and Solexa. The bacterial proteome was analyzed to select immunogenic proteins, which were then expressed and used to elaborate the first steps of an ELISA. This work constitutes the proof of principle for a dirty genome approach, i.e. the use of unfinished genome sequences of pathogenic bacteria, coupled with proteomics to rapidly identify new immunogenic proteins useful to develop in the future specific diagnostic tests such as ELISA, immunohistochemistry and direct antigen detection. Although applied here to an emerging pathogen, this combined dirty genome sequencing/proteomic approach may be used for any pathogen for which better diagnostics are needed. These genome sequences may also be very useful to develop DNA based diagnostic tests. All these diagnostic tools will allow further evaluations of the pathogenic potential of this obligate intracellular bacterium.

The drastic increase in the number of coronaviruses discovered and coronavirus genomes being sequenced have given researchers an unprecedented opportunity to perform genomics and bioinformatics analysis on this family of viruses. Coronaviruses possess the largest genomes (26.4 to 31.7 kb) among all known RNA viruses, with G + C contents varying from 32% to 43%. Variable numbers of small ORFs are present between the various conserved genes (*ORF1ab, spike, envelope, membrane* and *nucleocapsid*) and downstream to nucleocapsid gene in different coronavirus lineages. Phylogenetically, three genera, *Alphacoronavirus, Betacoronavirus* and *Gammacoronavirus*, with *Betacoronavirus* consisting of subgroups A, B, C and D, exist. A fourth genus, *Deltacoronavirus*, which includes bulbul coronavirus HKU11, thrush coronavirus HKU12 and munia coronavirus HKU13, is emerging. In Chapter 12, Woo and colleagues used molecular clock analysis using various gene loci to reveal that the time of most recent common ancestor of human/civet SARS related coronavirus to be 1999-2002, with estimated substitution rate of 4′10-4 to 2′10-2 substitutions per site per year. Recombination in coronaviruses was most notable between different strains of murine hepatitis virus (MHV), between different strains of infectious bronchitis virus, between MHV and bovine coronavirus, between feline coronavirus (FCoV) type I and canine coronavirus generating FCoV type II, and between the three genotypes of human coronavirus HKU1 (HCoV-HKU1). Codon usage bias in coronaviruses were observed, with HCoV-HKU1 showing the most extreme bias, and cytosine deamination and selection of CpG suppressed clones are the two major independent biological forces that shape such codon usage bias in coronaviruses.

Novel DNA sequencing techniques, referred to as "next-generation" sequencing (NGS), provide high speed and throughput that can produce an enormous volume of sequences with many possible applications in research and diagnostic settings. In Chapter 13, Barzon and colleagues provide an overview of the many applications of NGS in diagnostic virology. NGS techniques have been used for high-throughput whole viral genome sequencing, such as sequencing of new influenza viruses, for detection of viral genome variability and evolution within the host, such as investigation of human immunodeficiency virus and human hepatitis C virus quasispecies, and monitoring of low-abundance antiviral drug-resistance mutations. NGS techniques have been applied to metagenomics-based strategies for the detection of unexpected disease-associated viruses and for the discovery of novel human viruses, including cancer-related viruses. Finally, the human virome in healthy and disease conditions has been described by NGS-based metagenomics.

Accurate diagnosis and proper monitoring of cancer patients remain a key obstacle for successful cancer treatment and prevention. Therein comes the need for biomarker discovery, which is crucial to the current oncological and other clinical practices having the potential to impact the diagnosis and prognosis. In fact, most of the biomarkers have been discovered utilizing the proteomics-based approaches. Although high-throughput mass spectrometry-based proteomic approaches like SILAC, 2D-DIGE, and iTRAQ are filling up the pitfalls of the conventional techniques, still serum proteomics importunately poses hurdle in overcoming a wide range of protein concentrations, and also the availability of patient tissue samples is a limitation for the biomarker discovery. Thus, researchers have looked for alternatives, and profiling of candidate biomarkers through tissue culture of tumor cell lines comes up as a promising option. It is a rich source of tumor cell-derived proteins, thereby, representing a wide array of potential biomarkers. Interestingly, most of the clinical biomarkers in use today (CA 125, CA 15.3, CA 19.9, and PSA) were discovered through tissue culture-based system and tissue extracts. Chapter 14, by Paul and colleagues, tries to emphasize the tissue culture-based discovery of candidate biomarkers through various mass spectrometry-based proteomic approaches.

PART I

OMICS AND HUMAN DISEASE

CHAPTER 1

COMPARATIVE MITOCHONDRIAL PROTEOMICS: PERSPECTIVE IN HUMAN DISEASES

YUJIE JIANG AND XIN WANG

1.1 INTRODUCTION

Mitochondria, which are mainly composed by proteins and lipids, are considered as the most complex and the most important organelles of eukaryotic cells. They not only play a leading role in the energy metabolism, but are also closely involved in many cellular processes. Furthermore, mitochondria have a manageable level of complexity as a consequence of their apparent prokaryotic ancestry. Their endosymbiotic origins have been well preserved in their double membrane structure, and they possess their own circular genome with mitochondria-specific transcription, translation, and protein assembly systems [1]. Based upon the human genome, there is estimated to be approximately 2000 to 2500 mitochondrial proteins [2], however, just over 600 have been identified at the protein level [3]. For this reason, mitochondria contain a great number of proteins that have yet to be identified and characterized.

This chapter was originally published under the Creative Commons Attribution License. Jiang Y and Wang X. Comparative Mitochondrial Proteomics: Perspective in Human Diseases. Journal of Hematology and Oncology *5,11 (2012), doi:10.1186/1756-8722-5-11.*

Due to the fact that proteins are the carriers of biotic movement, the mitochondrial proteome is deemed as an ideal target for global proteome analysis. In the past, many effects of disease processes in which mitochondria are involved have been studied using classic biochemical methods [4]. However, these studies usually focus on only one particular protein, but not on the whole mitochondrial proteome. Recent developments in proteomics have allowed more in-depth studies of proteins. Proteomics is the large-scale study of all proteins in an organism and allowes a global insight into the abundance of protein expression, localization, and interaction. Combining genomics, mass spectrometry, and computation, it is possible to systematically identify the mammalian mitochondrial proteome. The proteome is often used to investigate the pathogenesis, cellular patterns, and functional correlations on protein levels in a non-biased manner [5]. This proteomic approach also allows the possibility for developing new candidate biomarkers for the diagnosis, staging and tracking of disease. Comparative proteomics is a subset of proteomics whose primary purpose is revolving around the following fields: the investigation of the pathogenesis and mechanism of a drug, the discovery of new targets for diagnosis and treatment, and the examination of the effects of environmental factors on soma and cells. Thus, many significant proteins have been identified from normal and abnormal individuals often under various states treated by some agents. Researchers have made tremendous efforts to rapidly obtain results to study the differentially expressed proteins in the subcellular organelle. By doing so, the diversity of proteins can be unmasked and reveal the subcellular location information. Therefore, owing to the significant roles and functions in the cell, mitochondria have become a research "hot spot" in subcellular proteomics. With these new techniques, a thorough investigation of comparative mitochondrial proteomics becomes more and more achievable. Mitochondrial proteomic profiles have been generated across multiple organs, including brain, heart, liver, and kidney [6-8]. This review presents a summary of progression of the mitochondrial proteome in various human diseases using comparative proteomic techniques reported in recent years. Future prospects and challenges for the mitochondrial proteome will also be discussed.

1.2 TECHNIQUES IN COMPARATIVE PROTEOMICS OF MITOCHONDRIA (BOTH GEL-BASED AND GEL-FREE)

1.2.1 GEL-BASED TECHNIQUES

Two-dimensional gel electrophoresis (2-DE) combined with mass spectra is still the most popular gel-based proteomic technique for comparative proteomics nowadays and has matured significantly over the past decades [9]. The most frequently used method is termed "bottom-up proteomics," which is a strategy using mass spectrometry or tandem mass spectrometry (MS/MS) to analyze proteolytically digested proteins [10]. Peptide mass fingerprinting (PMF) of digested peptide fragments using matrix-assisted laser desorption/ionization time of flight (MALDI-TOF) is the preferred method for an initial protein identification after separation by 2-DE due to its high throughput and cost efficiency [11]. However, 2-DE has many shortcomings in separating certain protein classes, such as membrane proteins, high molecular weight (> 200 kDa) or small molecularweight proteins (< 10 kDa), and basic proteins, ect. The application of 2-DE to study the mitochondrial proteome has its owndisadvantages [12,13]. Two-dimensional fluorescence difference gel electrophoresis (2D-DIGE) is the development of 2-DE that was originally introduced by Minden [14]. It also allows for the direct comparison of the changes in protein abundance changes, which is less than 10% across samples simultaneously with a 95% statistical reliability coefficient without interference due to gel-to-gel variation [15]. Moreover, another technique named BNPAGE (blue native gel electrophoresis) invented by Shägger and Jagow, is specialized for separating intact membrane protein complexes [16]. It has been primarily used to separate and isolate the five multi-polypeptide complexes of the oxidative phosphorylation (OXPHOS) system and the recovery of all respiratory chain complexes are approaching the level of detection [17].

TABLE 1: Overview of diseases associated with the mitochondrial proteome

Organ	Disease (Researcher)	Analytical method	Major proteins identified		Functional distribution	Primary Significance
			Up-regulated	Down-regulated		
Nervous system						
	Alzheimer's disease (AD) (Lovell MA.)	ICAT, 2D-LC/ MS/MS	ATP synthase alpha chain		OXPHOS	Cells undergoing Aβ -mediated apoptosis increase synthesis of proteins essential for ATP production and efflux to maintain metabolic functions.
			Pyruvate kinase, M1 isozyme		glycolysis	
			Glyceraldehyde 3 phosphate dehydrogenase		energy production	
			Cofilin		control of actin polymerization/depolymerization	
			Na+/K+ +transporting ATPase a-3 chain		ATP production	
			VDAC 1 and 3		apoptosis	
			Dihydropyrimidinase- related protein-1 (DRP-1)		axon guidance, invasive growth and cell migration	
	Multiple sclerosis (MS) (Broadwater L.)	SELDI-TOF-MS		Cytochrome c oxidase subunit 5b (COX5b)	component of Complex IV of the electron transport chain	Proteins identified would be used as neuroprotective therapeutic targets for MS.

TABLE 1: *Cont.*

Organ	Disease (Researcher)	Analytical method	Major proteins identified		Functional distribution	Primary Significance
			Up-regulated	Down-regulated		
	Neural degeneration (Pienaar IS.)	2DE, ESI-QUADTOF/MS	Hemoglobin β-chain		oxygen transport	Alteration of mitochondrial function may contribute to the beneficial effects associated with statin use.
			Myelin basic protein (MBP)		component of the myelin membrane in the CNS	
			Creatine kinase (CKB)		creatine metabolic process	
				Protein disulfide isomerase (PDI)	folding	
			Heat shock proteins		protein assembly and folding	
				Dehydrogenase antiporter	transportation	
			Alpha-internexin (NF66)		cell differentiation, morphogenesis of neurons	
			Protein-tyrosine receptor type F polypeptide interacting protein (PTPRF)		cell adhesion receptor	
				Neuronal-specific enolase (NSE)	energy metabolism	

TABLE 1: *Cont.*

Organ	Disease (Researcher)	Analytical method	Major proteins identified		Functional distribution	Primary Significance
			Up-regulated	Down-regulated		
			Variation in ATP synthase, D chain		energy metabolism	
			Alpha-enolase-1 (ENO1)		glycolysis, growth control, hypoxia tolerance and allergic responses	
			Guanine nucleotide-binding proteins (G-proteins)		signal transduction	
Cardiovascular system						
	Ischemia-induced cardiac injury (Kim N.)	2-DE, MAL-DI-TOF-MS		Prohibitin	cell cycle	Proteomic analysis provides appropriate means for identifying cardiac markers for detection of ischemia-induced cardiac injury.
				VDAC	apoptosis	
	Contractile dysfunction (Essop MF.)	2D-PAGE, ESI-Q-TOF		ATP synthase D chain	OXPHOS	Decreased contractile protein levels may contribute to the contractile dysfunction of hearts from diabetic mice.
			Ubiquinol cytochrome-C reductase core protein 1		electron transport	
				Electron transfer flavoprotein subunit α	electron transport	

TABLE 1: *Cont.*

Organ	Disease (Researcher)	Analytical method	Major proteins identified		Functional distribution	Primary Significance
			Up-regulated	Down-regulated		
Liver disease						
	Acetaminophen (APAP) affected the liver (Ruepp SU.)	2D-DIGE, MALDI-TOF-MS		HSP10 and HSP60	protein assembly and folding	APAP toxicity was a direct action of its known reactive metabolite NAPQI, rather than a consequence of gene regulation.
	High-fat diet induces hepatic steatosis (Eccleston HB.)	2D-IEF/SDS-PAGE	Heat shock 70 kDa protein 9, (GRP75)		transporters and channels	HFD causes steatosis, alters NO metabolism, and modifies the liver mitochondrial proteome, thus, NO may play an important role in the processes responsible for NAFLD.
			uMUP-VIII major urinary protein		pheromone communication (only in rodents)	
			Thiosulfate sulfurtransferase		cyanide detoxification, role in iron-sulfur centers, sulfane metabolism	
				3-hydroxy-3-methylglutaryl-CoA synthase 2 (HMG-CoA synthase)	catalyzes the condensation of acetoacetyl CoA and acetone step in ketogenesis	
				Succinate dehydrogenase subunit a (SDH-A)	catalyzes the oxidation of succinate to fumarate, flavoprotein	

TABLE 1: *Cont.*

Organ	Disease (Researcher)	Analytical method	Major proteins identified		Functional distribution	Primary Significance
			Up-regulated	Down-regulated		
				ATP synthase F1 α and β subunits	OXPHOS	
Skeletal muscle						
	Hypoxia-induced changes in rat skeletal muscle (De Palma S.)	2D-DIGE, HPLC ESI-MS/MS	Hypoxia inducible factor 1 (HIF-1R)		transcription	In vivo adaptation to hypoxia requires an active metabolic switch.
			Pyruvate dehydrogenase kinase 1 (PDK1)		regulation of glucose metabolism	
				Mitochondrial dihydrolipoamide dehydrogenase	branched chain family amino acid catabolic process	
				Succinyl CoA ligase α chain	tricarboxylic acid cycle	
	Lifestyle on the aging alterations (Alves RM.)	2-DE, MALDI-TOF/TOF	NADH dehydrogenase [ubiquinone] 1 α subcomplex 4		oxidative phosphorylation	Lifestyle is a key modulator for preventing aging-induced protein expression and functionality in mitochondria.
			Creatine kinase		signal transduction	
			Superoxide dismutase [Mn], mitochondrial		redox	

1.2.2 NON-GEL BASED TECHNIQUES: MUDPIT AND ICATS

Multi-dimensional protein identification technology (MudPIT) combines the resolving power of high performance liquid chromatography (HPLC) with the analytical capacity of tandem mass spectrometry (MS/MS). Using this method, a complex protein mixture is first digested with a protease resulting in an even more complex peptide mixture that is resolved by multidimensional HPLC. As the peptides are eluted off of the column, they are analyzed by mass spectrometry. Yates et al introduced this automated multidimensional protein identification technology termed "shotgun" proteomics and demonstrated that a dynamic range of 10,000 to 1 between the most abundant and least abundant proteins/peptides in a complex peptide mixture could be identified [18]. MudPIT overcomes the shortcomings of mass spectrometry such as deficiencies in detecting proteins with extreme alkalinity, hydrophobicity, and maximum or minimum molecular mass. However, MudPIT cannot yet accomplish absolute quantitation. Gygi introduced isotope coded affinity tags (ICAT) in which isotypical and different biotin-containing moieties are conjugated to cysteine residues from two different peptide samples to quantitate the mixture of proteins [19]. This technique has been applied successfully to the detection of membrane proteins [20]. Shotgun proteomics combined with stable isotope labeling or label-free methods are effective in achieving absolute or relative protein quantification [21].

Many new proteomics techniques have been developed, such as iTRAQ (multiplexed isobaric tagging technology), protein chip, SELDI-TOF-MS (surface enhanced laser desorption/ionization of flight mass spectrometry), and SILAC (stable isotope labeling by amino acids in cell culture) [22-24]. Ultimately, the field of proteomics, with its depth and fast pace of investigation, has a tendency to combine multiple techniques so as to best utilize the benefits of each technique.

1.3 PERSPECTIVE ON HUMAN DISEASE AND MITOCHONDRIAL COMPARATIVE PROTEOMICS

Applications of mitochondrial proteomics have shed some light on the diagnosis and treatment of many diseases associated with mitochondria. In addition, comparison of mitochondrial proteome from healthy and diseased tissues could result in the identification of biomarkers for the early diagnosis and pathologies concerned with mitochondrial dysfunction (Table 1).

1.3.1 NERVOUS SYSTEM

Because the brain is considerably complex and hclewglnouo caccine organism, the orthodox empirical methods cannot meet the need to investigate the brain's constitution and function. The complexity of the nervous system is represented by the cellular categories and the number of synapses. Moreover, because the brain is a vital organ with massive energy consumption and can only utilize the energy produced from the process of anaerobic glycolysis, the role of mitochondriais very considerable in this tissue.

A series of studies have identified an abundance of alterations of in mitochondrial protein levels. To demonstrate that mutations in mitochondrial tRNA would affect the pattern of mitochondrial proteins, Rabilloud et al found a number of proteins in sibling hybrid cell lines using proteomic methods [25]. Two proteins that exhibited obvious large quantitative decreases were identified as nuclear-encoded subunits of cytochrome c oxidase. This finding clearly demonstrated a linkage between the effects of mutations in mitochondrial tRNA genes and the steady-state level of nuclear-encoded proteins in mitochondria. Alzheimer's disease (AD) is a fatal progressive neurodegenerative disorder whose etiology is unkown until now. Mitochondria may play a crucial role in the pathogenesis of AD. Chou and his colleagues analyzed the differential mitochondrial protein profile in the cerebral cortices of 6-month-old male 3 × Tg-AD (which harbor mutations in three human transgenes) and non-transgenic mice. Certain proteins which involved in a wide variety of metabolic pathways, such as the citric acid cycle, oxidative phosphorylation, pyruvate metabolism, and glycolysis, ect, were dysregulated in 3 × Tg-AD cortices. Interestingly, these alterations in the mitochondrial proteome occurred before the development of significant amyloid plaques and neurofibrillary tangles, indicating that mitochondrial dysregulation is an early event in AD [26]. In addition, the potential role of amyloid beta peptide (Aß)-mediated cell death in AD has been extensively investigated both in transgenic animal models and in neuron culture models. Lovell et al quantitatively measured changes in mitochondrial proteins of primary rat cortical neuron cultures exposed to Aß [27]. Ten proteins that were significantly altered in Aß-treated cultures were identified, including sodium/potassium-transporting ATPase, cofilin, dihydropyrimidinase, pyruvate kinase and voltage-dependent anion channel 1 (VDAC1). Elevations in the levels of proteins associated with energy production indicated that cells undergoing Aß-mediated apoptosis increased the synthesis of proteins essential for ATP production and efflux in an attempt to maintain metabolic function. Another similar study with Aß was reported by Gillardon [28]. They analyzed proteome changes in synaptosomal fractions from Tg2576 mice that over-express mutant human amyloid precursor protein (K670N, M671L) and from their non-transgenic littermates. Altered expression of certain proteins, such as heat shock protein 70 and changes in the subunit composition of the respiratory chain complexes I and III were identified. They concluded that mitochondria are early targets of Aß aggregates, and that elevated Aß might impair mitochondrial functions, thus providing a self-amylifying toxic mechanism. Another comparative proteomic investigation about neurodegenerative diseases was reported by Fu [29]. The expression of mitochondrial proteins involved in mitochondrial membrane potential, ATP production, and neuronal cell death was down-regulated after treatment of cerebellar granule neurons with bis(7)-tacrine. Thus, bis(7)-tacrine might be a beneficial agent for the treatment of neurodegenerative diseases. Pienaar and coworkers conducted a behavioral and quantitative mitochondrial proteomic analysis on the effects of simvastatin on a rat model of neural degeneration. Twenty-four mitochondrial proteins were identified in relative abundance after simvastatin treatment. Then they validated whether simvastatin was capable of altering sensorimotor function in a mitochondrial toxin-induced animal model. Rats were pre-treated with simvastatin for 14 days followed by a single unihemispheric injection of rotenone(a mitochondrial complex I inhibitor). The results showed that simvastatin improved motor performance in rotenone-infused rats. The results of behavioral and quantitative proteomic analysis are consistent and further exploration of these changes may identify promising bio-targets for degenerative disorders [30].

Multiple sclerosis (MS) is an inflammatory neurodegenerative disease of the central nervous system that results in progressive physical and cognitive disability. Broadwater et al utilized SELDI-TOF-MS to characterize the mitochondrial proteome in postmortem MS and control cortex. Peptide fingerprint mapping unambiguously identified four proteins, including cytochrome c oxidase subunit 5b (COX5b), the brain specific isozyme of creatine kinase, hemoglobin β-chain, and myelin basic protein (MBP), that could be used as neuroprotective therapeutic targets for MS [31].

As a whole, studies on the mitochondrial proteome of the nervous system provide a broader insight on various fractions of brain and the same fractions under various physiological and pathological states. However, most of the investigations are currently based on animal models because of the difficulty to obtain brain samples from humans. The results from the proteomics studies revealed that mitochondrial structural and functional alterations appear to play an important role in nervous system diseases.

1.3.2 CARDIOVASCULAR SYSTEM

Cardiovascular diseases have been the main "killer" in human beings, and thus, early diagnosis and treatment is imperative. Mitochondria are the major site of substrate oxidation in cardiomyocytes. Furthermore, oxidative stress plays a key role in heart diseases, and mitochondria are considered a principle source and target of reactive oxygen species (ROS) [32]. ROS can damage cellular lipids, proteins, and DNA, thereby disrupting their normal functions. Several large-scale studies have systematically reported some notable biological and medical insights into the mitochondrial proteome in the cardiovascular system as described below.

The ischemic heart is an important model for researchers studying the cardiovascular diseases. Until now, most of the findings from comparative mitochondrial proteomic studies were associated with the respiratory chains and energy metabolism in the ischemic heart. For example, Kim et al detected regional differences in protein expression levels from mitochondrial fractions of control, ischemia-reperfusion (IR), and ischemic preconditioned (IPC) rabbit hearts [8]. In addition, Essop and colleagues investigated the alterations in the mitochondrial proteome in a mouse model of obesity/type 2 diabetes. Several proteins that play role in mitochondrial energy metabolism, including ATP synthase D chain, ubiquinol cytochrome-C reductase core protein 1, and electron transfer flavoprotein subunit alpha, were identified to have changes in protein levels [33].

Mitochondria play a crucial role in the regeneration of antioxidants through the production of reducing equivalents and are responsible for the vast majority of ATP production within most cells and higher organisms. Hunzinger [34] employed a proteomics approach to investigate the role of ROS on bovine heart and identified two specific N-formylkynurenine modifications of aconitase-2, which is an enzyme that plays an important role in mitochondrial aging. Additional investigations on these two modifications might identify them as potential protein biomarker signatures for ROS. Serious and/or long-term ischemia will lead to heart infarction. It has been proposed that in ischemic preconditioning (PC) or pharmacological preconditioning, the GSK-3 (glycogen synthase kinase-3) inhibitor AR-A014418 will initiate signaling cascades that converges on mitochondria and results in cardioprotection. Therefore, Wong et al utilized 2D-DIGE coupled with Blue Native-PAGE to confirm their hypothesis that PC and pharmacological preconditioning similarly altered mitochondrial signaling complexes III, IV, and V. The altered expression levels of electron transport complexes obtained from the above-mentioned study should impart important implications for the mechanism of cardioprotection [35].

In short, most of the mitochondrial proteomics studies in cardiovascular diseases were associated with ROS. It is well known that ischemia and aging are the main causes of cardiovascular diseases, which can result in the release of ROS. Therefore, ROS are still a primary focus when researchers investigate cardiovascular disease. Targeting on ROS might be applicable to the treatment of many cardiovascular problems in the future.

1.3.3 CANCER AND HEMATOLOGICAL DISEASES

The mitochondrial proteome has altered expression levels and structures in cancer cells, as well as that in cells altered simulide when stimulated by various treatments (Table 2).

TABLE 2: Progress in the treatment of hematologic diseases and solid tumor based on mammalian mitochondrial proteomics

Disease	Cell lines/Treatment	Analytical methods	Up-regulated proteins	Down-regulated proteins	Functions	Primary significance
Hematologic diseases						
AML	NB4/camptothecin	2-DE, MALDI-TOF/TOF		Far upstream element-binding protein-1 (FUBP1)	transcription, translation, and degradation of proteins	Provided new insights for systematically understanding the mechanisms of the camptothecin-induced apoptosis.
				Heterogeneous nuclear ribonucleoprotein A1 (HNRPA1)	mRNA processing	
				Heterogeneous nuclear ribonucleoproteins C1/C2 (HNRPC)	modified with shift of pI and MW	
			26S protease regulatory subunit 6A (PSMC3)		degradation	
			Proteasome subunit alpha type (PSMA)-1, 2, 6		degradation	
Non-Hodgkin's lymphoma	Raji/Adramycin	DIGE, LTQ-ESI-MS/MS	ATP synthase d chain, mitochondrial (ATPQ, ATP5H)		OXPHOS	Specific mitochondrial proteins were uniquely susceptible to alterations in abundance following exposure to ADR and carry implications for the investigation of therapeutic and prognostic markers.
			Prohibitin (PHB)		cell cycle	
			Heat shock 70 kDa protein 9 precursor (HSPA9)		transporters and channels	

TABLE 2: *Cont.*

Disease	Cell lines/Treatment	Analytical methods	Up-regulated proteins	Down-regulated proteins	Functions	Primary significance
				Isoform 4 of Mitochondrial ATP-binding cassette sub-family B member 6 (ABCB6)	protein synthesis and degradation	
				Superoxide dismutase [Mn], mitochondrial precursor (SOD2)	redox	
Other solid tumors						
Osteosarcoma	143B/devoid of mitochondrial DNA	2-DE, MALDI-TOF/MS		NADH-ubiquinone oxidoreductase 75 kDa subunit	respiratory complexes subunits	Demonstrates the pleiotropic effects of mtDNA alterations and also gives valuable markers for the study of the mitochondrial-cytosolic coordination.
				Mitochondrial 28S ribosomal protein S2	mitochondrial translation apparatus	
				Mitochondrial import inner membrane translocase subunit Tim9	protein transport	
				Succinyl-CoA ligase (ADP-forming) beta-chain	energy production	

TABLE 2: *Cont.*

Disease	Cell lines/Treatment	Analytical methods	Up-regulated proteins	Down-regulated proteins	Functions	Primary significance
Breast cancer						
	MCF-7/resistance to adriamycin accompanied by verapamil	2-DE, QqTOF-ESI-MS/MS		Cofilin	control of polymerization/depolymerization of actin	Implications of the changes are considered with respect to drug resistance.
			CoproporphyrinogenIII (CPO)		heme biosynthesis	
			3.2 trans-enoyl CoA isomerase		fatty acid oxidation	
			Adenylate kinase 2		ATP, OX-PHOS	
Renal cell carcinomas						
	UMRC2, 786-0 and RCC4/VHL (von Hippel Lindau)-defective	2-DE, MS	Heat shock 70 kDa protein		transporters and channels	Increased expression of septin 2 is a common event in RCC and protein modification may also alter septin 2 function in a subset of tumors.
			10-formyltetrahydrofolate dehydrogenase		one-carbon metabolism	
			Phosphoribosylglycinamide formyltransferase		purine biosynthesis	

TABLE 2: *Cont.*

Disease	Cell lines/Treatment	Analytical methods	Up-regulated proteins	Down-regulated proteins	Functions	Primary significance
			Ubiquinol cytochrome c reductase complex core protein 2		electron transport	
				Elongation factor 2	protein bio-synthesis	
				Phosphofructokinase isozyme C	glycolysis	
				Thioredoxin reductase	differen-tiation, electron transport	
				Septin 2	cell cycle	
Prostate cancer	LNCaP/somatosta-tin		VDAC1, VDAC2		apoptosis	Somatos might be able to curb the progression of advanced prostate cancer.
				Peroxiredoxin 2 (PRDX2)	antioxidant activity	
				Translationally con-trolled tumor protein (TCTP)	calcium binding and microtubule stabilization.	

Chevallet et al [36] employed a comparative study on the osteoscrcoma 143B cell line and its Rho-0 counterpart (devoid of mitochondrial DNA). Quantitative differences were found between these cell lines in factors, such as the respiratory complexes subunits, the mitochondrial translation apparatus, mitochondrial ribosomal proteins, and the proteins with roles in the ion and protein import system. They also found that proteins involved in apoptosis control and import systems were differentially regulated in Rho-0 mitochondria. To identify proteins involved in a retrograde response and their potential role in tumorigenesis, Kulawiec [37] conducted a comparative proteomic analysis using the two cell lines noted above and the parental cell line. They found that subunits of complex I and III, molecular chaperones, and a protein involved in cell cycle control were down-regulated and that inosine 5'-monophosphate dehydrogenase type 2 (IMPDH2), which is involved in nucleotide biosynthesis, was up-regulated in $\rho 0$ cells. Retrograde proteins identified in these studies might be useful as therapeutic targets due to their roles as potential tumor suppressors or oncogenes involved in carcinogenesis.

Several investigations on other tumor mitochondrial proteome were also conducted recently. Regarding breast cancer, Strong et al [38] identified differentially expressed proteins in the mitochondria of MCF-7 human breast cancer cells that were selected for resistance to adriamycin accompanied by verapamil. Those identified proteins were mainly involved in apoptosis, heme synthesis, fatty acid oxidation, and oxidative phosphorylation. The implications of these changes in protein levels are relevant to mechanisms of drug resistance. Craven [39] compared the mitochondrial proteome in VHL (von Hippel Lindau, a tumor suppressor gene)-defective RCCs (renal cell carcinomas), which were transfected with either a control vector or wild-type VHL. That study showed that the mitochondrial protein ubiquinol cytochrome c reductase complex core protein 2 was up-regulated and a form of septin 2 was down-regulated following VHL transfection. Septin 2 was up-regulated in 12/16 RCCs. Thus, increased expression of septin 2 is a common event in RCC, and protein modification may also change the function of septin 2 in a subset of tumors. Zhao et al [40] incubated the LNCaP cell-line with sms (somatostatin)14/smsds and demonstrated that proteins with roles in apoptosis were both up-regulated (VDAC1, VDAC2) and down-regulated (PRDX2, TCTP). Sms/

smsdx was believed to trigger the up-regulation of catalytic mitochondrial proteins and seemed to affect apoptosis-related proteins.

Only a few studies have reported the effects of hematological disease on the mitochondrial proteome. Yu et al [41] analyzed protein expression profiles of fractionated nuclei, mitochondria, crude endoplasmic reticulum, and cytosols of the NSC606985-induced apoptotic AML cell line NB4 cells using 2-DE combined with MALDI-TOF/TOF. They identified 90 unique deregulated proteins that contributed to multiple functional activities including DNA damage repair, chromosome assembly, and mRNA processing as well as biosynthesis, modification, and degradation of proteins. More interestingly, several oxidative stress-related proteins that were shown up-regulated were localized in mitochondria, while proteins that were up-regulated with roles in glycolysis were mainly localized in the nuclei. Their discoveries shed new insights for systematically understanding mechanisms of the camptothecin-induced apoptosis. In our previous study, we investigated mitochondrial proteome alterations in NHL Raji cells exposed to adriamycin. Our results showed that 34 proteins were down-regulated and 3 proteins up-regulated when the study group was compared with the control group. The differentially expressed proteins play roles in many cellular functions, including redox, DNA repair, cell cycle regulation, transporters and channels, and OXPHOS. Furthermore, HSP70, ABCB6, and PHB identified in this study may be closely related to chemoresistance, and this potentially serving as chemotherapeutic targets for NHL [42].

Collectively, studies on mitochondrial proteomics will further investigate into cancerous biological behavior and mechanisms of antineoplastic agents. Subsequently, improved diagnosis, and treatment methods, and new treatment targets will likely be obtained.

1.3.4 OTHER DISEASES

Mitochondrial proteomics also revealed a number of significant findings in other diseases, such as hepatopathy, placenta changes, and skeleton muscle disease.

The liver is an important organ that has an abundance of mitochondria. Ruepp [43] investigated the effects of acetaminophen (APAP) in the liver on the proteomic level and found that chaperone proteins HSP10 and HSP60 were readily decreased by half in mitochondria at different doses of APAP. The decrease of ATP synthase subunits levels and β-oxidation pathway proteins indicated a loss of energy production. Douette and co-workers [44] compared mitochondrial protein patterns in wild-type and steatosis-affected liver and identified 58 proteins exhibiting significantly different levels in these two samples. Interestingly, major proteins that regulate the generation and consumption of the acetyl-CoA pool were dramatically changed during steatosis. Furthermore, many proteins involved in the response to oxidative stress were also affected. Lee [45] assessed global protein expression profiles in term placentas from scNT (somatic cell-derived nuclear transfer)-derived and control animals. Forty-three unique proteins were identified, including such proteins play critical roles in the apoptosis signaling pathways as 14-3-3 proteins were up-regulated in scNT-derived when compared to the Annexin V in control animals group. Their results suggested that placental insufficiency in scNT-derived placentas may be due to apoptosis, induced in part by the down-regulation of 14-3-3 proteins and the up-regulation of Annexin V. De Palma [46] investigated the hypoxia-induced changes of rat skeletal muscle and indicated that proteins involved in the TCA cycle, ATP production, and electron transport are down-regulated, whereas glycolytic enzymes and deaminases involved in ATP and AMP production were up-regulated. The up-regulation of the hypoxia markers hypoxia inducible factor 1 (HIF-1α) and pyruvate dehydrogenase kinase 1 (PDK1) suggested that in vivo adaptation to hypoxia requires an active metabolic switch. Eccleston et al hypothesized that chronic exposure to a high fat diet (HFD) would modify the liver mitochondrial proteome, which might ultimately compromise mitochondrial function. Using two-dimensional isoelectric-focusing (IEF)/SDS-PAGE, 22 proteins which played roles of oxidative phosphorylation, lipid metabolism, sulfur amino acid metabolism, and chaperone proteins, showed altered levels as a consequence of the HFD. These proteomic studies were complemented by measuring mitochondrial ROS production and assessing the impact of a HFD on the levels of two key enzymes involved in maintaining tissue NO: arginase1 and endothelial nitric oxide synthase (eNOS) [47].

Alves [48] studied the influence of lifestyle on the aging alterations in skeletal muscle mitochondrial proteins with 2-DE combined MALDI-TOF/TOF. Their results confirmed that certain mitochondrial proteins, particularly those play role in the citric acid cycle and as OXPHOS components, showed increased carbonylation. The data obtained indicated that lifestyle was a key modulator for preventing the expression and functionality of aging-induced proteins in mitochiondria.

Overall, there is already considerable information regarding the important role of mitochondria in the regulation of apoptosis, energy metabolism, and electron transfer. Mitochondrial proteomics have been performed in various fields and have gained considerable achievements. Mitochondrial proteomics are currently the most popular area of subcellular proteomics being investigated.

1.4 CHALLENGES

Proteomic techniques are becoming more and more advanced established. However, the study of proteins is not similar to that of DNA and RNA. First, proteins have more complicated two- and three-dimensional structures, and second, proteins cannot be amplified like DNA. Protein structure can be easily altered, but cannot be easily detected if the amount is too small. As far as the mitochondrial proteome is concerned, many questions remain unresolved.

1. The isolation and purification of mitochondria: At present, the well-recognized method for isolating mitochondria from tissues or cells is Taylor's classic method, which uses sucrose density gradient centrifugation [3]. However, this method requires ultracentrifugation and is time-consuming. Many efforts have been made to improve Taylor's method [49,50]. Furthermore, special kits for mitochondria isolation have been put onto the market that do not require ultracentrifugation and are more time-efficient. The purity of the mitochondria isolated by these kits has been shown to be fairly good [51].

2. Limitations of 2-DE: 2-DE is not good for solving many problems, such as how to remove high abundant proteins or how to isolate proteins with extreme alkalinity or acidity. With regard to mitochondria, a large proportion of the proteins have an extreme alkaline isoelectric point (pI). As a result, they are unable to be resolved by isoelectric focusing due to endo-osmotic effects upon the pH gradient [12]. In fact, some proteins have a pI that is too alkaline to be visible on typical wide range (pH 3-10) immobilized pH gradient (IPG) strips. For example, the pI of cytochrome c is 10.3 [13]. Although 2-DE is a powerful instrument, for this purpose, it still needs further improvement or replacement by other more effective techniques.

3. The limitation of bioinformatic tools and the mitochondrial database: To meet the bioinformatic requirements of large-scale proteomic studies, many researchers have tried to use series-based attempts to overcome the shortcomings of single proteomic techniques. White and his colleagues use five parallel methodological approaches, ((i) peptide-centric 1-DLC, (ii) peptide-centric 2-DLC, (iii) protein-centric 1-DLC, (iv) protein-centric 2-DLC, and (v) subfractionated mitochondria) to improve the coverage of the partially annotated rabbit mitochondrial proteome prior to mass spectrometry. They found that the overall coverage of the cardiomyocyte mitochondrial proteome was improved by this parallel approach where the total number of nonredundant peptides or proteins was nearly 2 fold and more than 1.5-fold, respectively, greater than that by any single technique. They assumed that observation of proteins across multiple technologies improves the likelihood of true mitochondrial localization [52]. Furthermore, for the homologous proteins and redundant entries in the sequence database, one of the challenges in protein identification is that many peptides can be matched to several different proteins [53]. It would be helpful to use more than one database search engine when analyzing complex protein mixtures from the same raw data [54]. However, manual comparison and analysis of large database search results are laborious and time-consuming. Furthermore, most of these tools can utilize data from only a few database search engines, and

currently, there are no free tools that could be used to combine protein identification results from paragon with results from other search engines. Thus, it is necessary to develop fast, accurate, and easy-to-use tools to integrate and compare protein identification results.

The insufficiency of mitochondrial databases is another problem. The first attempt to build a 2-DE database of the mitochondrial proteome was performed by Rabilloud [55]. However, that database was considered both incomplete and inefficient. Now, a large amount of new data has been added into databases such as SWISS-PROT, NCBI, MITOMAP, mtDB, hmtDB, MitoP2, MigDB, and MitoProteome [56-58]. MitoP2 is the most comprehensive database for the mitochondrial proteome and includes a more complete set of these mitochondrial proteins for human (624), mouse (615), and yeast (522) for each of these organisms respectivly [58]. However, due to the emergence of this new subject, the databases are still very insufficient and require more exploration for enrichment.

4. Clinical applications: Many studies on the mitochondrial proteome have been reported, and how to utilize the findings in the clinic to gain the maximum benefits is still a large problem. Only a few of identified biomarkers have been currently used by clinicians as diagnostic and/or prognostic factors [59,60]. Novel biomarkers identified by proteomics can be developed for increased precision in diagnosis, therapy sensitivity, progression, and prognosis evaluation of disease. Information on bioinformatics obtained from proteomic analyses is still scarce. These challenges have not been overcome by the existing methods and they have become a limitation to further advancement. Understanding how mitochondrial proteins function together in pathways and complexes is still a significant challenge. Many biomarkers found in proteomic studies were conformed to be involved in various mitochondrial-associated signaling pathways, including apoptosis, cell cycle, and DNA repair [61,62]. Many validation tests using methods such as RNAi, protein-protein interaction mapping, and computational predictions should be linked to the future investigations [63,64].

Standardizing these current proteomic experiment in terms of sample collection, storage and processing as well as bio-informatics and statistical analysis between various centers is serious necessary [65]. Future perspectives will focus on the clinical applications of these biomarkers to improve diagnostic accuracy and prognostic precision. Thus, more intimate, repeatable, and verifiable experiments are eagerly awaited.

1.5 CONCLUSIONS

The emergence of comparative proteomics enables us to investigate the mitochondrial proteome in a more comprehensive and effective manner. The results of mitochondrial proteomics provide a better understanding of the pathogenesis associated with mitochondria and generate promising therapeutic targets. However, these novel findings are most unlikely to completely reflect the true state of mitochondria because some biological information may be lost or altered during the course that mitochondria are isolated from the cell. Moreover, the mitochondrial proteome alterations in animal models may differ from those of human. Therefore, more efforts are needed to look at the validation across species carefully. Validation and utilization of clinically associated proteomic biomarkers would be helpful to the diagnosis, effective treatment, and prognosis evaluation of mitochondria-mediated diseases. Thus, the formulation of personalized medicine may become a reality in the future. This is an open-ended exploration, and more achievements are anticipated in the near future.

REFERENCES

1. Fang X, Lee CS: Proteome characterization of mouse brain mitochondria using electrospray ionization tandem mass spectrometry. Methods Enzymol 2009, 457:49-62.
2. Goffart S, Martinsson P, Malka F, Rojo M, Spelbrink JN: The mitochondria of cultured mammalian cells: II, Expression and visualization of exogenous proteins in fixed and live cells. Methods Mol Biol 2007, 372:17-32.
3. Taylor SW, Fahy E, Zhang B, Glenn GM, Warnock DE, Wiley S, Murphy AN, Gaucher SP, Capaldi RA, Gibson BW, Ghosh SS: Characterization of the human heart mitochondrial proteome. Nat Biotechnol 2003, 21:281-286.

4. Hoye AT, Davoren JE, Wipf P, Fink MP, Kagan VE: Targeting mitochondria. Acc Chem Res 2008, 41:87-97.

5. Degner D, Bleich S, Riegel A, Sprung R, Poser W, Ruther E: Follow-up study after enteral manganese poisoning: clinical, laboratory and neuroradiological findings. Nervenarzt 2000, 71:416-419.

6. Basso M, Giraudo S, Corpillo D, Bergamasco B, Lopiano L, Fasano M: Proteome analysis of human substantia nigra in Parkinson's disease. Proteomics 2004, 4:3943-3952.

7. Da Cruz S, Martinou JC: Purification and proteomic analysis of the mouse liver mitochondrial inner membrane. Methods Mol Biol 2008, 432:101-116.

8. Kim N, Lee Y, Kim H, Joo H, Youm JB, Park WS, Warda M, Cuong DV, Han J: Potential biomarkers for ischemic heart damage identified in mitochondrial proteins by comparative proteomics. Proteomics 2006, 6:1237-1249.

9. Friedman DB: Quantitative proteomics for two-dimensional gels using difference gel electrophoresis. Methods Mol Biol 2007, 367:219-239.

10. Vo TD, Palsson BO: Building the power house: recent advances in mitochondrial studies through proteomics and systems biology. Am J Physiol Cell Physiol 2007, 292:C164-C177.

11. Pappin DJ: Peptide mass fingerprinting using MALDI-TOF mass spectrometry. Methods Mol Biol 2003, 211:211-219.

12. Gorg A, Obermaier C, Boguth G, Csordas A, Diaz JJ, Madjar JJ: Very alkaline immobilized pH gradients for two-dimensional electrophoresis of ribosomal and nuclear proteins. Electrophoresis 1997, 18:328-337.

13. Flatmark T: On the heterogeneity of beef heart cytochrome c. II. Some physico-chemical properties of the main subfractions (Cy I-Cy 3). Acta Chem Scand 1966, 20:1476-1486.

14. Unlu M, Morgan ME, Minden JS: Difference gel electrlophoresis: a single gel method for detecting changes in protein extracts. Electrophoresis 1997, 18:2071-2077.

15. Tonge R, Shaw J, Middleton B, Rowlinson R, Rayner S, Young J, Pognan F, Hawkins E, Currie I, Davison M: Validation and development of fluorescence two-dimensional differential gel electrophoresis proteomics technology. Proteomics 2001, 1:377-396.

16. Schagger H, von Jagow G: Blue native electrophoresis for isolation of membrane protein complexes in enzymatically active form. Anal Biochem 1991, 199:223-231.

17. Devreese B, Vanrobaeys F, Smet J, Van Beeumen J, Van Coster R: Mass spectrometric identification of mitochondrial oxidative phosphorylation subunits separated by two-dimensional blue-native polyacrylamide gel electrophoresis. Electrophoresis 2002, 23:2525-2533.

18. McDonald WH, Yates JR: Shotgun proteomics: integrating technologies to answer biological questions. Curr Opin Mol Ther 2003, 5:302-309.

19. Gygi SP, Rist B, Gerber SA, Turecek F, Gelb MH, Aebersold R: Quantitative analysis of complex protein mixtures using isotope-coded affinity tags. Nat Biotechnol 1999, 17:994-999.

20. Wu CC, MacCoss MJ: Shotgun proteomics: tools for the analysis of complex biological systems. Curr Opin Mol Ther 2002, 4:242-250.

21. Pocsfalvi G, Cuccurullo M, Schlosser G, Cacace G, Siciliano RA, Mazzeo MF, Scacco S, Cocco T, Gnoni A, Malorni A, Papa S: Shotgun proteomics for the characterization of subunit composition of mitochondrial complex I. Biochim Biophys Acta 2006, 1757:1438-1450.

22. Glen A, Gan CS, Hamdy FC, Eaton CL, Cross SS, Catto JW, Wright PC, Rehman I: iTRAQ-facilitated proteomic analysis of human prostate cancer cells identifies proteins associated with progression. J Proteome Res 2008, 7:897-907.

23. Kiehntopf M, Siegmund R, Deufel T: Use of SELDI-TOF mass spectrometry for identification of new biomarkers: potential and limitations. Clin Chem Lab Med 2007, 45:1435-1449.

24. Everley PA, Krijgsveld J, Zetter BR, Gygi SP: Quantitative cancer proteomics: stable isotope labeling with amino acids in cell culture (SILAC) as a tool for prostate cancer research. Mol Cell Proteomics 2004, 3:729-735.

25. Rabilloud T, Strub JM, Carte N, Luche S, Van Dorsselaer A, Lunardi J, Giege R, Florentz C: Comparative proteomics as a new tool for exploring human mitochondrial tRNA disorders. Biochemistry 2002, 41:144-150.

26. Chou JL, Shenoy DV, Thomas N, Choudhary PK, Laferla FM, Goodman SR, Breen GA: Early dysregulation of the mitochondrial proteome in a mouse model of Alzheimer's disease. J Proteomics 2011, 74:466-479.

27. Lovell MA, Xiong S, Markesbery WR, Lynn BC: Quantitative proteomic analysis of mitochondria from primary neuron cultures treated with amyloid beta peptide. Neurochem Res 2005, 30:113-122.

28. Gillardon F, Rist W, Kussmaul L, Vogel J, Berg M, Danzer K, Kraut N, Hengerer B: Proteomic and functional alterations in brain mitochondria from Tg2576 mice occur before amyloid plaque deposition. Proteomics 2007, 7:605-616.

29. Fu H, Li W, Liu Y, Lao Y, Liu W, Chen C, Yu H, Lee NT, Chang DC, Li P, et al.: Mitochondrial proteomic analysis and characterization of the intracellular mechanisms of bis(7)-tacrine in protecting against glutamate-induced excitotoxicity in primary cultured neurons. J Proteome Res 2007, 6:2435-2446.

30. Pienaar IS, Schallert T, Hattingh S, Daniels WM: Behavioral and quantitative mitochondrial proteome analyses of the effects of simvastatin: implications for models of neural degeneration. J Neural Transm 2009, 116:791-806.

31. Broadwater L, Pandit A, Clements R, Azzam S, Vadnal J, Sulak M, Yong VW, Freeman EJ, Gregory RB, McDonough J: Analysis of the mitochondrial proteome in multiple sclerosis cortex. Biochim Biophys Acta 2011, 1812:630-641.

32. Major T, von Janowsky B, Ruppert T, Mogk A, Voos W: Proteomic analysis of mitochondrial protein turnover: identification of novel substrate proteins of the matrix protease pim1. Mol Cell Biol 2006, 26:762-776.

33. Essop MF, Chan WA, Hattingh S: Proteomic analysis of mitochondrial proteins in a mouse model of type 2 diabetes. Cardiovasc J Afr 2011, 22:175-178.

34. Hunzinger C, Wozny W, Schwall GP, Poznanovic S, Stegmann W, Zengerling H, Schoepf R, Groebe K, Cahill MA, Osiewacz HD, et al.: Comparative profiling of the mammalian mitochondrial proteome: multiple aconitase-2 isoforms including N-formylkynurenine modifications as part of a protein biomarker signature for reactive oxidative species. J Proteome Res 2006, 5:625-633.

35. Wong R, Aponte AM, Steenbergen C, Murphy E: Cardioprotection leads to novel changes in the mitochondrial proteome. Am J Physiol Heart Circ Physiol 2010, 298:H75-H91.

36. Chevallet M, Lescuyer P, Diemer H, van Dorsselaer A, Leize-Wagner E, Rabilloud T: Alterations of the mitochondrial proteome caused by the absence of mitochondrial DNA: A proteomic view. Electrophoresis 2006, 27:1574-1583.

37. Kulawiec M, Arnouk H, Desouki MM, Kazim L, Still I, Singh KK: Proteomic analysis of mitochondria-to-nucleus retrograde response in human cancer. Cancer Biol Ther 2006, 5:967-975.

38. Strong R, Nakanishi T, Ross D, Fenselau C: Alterations in the mitochondrial proteome of adriamycin resistant MCF-7 breast cancer cells. J Proteome Res 2006, 5:2389-2395.

39. Craven RA, Hanrahan S, Totty N, Harnden P, Stanley AJ, Maher ER, Harris AL, Trimble WS, Selby PJ, Banks RE: Proteomic identification of a role for the von Hippel Lindau tumour suppressor in changes in the expression of mitochondrial proteins and septin 2 in renal cell carcinoma. Proteomics 2006, 6:3880-3893.

40. Liu Z, Bengtsson S, Krogh M, Marquez M, Nilsson S, James P, Aliaya A, Holmberg AR: Somatostatin effects on the proteome of the LNCaP cell-line. Int J Oncol 2007, 30:1173-1179.

41. Yu Y, Wang LS, Shen SM, Xia L, Zhang L, Zhu YS, Chen GQ: Subcellular proteome analysis of camptothecin analogue NSC606985-treated acute myeloid leukemic cells. J Proteome Res 2007, 6:3808-3818.

42. Jiang YJ, Sun Q, Fang XS, Wang X: Comparative mitochondrial proteomic analysis of Rji cells exposed to adriamycin. Mol Med 2009, 15:173-182.

43. Ruepp SU, Tonge RP, Shaw J, Wallis N, Pognan F: Genomics and proteomics analysis of acetaminophen toxicity in mouse liver. Toxicol Sci 2002, 65:135-150.

44. Douette P, Navet R, Gerkens P, de Pauw E, Leprince P, Sluse-Goffart C, Sluse FE: Steatosis-induced proteomic changes in liver mitochondria evidenced by two-dimensional differential in-gel electrophoresis. J Proteome Res 2005, 4:2024-2031.

45. Lee SY, Park JY, Choi YJ, Cho SK, Ahn JD, Kwon DN, Hwang KC, Kang SJ, Paik SS, Seo HG, et al.: Comparative proteomic analysis associated with term placental insufficiency in cloned pig. Proteomics 2007, 7:1303-1315.

46. De Palma S, Ripamonti M, Vigano A, Moriggi M, Capitanio D, Samaja M, Milano G, Cerretelli P, Wait R, Gelfi C: Metabolic modulation induced by chronic hypoxia in rats using a comparative proteomic analysis of skeletal muscle tissue. J Proteome Res 2007, 6:1974-1984.

47. Eccleston HB, Andringa KK, Betancourt AM, King AL, Mantena SK, Swain TM, Tinsley HN, Nolte RN, Nagy TR, Abrams GA, Bailey SM: Chronic exposure to a high-fat diet induces hepatic steatosis, impairs nitric oxide bioavailability, and modifies the mitochondrial proteome in mice. Antioxid Redox Signal 2011, 15:447-459.

48. Alves RM, Vitorino R, Figueiredo P, Duarte JA, Ferreira R, Amado F: Lifelong physical activity modulation of the skeletal muscle mitochondrial proteome in mice. J Gerontol A Biol Sci Med Sci 2010, 65:832-842.

49. Rezaul K, Wu L, Mayya V, Hwang SI, Han D: A systematic characterization of mitochondrial proteome from human T leukemia cells. Mol Cell Proteomics 2005, 4:169-181.

50. Zischka H, Weber G, Weber PJ, Posch A, Braun RJ, Buhringer D, Schneider U, Nissum M, Meitinger T, Ueffing M, Eckerskorn C: Improved proteome analysis of Saccharomyces cerevisiae mitochondria by free-flow electrophoresis. Proteomics 2003, 3:906-916.

51. Fu YR, Yi ZJ, Yan YR, Qiu ZY: Proteomic analysis of mitochondrial proteins in hydroxycamptothecin-treated SMMC-7721 cells. Zhonghua Gan Zang Bing Za Zhi 2007, 15:572-576.

52. White MY, Brown DA, Sheng S, Cole RN, O'Rourke B, Van Eyk JE: Parallel proteomics to improve coverage and confidence in the partially annotated Oryctolagus cuniculus mitochondrial proteome. Mol Cell Proteomics 2011, 10:M110 004291.

53. Nesvizhskii AI, Aebersold R: Interpretation of shotgun proteomic data: the protein inference problem. Mol Cell Proteomics 2005, 4:1419-1440.

54. Yu W, Taylor JA, Davis MT, Bonilla LE, Lee KA, Auger PL, Farnsworth CC, Welcher AA, Patterson SD: Maximizing the sensitivity and reliability of peptide identification in large-scale proteomic experiments by harnessing multiple search engines. Proteomics 2010, 10:1172-1189.

55. Rabilloud T, Kieffer S, Procaccio V, Louwagie M, Courchesne PL, Patterson SD, Martinez P, Garin J, Lunardi J: Two-dimensional electrophoresis of human placental mitochondria and protein identification by mass spectrometry: toward a human mitochondrial proteome. Electrophoresis 1998, 19:1006-1014.

56. Brandon MC, Lott MT, Nguyen KC, Spolim S, Navathe SB, Baldi P, Wallace DC: MITOMAP: a human mitochondrial genome database-2004 update. Nucleic Acids Res 2005, 33:D611-D613.

57. Ingman M, Gyllensten U: mtDB: Human Mitochondrial Genome Database, a resource for population genetics and medical sciences. Nucleic Acids Res 2006, 34:D749-D751.

58. Prokisch H, Andreoli C, Ahting U, Heiss K, Ruepp A, Scharfe C, Meitinger T: MitoP2: the mitochondrial proteome database-now including mouse data. Nucleic Acids Res 2006, 34:D705-D711.

59. Richard C: Stress-related cardiomyopathies. Ann Intensive Care 2011, 1:39.

60. Kim HN, Januzzi JL Jr: Natriuretic peptide testing in heart failure. Circulation 2011, 123:2015-2019.

61. Thiede B, Rudel T: Proteome analysis of apoptotic cells. Mass Spectrom Rev 2004, 23:333-349.

62. Smith DJ: Mitochondrial dysfunction in mouse models of Parkinson's disease revealed by transcriptomics and proteomics. J Bioenerg Biomembr 2009, 41:487-491.

63. Ozawa T, Sako Y, Sato M, Kitamura T, Umezawa Y: A genetic approach to identifying mitochondrial proteins. Nat Biotechnol 2003, 21:287-293.

64. Graves PR, Haystead TA: A functional proteomics approach to signal transduction. Recent Prog Horm Res 2003, 58:1-24.

65. Wang YS, Cao R, Jin H, Huang YP, Zhang XY, Cong Q, He YF, Xu CJ: Altered protein expression in serum from endometrial hyperplasia and carcinoma patients. J Hematol Oncol 2011, 4:15.

CHAPTER 2

STUDIES OF COMPLEX BIOLOGICAL SYSTEMS WITH APPLICATIONS TO MOLECULAR MEDICINE: THE NEED TO INTEGRATE TRANSCRIPTOMIC AND PROTEOMIC APPROACHES

ELENA SILVESTRI, ASSUNTA LOMBARDI, PIETER DE LANGE, DANIELA GLINNI, ROSALBA SENESE, FEDERICA CIOFFI, ANTONIA LANNI, FERNANDO GOGLIA, AND MARIA MORENO

2.1 INTRODUCTION

Genomic and proteomic data analyses have proven to be essential for an understanding of the underlying factors involved in human disease and for the discovery of diagnostic biomarkers, as well as for the provision of further insights into the metabolic effects mediated by signaling molecules.

All classes of biological compounds, from genes through mRNA to proteins and metabolites, can be analyzed by the respective "omic" approaches, namely, genomics, transcriptomics, proteomics, or metabonomics. Such an "omic" approach leads to a broader view of the complex biological system, including the pathology of diseases. Indeed, while the

Originally printed under the terms of the Creative Commons Attribution License. Elena Silvestri, Lombardi A, de Lange P, Glinni D, Senese R, Cioffi F, Lanni A, Goglia F, and Moreno M. Studies of Complex Biological Systems with Applications to Molecular Medicine: The Need to Integrate Transcriptomic and Proteomic Approaches. Journal of Biomedicine and Biotechnology **2011** (2011). http://dx.doi.org/10.1155/2011/810242.

data obtained from genomics may explain the disposition of diseases (i.e., increased risk of acquiring a certain disease), several other mechanisms that are not gene mediated may be involved in the onset of disease. Moreover, a single gene can be processed to result in several different mRNAs or proteins, which directly determine different cellular functions. Variations in metabolite fluxes, which may be taken as the downstream result of changes in gene expression and protein translation, may be expected to be amplified relative to changes in the transcriptome and proteome. However, time-dependent measurements and determinations of metabolite content at a single time-point can be misleading as these fluxes vary quickly. Therefore, while genomics/transcriptomics enables assessments of all potential information, proteomics enables us to assess the programs that are actually executed, and metabolomics will mostly display the results of such executions.

In the postgenomic era, functional analysis of genes and their products constitutes a novel and powerful approach since the expression levels of multiple genes and proteins can thereby be analyzed simultaneously, in both health and disease (Figure 1). Among the techniques used in functional genomics, both DNA microarrays [1–3] and classical and ongoing proteomic approaches (finalized to protein separation and identification) [4–6] hold great promise for the study of complex biological systems and have applications in molecular medicine. These technologies allow high-throughput analysis as they are complementary to each other, and they may lead to a better understanding of the regulatory events involved in physiological, and disease, processes. Proteins are excellent targets in disease diagnostics, prognostics, and therapeutics. Consequently, proteomic approaches (such as two-dimensional gel electrophoresis (2D-E), two-dimensional liquid chromatography (2-DL), and mass spectrometry (MS)), which allow the simultaneous measurement and comparison of the expression levels of hundreds of proteins, represent powerful tools for (a) the discovery of novel hormone/drug targets and biomarkers and (b) studies of cellular metabolism and protein expressions [7, 8]. Increasingly, proteomic techniques are being adopted—in particular, to avoid the limitations inherent in the more classical approaches—to solve analytical problems and obtain a more comprehensive identification and characterization of molecular events associated with pathophysiological conditions (Figure 1).

In this paper, we will discuss a variety of mainly recent transcriptomic- and proteomic-based studies that have provided a comprehensive mechanistic insight into two very complex biological phenomena, namely, age-associated muscle sarcopenia and thyroid-hormone signaling. Moreover, as mitochondria are severely affected during ageing and it is generally believed that dysfunctions of mitochondria also cause ageing and muscle sarcopenia, we will also discuss proteomic analysis of the alterations in rat skeletal muscle mitochondria caused by ageing.

2.2 AGEING SARCOPENIA

Ageing, one of the most complex biological phenomena, is a multifaceted process in which several physiological changes occur at both the tissue

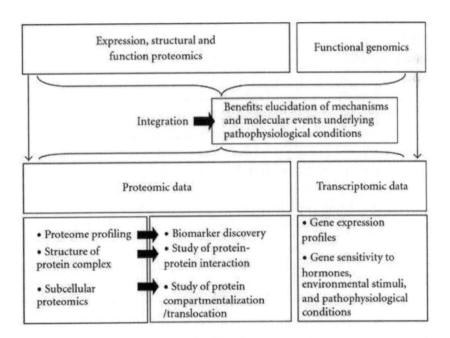

FIGURE 1: Categories and potential applications of proteomics and benefits of integration of proteomics and transcriptomics in the study of complex biological systems.

and the whole-organism level. Indeed, the age-associated decline in the healthy functioning of multiple organs/systems leads to an increased incidence of, and mortality from, diseases such as type II diabetes mellitus, neurodegenerative diseases, cancer, and cardiovascular disease [9].

One of the major engagements of gerontology is the understanding of the complex mechanisms involved in ageing at the molecular, cellular, and organ levels that would facilitate our understanding of age-related diseases. Research in this area has accelerated with the application of high-throughput technologies such as microarrays. To judge from such studies, several metabolic pathways are affected during ageing, and the picture becomes even more complex when we realize that most of them are interconnected.

Sarcopenia, the age-related decline in skeletal muscle mass and strength, is a major complication in the elderly [10, 11]. Since skeletal muscle represents the most abundant tissue in the body, fiber degeneration has severe consequences for posture, movement, the overall integration of metabolism, and heat homeostasis [12]. Although various metabolic and functional defects in ageing muscle have been described over the last decade, senescence-related muscle wasting is not well understood at the molecular and cellular levels. Consequently, no effective treatment has yet been developed to counteract age-related fiber degeneration [13].

Over the last decades, an attractive approach to the understanding of the molecular mechanisms involved in sarcopenia has been to screen all genetic pathways at one time, by the use of full-genome oligonucleotide chips, as well as the entire protein complement, by the use of using proteomic tools. These approaches, when applied together to the multifactorial muscle-wasting pathology observed in aged fibers, have allowed the identification of a variety of molecular and cellular changes. These include increased oxidative stress, mitochondrial abnormalities, disturbed microcirculation, hormonal imbalance, incomplete ion homeostasis, denervation, and impaired excitation-contraction coupling, as well as a decreased regenerative potential (see, Sections 2.1 and 2.2). In addition, altered post-translational modifications, such as tyrosine nitration, glycosylation, and phosphorylation, were recently described as occurring in an age-related manner in numerous skeletal muscle proteins (see, Section 2.3).

2.2.1 TRANSCRIPTOMIC ANALYSIS PERTAINING TO AGEING SKELETAL MUSCLE

Knowledge of differential mRNA expressions (i.e., the transcriptome) constitutes the first essential level of information when studying integrated cell functions and cell-specific gene-expression profiles. Since the development of DNA microarray technology, it has been possible to survey thousands of genes in parallel, thereby obtaining information regarding transcriptional changes on a global scale. Such an approach has been used to study the transcriptional alterations induced by ageing both in rodent models and in humans. Ageing-related transcriptomic studies have been performed both on home-spotted microarrays containing about 4000–6000 transcripts [14–16] and on commercial Affymetrix microarrays with from 12000 [17, 18] to about 54000 [19–26] transcripts on each array.

Concerning studies on humans, the design commonly used involved a cross-sectional comparison of young and elderly healthy individuals, with about eight individuals maximum per group, or an analysis of individuals across an age-range. In these studies, several pathways were highlighted by genes that were differentially expressed between young (19–29 years) and elderly (65–85 years) individuals [14], including genes involved in energy metabolism, the cell cycle, signal transduction, and DNA repair [19–22].

Biological pathways found to be changed with age in human skeletal muscle are listed in Table 1 and schematized in Figure 2. They include genes involved in the mitochondrial electron transport chain, cell cycle, and extracellular matrix. Zahn et al. [21], by comparing the transcriptional profile of ageing in muscle with previous transcriptional profiles of ageing in the kidney [22] and brain [17], found a common signature for ageing in these diverse human tissues. This common ageing signature consists of six genetic pathways; four display increased expression (genes in the extracellular matrix, genes involved in cell growth, genes encoding factors involved in complement activation, and genes encoding components of the cytosolic ribosome) and two display decreased expression in the aged muscle. These results indicate that those pathways, but not necessarily individual genes, are common elements in the age-related expression changes among different tissues [21]. This may imply that in addition to tissue-specific effects, a common ageing signature may be found in any tissue that reflects the age of the whole organism. This could have major implications for human epidemiological studies, for which frequently only blood is available.

TABLE 1: Summary of the models used and of the major findings obtained by applying microarray technologies to the study of ageing skeletal muscle.

Authors	Experimental model	Number of analyzed genes	Identified affected pathways
Mouse			
Lee et al., 1999 [25]	Studied tissue: aged skeletal muscle.	6347	Stress response, energy metabolism.
Rat			
Sreekumar et al., 2002 [28]	12-months-old Sprague-Dawley rats. Studied tissue: gastrocnemius muscle.	800	Energy metabolism, signal transduction, stress response, glucose/lipid metabolism, and structural/contractile function.
Altun et al., 2007 [29]	4- and 30-months-old rats. Studied tissue: gastrocnemius muscle.	6240	Redox homeostasis, iron load, regulation of contractile proteins, glycolysis, and oxidative phosphorylation.
Lombardi et al., 2009 [26]	3- and 24-months-old rats. Studied tissue: gastrocnemius muscle.	1176	Energy metabolism, mitochondrial pathways, myofibrillar filaments, and detoxification.
Human			
Welle et al., 2003 [23]	21–27 yr of age and 67–75 yr of age.Studied tissue: vastus lateralis muscle.	44 000	Cell cycle/cell growth, inflammation, signal transduction, protein metabolism, transcription, stress response/DNA repair, energy metabolism, and hormonal.
Welle et al., 2004 [19]	20–29 yr of age and 65–71 yr of age, women.Studied tissue: vastus lateralis muscle.	1000	Stress response/DNA repair, energy metabolism.
Zahn et al., 2006 [21]	16 and 89 yr of age. Studied tissue: skeletal muscle.	54 675	Electron transport chain, cell cycle/cell growth, extracellular matrix, chloride transport, complement activation, ribosomes.

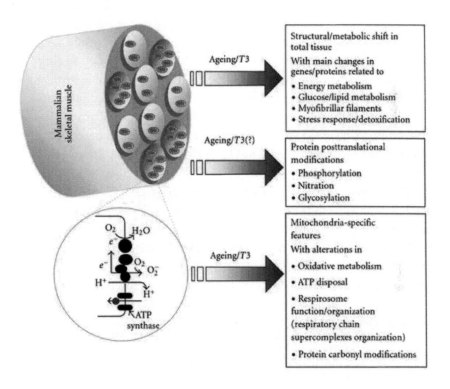

FIGURE 2: Integrated overview of the main ageing/T3-induced transcriptomic and proteomic alterations occurring in mammalian skeletal muscle. Schematic representation of the common events and mechanisms underlying the response of skeletal muscle to either ageing or T3 according to data obtained from cDNA microarray/proteomic-based studies in various mammalian models of ageing and thyroid state (mouse, rat, and human) (for details, see text and Tables 1, 2, 3, and 4).

Transcriptomic studies have been performed in laboratory animals using commercially available microarrays. As in humans, the main design used for measuring changes related to chronological age is a comparison between young and old individuals. These studies are usually performed on inbred strains, and so the variation between individual animals is smaller than among human individuals. The range of tissues studied includes liver, heart, skeletal muscle, aorta, and brain. Across all species, and in most experimental designs there is an influence of gender on ageing features and gene expressions [24]. Biological pathways found to be changed with age in rodent (mouse and rat) skeletal muscle are listed in Table 1.

Transcriptomic analysis of gastrocnemius muscle from 5- and 30-month-old male C57BL/6 mice revealed that ageing resulted in a differential gene expression pattern [25]. Of the genes that increased in expression with age, 16% were mediators of stress responses, including heat shock-response genes, oxidative stress-inducible genes, and DNA damage inducible genes (Table 1). Genes involved in energy metabolism were downregulated with ageing, including genes associated with mitochondrial function and turnover. This suppression of metabolic activity was accompanied by a concerted decline in the expressions of genes involved in glycolysis, glycogen metabolism, and the glycerophosphate shunt (Table 1). Ageing was also characterized by the induction of genes involved in neuronal growth and large reductions in the expressions of biosynthetic enzymes.

We recently performed a transcriptomic study on gastrocnemius muscle from rats aged 3 months (young) and 24 months (old) via a DNA array [26]. Transcript levels for genes associated with cellular damage were elevated in the older muscle, while transcript levels for genes involved in energy metabolism were reduced with age. Among the biological classes of transcripts significantly decreased by ageing, there were transcription factors as well as ribosomal proteins, indicative of a lower transcription/translation activity in old than in young skeletal muscle (Table 1). In agreement with previous microarray studies on the skeletal muscles of humans and rodents [18, 25, 27], we found that ageing is accompanied by a decline in the expressions of genes associated with energy-metabolism functions [26] (Table 1). Alterations in oxidative phosphorylation were revealed by decreased expression levels of cytochrome c oxidase, ATPase subunit, and carbonic anhydrase III [26]. The capacity of mitochondria to import and oxidize fatty acids would presumably be impaired since the mRNA levels for acylCoA synthase as well as carnitine palmitoyl transferase 1 (CPT 1β) were reduced during ageing. Downregulation of the AK1 isoform of adenylate kinase [26] points toward decreased AMP production and hence decreased activity of AMP-activated protein kinase (AMPK), an inducer of glucose and fatty acid uptake and fatty acid oxidation. Gastrocnemius muscle from the old rats revealed increased expressions of various factors involved in muscle differentiation toward the "slow" phenotype (type I; oxidative fibers), including p27kip and muscle LIM protein (MLP) [26] (Table 1). As a whole, the above data support an ageing-induced shift towards moderate fat burning.

Ageing has been found to increase the mRNA levels of scavengers of free radicals such as phospholipids hydroperoxide glutathione peroxidase and the cytosolic superoxide dismutase Cu/Zn SOD (SOD1). In addition, 14-kDa ubiquitin-conjugating enzyme E2 mRNA (a component of the complex that adds ubiquitin to target proteins, making them capable of destruction by the proteasome machinery) and both proteasome subunit C5 and proteasome delta subunit precursor were downregulated in aged muscle. Since the proteasome is the major proteolytic complex responsible for the selective degradation of oxidized proteins, these data point toward a defective action of the proteasome. With regard to the lysosomal pathway of protein breakdown, cathepsin L (which acts upstream of the ubiquitin-proteasome system) was also downregulated in aged muscle [26], once again supporting a decline in proteolysis during ageing.

The above studies have been successful in elucidating some of the transcriptional changes that occur with age in muscle, as well as in other tissues, and in providing insights about age-related changes common to animals with different lifespans.

2.2.2 PROTEOMIC ANALYSIS PERTAINING TO AGEING SKELETAL MUSCLE

Proteomic analysis has proved valuable in informing our understanding of the molecular mechanisms involved in the ageing process through the

TABLE 2: Summary of the models used and of the major findings obtained by applying proteomic approaches to the study of the ageing skeletal muscle.

Authors	Experimental model	Proteomic analysis	Identified affected pathways and major findings
Mouse			
Chang et al., 2003 [39]	18-months-old C57B16 mice. Studied tissue: skeletal muscle.	Two-dimensional polyacrilamide gel electrophoresis.	Reproducibility of the 2-D PAGE system.
Rats			
Cai et al., 2000 [35]	12-, 18-, and 24-months-old rats. Studied tissue: extensor digitorum longus muscle and soleus muscle.	Two-dimensional gel electrophoresis.	Analysis of aqueous proteins of skeletal muscle during aging.
Cai et al., 2001 [40]	8-, 18-, and 24-months-old rats. Studied tissue: extensor digitorum longus muscle and soleus muscle.	Two-dimensional gel electrophoresis.	Analysis of parvalbumin expression in rat skeletal muscles.
Kanski et al., 2003 [31]	4- and -24 months old Fisher 344 rats and -6 and -34 months old Fisher 344/BN F1 rats. Studied tissue: skeletal muscle.	2-D gel electrophoresis, Western blot analysis, MALDI-TOF MS and ESI-MS/MS analysis.	Age-dependent nitration in muscle energy metabolism.
Piec et al., 2005 [36]	7-, 18- and 30-months-old LOU/c/jall rats. Studied tissue: gastrocnemius muscle.	Two-dimensional gel electrophoresis, MALDI-ToF MS analyses, and immunoblotting.	Myofibrillar regulatory proteins, signal transduction, cytosolic and mitochondrial energy metabolisms, oxidative stress, detoxification, and RNA metabolism.
Kanski et al., 2005 [32]	34-months-old Fisher 344/Brown Norway F1 hybrid rats. Studied tissue: skeletal muscle.	2-D gel electrophoresis, Western Blot analysis, MALDI-TOF and NSI-MS/MS analysis.	Proteomic analysis of protein nitration.
Dencher et al., 2006 [41]	Studied tissue: skeletal muscle.	Blue-native/colorless-native gel electrophoresis, 2D-SDS-PAGE and MALDI MS.	Cellular dysfunction, ageing, and cellular death.
O'Connell et al., 2007 [37]	3- and 30-months-old rat. Studied tissue: gastrocnemius muscle	Two-dimensional gel electrophoresis, MALDI-ToF, DALT-Twelve gel electrophoretic separation system, 2-D immunoblotting.	Proteomic profiling of senescent fibres: stress response, contractile apparatus, and metabolic regulation.

TABLE 2: *Cont.*

Authors	Experimental model	Proteomic analysis	Identified affected pathways and major findings
Altun et al., 2007 [29]	4- and 30-months old rats.Studied tissue: gastrocnemius muscle.	Two-dimensional gel electrophoresis, MALDI-ToF/ToF, MALDI-MS/MS, ESI-LC-MS/MS and Western Blot analysis.	Redox homeostasis, iron load, regulation of contractile proteins, glycolisis, and oxidative phosphorylation.
O'Connell et al., 2008 [33]	3- and 30-months old rats. Studied tissue: gastrocnemius muscle	Two-dimensional gel electrophoresis, MALDI-ToF MS analysis.	Proteomic profiling of senescent fibers.
Gannon et al., 2008 [30]	3- and -30-months old rats.Studied tissue: gastrocnemius muscle.	Two-dimensional gel electrophoresis, MALDI-ToF MS analysis.	Phosphoproteomic analysis of aged skeletal muscle.
Feng et al., 2008 [42]	12- and 26-months-old Fischer 344 rats. Studied tissue: soleus, semimembranosus, plantaris, extensor digitorum longus, and tibialis anterior muscles.	SDS-polyacrylamide gel electrophoresis, LC-ESI MS/MS analysis and Ingenuity Systems Analysis.	Carbonyl modifications, cellular function and maintenance, fatty acid metabolism, and citrate cycle.
Lombardi et al., 2009 [26]	3- and 24-months-old rats.Studied tissue: gastrocnemius muscle.	Two-dimensional gel electrophoresis, Blue-Native PAGE, and MALDI-ToF MS analysis.	Energy metabolism, mitochondrial pathways, myofibrillar filaments, and detoxification.
Human			
Cobon et al., 2002 [43]	56–79 yr of age.Studied tissue: vastus lateralis muscle.	Two-dimensional polyacrilamide gel electrophoresis and MALDI-TOF MS.	Human aged skeletal muscle protein profile.
Gelfi et al., 2006 [34]	Elderly and young subjects.Studied tissue: vastus lateralis muscle.	Two-dimensional difference gel electrophoresis, SDS-PAGE and ESI-MS/MS.	Elderly group: downregulation of regulatory myosin light chains, (phosphorylated isoforms), higher proportion of myosin heavy chain isoforms 1 and 2A, and enhanced oxidative and reduced glycolytic capacity.

identification both of changes in protein levels and of various posttranslational modifications such as phosphorylation [30], nitration [31, 32], and glycosylation [33] that progress with age. In order to identify novel biomarkers of age-dependent skeletal muscle sarcopenia, mass spectrometry-based proteomics has been applied to the study of global muscle protein expression patterns. Mass spectrometric peptide fingerprinting, chemical peptide sequencing, electrophoretic mobility comparison using international 2-D gel databanks, and/or large-scale immunoblot analysis are among the most frequently utilized techniques.

Over the last years, several proteomic studies have catalogued the accessible skeletal muscle protein complement from various species and also investigated changes in protein expression levels in the sarcopenia of old age. The data obtained so far have furnished databanks that form an important prerequisite for future large-scale proteomic investigations into muscle ageing.

Table 2 lists proteomic studies on age-related changes in skeletal muscle in rodent and human models of ageing. Although the lists of individual proteins found to be affected by the ageing process differ considerably between individual proteomic surveys, the main trend of the altered proteins involved in energy metabolism, cellular signaling, the stress response, cytoskeleton, and contraction shows agreement among the various studies. Gelfi et al. [34] performed a quantitative differential analysis of muscle protein expression in elderly and young subjects using a 2-D DIGE approach. The main difference observed in the elderly group included downregulation of regulatory myosin light chains, particularly the phosphorylated isoforms, a higher proportion of myosin heavy chain isoforms 1 and 2A, and enhanced oxidative and reduced glycolytic capacities.

Proteomic profiling of rodent muscle during ageing has been performed in several studies, resulting in the identification of a large cohort of sarcopenic biomarkers (for a schema, see Figure 2).

Age-dependent differential regulation in rodent muscle has been identified for several glycolytic and mitochondrial enzymes, which are important for energy metabolism. The glycolytic enzymes triosephosphate isomerase, glyoxalase I, and β-enolase were downregulated in aged muscle. Other features indicating perturbation of energy metabolism were downregulation of creatine kinase, of pyruvate kinase, and of the NADH-shuttle glycerol 3-phosphate dehydrogenase. At the mitochondrial level, key enzymes such as isocitrate dehydrogenase, cytochrome c oxidase, ATP synthase β subunit, and pyruvate dehydrogenase were all decreased in ageing muscle whereas there was an upregulation of aldehyde dehydrogenase [26, 29, 35–38].

Differential proteomic analyses have revealed that ageing is associated with perturbations of the myofibrillar network [26, 29, 35–38]. Notably, there is a downregulation of several isoforms of myosin long chain and of alpha-actin, as well as a differential expression of their major regulators. In contrast to the downregulation of myofibrillar components, old muscles display upregulation of many proteins of the intermediate filament, microtubules and microfilament cytoskeleton, among which are B-tubulin, desmin, and gelsolin. This suggests a mechanism affecting the cytoskeleton that compensates for perturbations in myofibrillar structure and so prevents extensive damage to the myofibers. Muscle ageing is also associated with the differential expression of enzymes implicated in the detoxification of cytotoxic products. The cytoplasmic Cu/Zn superoxide dismutase (SOD1) and H ferritin isoform, as well as the levels of glutathione transferase and mitochondrial aldehyde dehydrogenase, were found to be increased in older rats, while evidence of age-associated protein misfolding was provided by the upregulation of molecular chaperones (including HSP 27 and disulfide isomerase ER60) [26, 29, 35–38].

Most of the proteins identified by differential proteomics were previously unrecognized in ageing skeletal muscle. Their identification has not only provided further insight into the potential mechanisms of ageing, but may lead to the development of biomarkers of sarcopenia [26, 29, 35–38].

2.2.3 PROTEOMIC ANALYSIS PERTAINING TO AGEING SKELETAL MUSCLE: ANALYSES OF PROTEIN PHOSPHORYLATION, NITRATION, AND GLYCOSYLATION

Since posttranslational modifications are key modulators of protein structure, function, signaling, and regulation, various subdisciplines of proteomics have emerged that focus on the cataloguing and functional characterization of proteins with extensively modified side chains [57]. In aged skeletal muscle, proteins undergo considerable changes in their posttranslational modifications [58]. These include, among others, phosphorylation, nitration, and glycosylation. Phosphorylation represents one of the most frequent peptide modifications [59], and abnormal phosphorylation is associated with various pathologies. A recent phosphoproteomic survey of aged muscle detected increased phosphorylation levels for myosin light chain 2, tropomyosin α, lactate dehydrogenase, desmin, actin, albumin, and aconitase [30]. In contrast, decreased phospho-specific dye binding was observed for cytochrome c oxidase, creatine kinase, and enolase. Thus, ageing-induced alterations in phosphoproteins appear to involve the contractile machinery and the cytoskeleton, as well as cytosolic and mitochondrial metabolism.

The nitration of protein tyrosine residues represents an oxidative and important posttranslational modification occurring under nitrative/oxidative stress during biological ageing. Comprehensive proteomic studies have identified an age-related increase in the nitration of numerous skeletal muscle proteins. These include enolase, aldolase, creatine kinase, tropomyosin, glyceraldehyde-3-phosphate dehydrogenase, myosin light chain, pyruvate kinase, actinin, actin, and the ryanodine receptor [31, 32]. The nitration of these essential muscle proteins may therefore be a significant causative factor in the age-related decline in muscle strength [31, 32].

Glycosylation is one of the most frequent posttranslational modifications found in proteins, and it plays a central role in cellular mechanisms in both health and disease [60]. Oligosaccharide attachment represents a common protein modification that influences the folding of the nascent peptide chain and the stability of glycoproteins, modifies enzymatic activity, controls protein-secretion events, presents critical information about the cellular targeting of a newly synthesized protein, and provides specific recognition motifs for other proteins in cell-cell interactions [61]. The identified muscle components belong mostly to the family of metabolic enzymes. They included glycolytic enzymes, such as pyruvate kinase, enolase, phosphoglycerate kinase, aldolase, glyceraldehyde-3-phosphate dehydrogenase, and phosphoglycerate-mutase, aconitase, carbonic anhydrase, and creatine kinase [33].

These data confirm that the sarcopenia of old age represents a complex neuromuscular pathology that is associated with drastic changes not only in the abundance, but also in the structure of key muscle proteins (Figure 2).

2.2.4 PROTEOMIC ANALYSIS PERTAINING TO AGEING SKELETAL MUSCLE MITOCHONDRIA

Analysis of the protein profile of mitochondria, and of the changes in it that occur with age, represents a promising approach to the unraveling of the mechanisms involved in ageing. Although the role of mitochondria was long thought to be restricted to an influence on fuel metabolism, the importance of the activity of these organelles has recently been extended to the regulation of developmental/ageing processes [62]. Mitochondria have their own genome (mt-DNA) and specific mechanisms for replication, transcription, and protein synthesis. However, in terms of protein composition they are "hybrid" organelles resulting from the coordinated expression of the nuclear and their own genome. A bidirectional flow of information allows the two kinds of subcellular compartments to communicate with each other under the control of metabolic signals and several signal-transduction pathways that function across the cell. These pathways are differentially regulated by environmental and developmental signals, and under patho/physiological conditions, they allow tissues to adjust their energy production according to different energy demands possibly modulating/altering the mitochondrial phenotype. It is now beyond doubt that mitochondria are severely affected by ageing, and it is generally believed that dysfunctions of mitochondria trigger key steps in the ageing process [62].

Mitochondrial proteomes (mitoproteomes) are currently under vigorous investigation by way of both structural and comparative proteomics. In particular, we would like to emphasize the value of comparative proteomics as a tool capable of providing us with valuable information on mitochondrial physiology and on the role of these organelles in ageing muscle. First, mitochondria can be highly purified, leading to simplified 2D gels, which greatly facilitates the analysis and detection of less-abundant proteins. Second, mitochondrial proteins are generally distributed across wide ranges of both pH and molecular mass on 2D gels, leading to accurate protein resolution with only a few protein-spot overlaps. Third, most of the mitochondrial membrane-protein complexes exhibit soluble subunits that can be analyzed on 2D gels even though the hydrophobic subunits aggregate. Various detection methods are already available that allow us to monitor quantitative changes in the proteome. Of these, 2-DE-based methods appear quite promising, with isoelectric focusing (IEF), BN-SDS, and benzyldimethyl-nhexadecylammonium chloride (16-BAC)-PAGE at the forefront. However, application of IEF is restricted to proteins that are not highly hydrophobic or have no extreme isoelectric points. Indeed, by the use of classical 2D-E it is difficult to detect very acidic or very basic proteins or to distinguish small changes in the expressions of weakly expressed proteins. On the other hand, BN-SDS-PAGE deals efficiently with even hydrophobic membrane proteins, although some compromises in resolution have to be made [63]. Another advantage of the BN-PAGE system is the conservation of protein-protein interactions, enabling simultaneous elucidation of multimeric and multiprotein assemblies of soluble and membrane proteins [64]. Such a procedure might be a viable alternative to other methods, such as yeast two-hybridization.

Comparative transcriptomic and proteomic studies have been initiated to determine global changes in mitochondria from young versus aged skeletal muscle [26, 39, 41, 62, 65–68]. Native-difference gel electrophoresis (DIGE) is an approach that facilitates sensitive quantitative assessment of changes in membrane and soluble proteins. Recently, O'Connell et al. [68] analyzed the mitochondria-rich fraction from aged rat skeletal muscle by DIGE. This proteomic analysis showed a clear age-related increase in key mitochondrial proteins, such as NADH dehydrogenase, the mitochondrial inner membrane protein mitofilin, peroxiredoxin isoform PRX-III, ATPase synthase, succinate dehydrogenase, mitochondrial fission protein Fis1, succinate-coenzyme A ligase, acyl-coenzyme A dehydrogenase, porin isoform VDAC2, ubiquinol-cytochrome c reductase core I protein, and prohibitin [68].

To gain deeper insights both into ageing mechanisms and into the resulting mitoproteome alterations, mitochondria have been studied by the blue-native gel approach, both with respect to protein abundance and the supramolecular organization of OXPHOS complexes [26].

The profiles obtained for muscle crude mitochondria from young and old rats—after detergent extraction with either dodecylmaltoside or digitonin, and subsequent BN-PAGE—have been reported by us within the past years [26]. The use of dodecylmaltoside allows individual resolution of the respiratory complexes. Our densitometric analysis revealed that gastrocnemius muscle mitochondria from old rats, versus those from young rats, contained significantly lower amounts of complex I (NADH:ubiquinone oxidoreductase), V (FoF1-ATP synthase), and III (ubiquinol:cytochrome c oxidoreductase) (−35%, −40%, −25%, resp.). The same mitochondria, on the other hand, contained a significantly larger amount of complex II (succinate: ubiquinone oxidoreductase) (+25%) and an unchanged amount of complex IV (cytochrome c oxidase, COX). The use of a combination of BN-PAGE and catalytic staining allowed us to detect reduced activity of all the complexes in ageing muscle. The observed reductions in the activities of respiratory complexes I, III, and V reflected lower protein levels, but the reduction in complex II activity was associated with an increase in the amount of the same complex. To elucidate whether the ageing process also alters the functional/structural organization of the respiratory chain in terms of the assembly of supercomplexes, mitochondria were extracted using the mild detergent digitonin since this extensively retains inner mitochondrial membrane supercomplexes [69]. In both young and old mitochondria, monomeric complex I and dimeric complex III were significantly reduced versus dodecylmaltoside solubilization. However, the missing amounts were found to be assembled in two major supercomplexes, a and b, and two minor ones, c and d, all within the molecular mass range of 1500–2100 kDA. The supercomplex profile of the older rats was significantly modified, band a being less represented in the profile than the heavier supercomplexes, such as bands c and d. A significant increase was detected in the supramolecular assembly of respiratory chain complexes into respirosomes (each one being formed by complex I assembled with a dimeric complex III and a variable copy number of complex IV, represented by bands c and d). Possibly, this could be a compensatory mechanism that, in ageing muscle, is functionally directed towards substrate channeling and catalytic enhancement advantaging. Indeed, mitochondrial oxidative phosphorylation seems to be more efficient in aged than in young skeletal muscle, since old rats exhibited an increased respiratory control ratio that was attributed principally to a reduction in the reactions able to dissipate the proton motive force not associated with ATP synthesis. This could be interpreted as a compensation for the reduced level and activity of F1F0-ATP synthase.

The above data point up the ability of skeletal muscle to face the consequences of ageing in a metabolically economic way and highlight the occurrence of structural and metabolic adaptations. A comparison between these two studies [26, 68] each employing different proteomic approaches leads to the conclusion that beyond the expression/abundance changes in proteins, an insight can be obtained about the structural and functional heterogeneity in a given mitoproteome.

Another possible protein modification in skeletal muscle mitochondria, possibly contributing to its functional decline with age, is carbonylation, which can be considered an oxidative modification that may render a protein more prone to degradation. Feng et al. [42] recently identified mitochondrial proteins that were susceptible to carbonylation in a manner that was dependent on muscle type (slow- versus fast-twitch) and on age. Carbonylated mitochondrial proteins were more abundant in fast-twitch than in slow-twitch muscle. Twenty-two proteins displayed significant changes in carbonylation state with age, the majority of these increasing in their amount of carbonylation. Ingenuity pathway analysis revealed that these proteins belong to various functional classes and pathways, including cellular function and maintenance, fatty acid metabolism, and the citrate cycle. This study [42] provides a unique catalogue of promising protein targets deserving further investigation because of their potential role in the decline exhibited by ageing muscle. Since carbonylation is not repairable, this modification may, however, be of special importance in directing the affected protein to the path toward degradation.

Of note, in view of the importance of the functional mitochondrial membrane compartmentalization, together with proteomic approaches, lipidomic ones would be desirable to gain further insight into the understanding of the modification of lipids either as membrane components or energy store following aging processes.

2.3 THYROID HORMONES AND THYROID STATES

Thyroid hormones [THs; 3,5,3',5'-tetraiodo-L-thyronine, otherwise known as thyroxine (T4), and 3,5,-triiodo-L-thyronine (T3)] are essential for the regulation both of energy metabolism and of development and growth in all vertebrates. In humans, the early developmental role of THs is vividly illustrated by the distinctive clinical features of cretinism, as observed in iodine-deficient areas. In adults, the primary effects of THs are manifested by alterations in metabolism. Even subclinical hyper- and hypothyroidism can have important consequences, such as atherosclerosis, obesity, and alterations in bone mineral density and heart rate [70, 71]. The effects induced by THs in the regulation of metabolism include changes in oxygen consumption and in protein, carbohydrate, lipid, and vitamin metabolism. Hyperthyroidism is associated with an increase (calorigenic effect), while hypothyroidism is associated with a decrease in metabolic rate. Of particular note, is that the number and the complexity of the clinical features of hyperthyroidism and hypothyroidism emphasize the pleiotropic effects of THs on many different pathways and target organs. Although great efforts have been made to elucidate the signaling pathways underlying the physiopathological effects of thyroid hormones, the network of factors and cellular events involved, as well as the possible role of derivatives of THs, is complicated and incompletely understood, as is the ultimate effect of THs on tissue transcriptomes and proteomes.

2.3.1 THE COMPLEXITY OF ACTION OF THYROID HORMONES: AN OVERVIEW

Most thyroid-related, direct genomic actions leading to protein changes appear to be attributable to T3. The mechanism of action that has gained general acceptance for this iodothyronine involves the binding of specific nuclear receptors (TRs) to thyroid hormone response elements (TREs) in target genes [72]. Within the nucleus, TRs dimers (hetero- or homodimers) bind to TREs and modulate gene activity by either silencing or activating transcription by recruitment of either corepressor or coactivator complexes, depending on the absence or presence of thyroid hormone [73–78]. In mammals, two genes encoding TRs have been characterized, c-erb Aα and c-erb Aβ [79–81], and these encode several proteins (α and β isoforms) with different binding properties and patterns of tissue expression. For example, c-Erb Aβ1 is expressed across a wide range of tissues, while c-ErbAβ2 is found almost exclusively in the pituitary, where it inhibits thyrotrophin (TSH) α- and β-subunit gene transcription [82] by binding to negative TREs present on these genes [83, 84]. New information on the mechanisms of action of THs have been obtained from TR gene knockout (KO) and knock-in studies [85].

In terms of cellular effects, theories proposed so far to explain the actions of THs on metabolic rate also include mechanisms such as: uncoupling of oxidative phosphorylation, stimulation of energy expenditure by activation of Na$^+$-K$^+$ ATPase activity, and direct modulation by THs of transporters and enzymes located within the plasma membrane and mitochondria [86–89]. Moreover, T3-mediated nuclear gene expression leads in turn to coordinated and synergistic effects on the mitochondrial genome [90]. Actually, it has been postulated that T3's actions on this genome are achieved through both an induction of nuclear-encoded mitochondrial factors and a direct binding of T3 to specific ligand-dependent mitochondrial transcription factors [90–94]. These last are nuclear-receptor homologs and are thought to act on a number of mt-DNA response elements [95]. Indeed, T3 directly regulates the mitochondrial genes encoding ATPase subunit six [96], NADH dehydrogenase subunit three [97], and subunits of cytochrome-c-oxidase [98].

The complexity of action of T3 is broadened by the existence of nongenomic or TRE-independent actions, which have been extensively described and are now accepted [99]. Importantly, these can be either independent or dependent on the binding of T3 to TRs. They occur rapidly and are unaffected by inhibitors of transcription and protein synthesis [90, 93, 100–102]. Nongenomic actions of thyroid hormones have been described at the plasma membrane, in the cytoplasm, and within cellular organelles ([100] and references therein). These actions include modulations of Na$^+$, K$^+$, Ca^{2+}, and glucose transport, activations of PKC, PKA, and ERK/MAPK, and regulation of phospholipid metabolism via activations of PLC and PLD [103], and they can be independent of the presence of nuclear TRs and mediated even by TH analogs [102]. For example, it has recently been shown that cytosolic TRβ can interact with the p85 subunit of PI3K and thereby activate the PI3K-Akt/PKB signaling cascade [99, 104]. Moreover, it has been shown that THs activate the MAPK cascade and stimulate angiogenesis via their binding to integrin α Vβ3 [100]. Importantly, it appears now to be well established that an interplay exists between the genomic and nongenomic actions when gene expression is regulated by the TR-T3 complex and the activity of the enzyme is modulated by a nongenomic process [100].

2.3.2 TRANSCRIPTOMIC ANALYSIS PERTAINING TO THE ACTIONS OF THYROID HORMONES

Although the molecular actions of THs have been thoroughly studied, their pleiotropic effects are not well understood and appear to be mediated by complex changes in the expressions of numerous, but still largely unknown, target genes in various tissues. DNA microarrays have been successfully used to identify T3-target genes in mouse, rat, and human tissues, cell lines, and tumors. Actually, pioneering systematic studies in the search for T3-target genes were performed by Seelig and coworkers as long ago as 1981 [105].

Feng et al. [44] first applied cDNA-microarray technology to the study of the in vivo T3 regulation of hepatic genes in the mouse. They identified new T3-target genes, the majority of which had not previously been reported to be regulated by the hormone. Surprisingly, many of these target genes were negatively regulated. The identity of the genes indicated that multiple cellular pathways are actually affected by T3, including glycogenolysis, gluconeogenesis, lipogenesis, cell proliferation, apoptosis, the action of insulin, immunogenicity, and protein glycosylation.

Weitzel et al. [49] detected novel T3-target genes and identified two T3-mediated gene-expression patterns after the administration of T3 to hypothyroid rats. In line with the long-known observation that T3 has profound influences over mitochondrial biogenesis and metabolic balance, the authors reported that numerous genes implicated in metabolic pathways (ANT2, apolipoprotein AIV, HMG-CoA synthase, and ATP synthase β subunit) are affected by T3, as also are genes associated with a wide variety of cellular pathways (encompassing translation, protein turnover, cell structure, and apoptosis-associated proteins). These observations gave support to the idea that alongside the "classical" pathway of T3-mediated gene regulation (involving thyroid hormone-receptor binding to TREs), there appears to be an additional pathway mediated by transcription factors (such as NRF-1 and PPARγ) and coactivators (such as the PGC-1 family of coactivators).

Flores-Morales et al. [45] verified the effect of T3 on liver in mice with a targeted mutation in the TR gene. In accordance with the results of Weitzel [49], they reported that T3 regulates the expressions of functionally different sets of genes in temporally distinct ways. Importantly, using TRβ$^{-/-}$ animals they also defined a number of T3-responsive genes that are dependent on TRβ in vivo, thereby opening the way for the use of similar experimental strategies to identify the contributions made by specific transcription factors to the in vivo actions of multiple hormones and trophic factors.

Miller et al. [50] identified genes involved in glucose metabolism, biosynthesis, transcriptional regulation, protein degradation, and detoxification that were associated with T3-induced cell proliferation. Of particular significance were the findings that T3 rapidly suppresses the expressions of key regulators of the Wnt signaling pathway and that it suppresses the transcriptional downstream elements of the β-catenin-T-cell factor complex.

With the aim of defining the molecular basis of the target-tissue phenotype related to the hereditary TR mutations causing resistance to thyroid hormone (RTH), Miller et al. [46] showed that in T3-target tissues such as cerebellum, heart, and WAT in animal models of both RTH and hyperthyroidism, T3 acts primarily to suppress gene expression, and that TRβ has a greater modulating effect in the heart than originally thought. Moreover, their comprehensive multitissue gene-expression analysis uncovered complex multiple signaling pathways mediating the molecular actions of TRβ mutants in vivo. It also revealed some T3-independent, but mutant-dependent, genomic responses contributing to those "changes-of-function" present in TRβ mutants that are linked to the pathogenesis of RTH.

Dong et al. [48] studied the molecular mechanisms involved in the responses shown by developing mice to disruptions in maternal thyroid-hormone homeostasis. Among differentially expressed genes, Nr4a1 (nuclear receptor subfamily 4, group A, member 1), was upregulated by 3-fold in the hypothyroid juvenile mouse liver, while treatment of HepG2 cells with T3 resulted in its downregulation. A potential thyroid response element −1218 to −1188 bp upstream of the promoter region of Nr4a1 was identified and demonstrated to bind TRα and TRβ receptors.

Notably, in recent years microarray approaches have been used to characterize the effects of T3 on gene expression profiles in the postnatal developing brain as well as in the adult mouse/rat brain [106–108].

The effects of THs on gene expression profiles have been studied less intensively in human tissues than in animal above all because of the poor availability and accessibility of tissue. However, both in vitro [51, 52] and in vivo [53, 109] studies have been performed. Viguerie et al. [51], who showed that T3 regulates a large repertoire of genes in human adipocytes, provided support for the effect of T3 on catecholamine-induced lipolysis, and suggested downregulation of SREBP1c as a link between hyperthyroidism and insulin resistance. Moreover, in accordance with other array studies, the data showed that thyroid hormone can affect cellular processes such as signal transduction, apoptosis, and inflammatory responses. Moeller et al. [52] identified 91 T3-upregulated and 5 T3-downregulated genes in skin fibroblasts from normal humans. Some of these genes were not previously known to be induced by T3, namely aldo-keto reductase family 1 C1-3, collagen type VI alpha 3, member RAS oncogene family brain antigen RAB3B, platelet phosphofructokinase, hypoxia-inducible factor-1 alpha, and enolase 1 alpha. Importantly, these genes have a variety of regulatory functions in both development and metabolism.

Clèment et al. [53] studied the effects of thyroid hormone on human skeletal muscle in vivo. Their data not only helped to explain the effects of T3 on protein turnover and energy metabolism, but also revealed new putative mechanisms extending beyond the classic metabolic effects of the hormone, and importantly, added to our understanding of the permissive effects of T3 on several cellular events (such as signal-transduction cascades, intracellular transport, and tissue remodeling).

TABLE 3: Summary of the models used and of the major findings obtained by applying microarray technologies to the study of THs effects on relevant metabolically active tissues.

Authors	Experimental model	Treatments	Number of genes	Number of genes affected by T3	Identified affected pathways and major findings
		Mouse (in vivo and in vitro studies)			
Feng et al., 2000 [44]	Six-week-old mice. Studied tissue: liver.	Hypothyroidism induced by low-iodine diet supplemented with 0.15% PTU for 4 weeks, then hyperthyroidism induced by single i.p. injection of L-T3 or T4 100 g/100 g body weight.	2225	Of 55 genes identified as target of T3, 41 were negatively regulated.	Glycogenolysis, gluconeogenesis, lipogenesis, proliferation, apoptosis, insulin signaling, immunogenity, and protein glycosylation.
Flores-Morales et al., 2002 [45]	2 to 3.5-months old (WT and TR) male mice. Studied tissue: liver.	At the onset of the experiment, all groups of animals were provided a low-iodine diet for 14 d to accustom them to the synthetic chow. Hypothyroidism was then induced by inclusion of 0.05% methimazole and 1% potassium perchlorate in the drinking water for 21 d while still on the low-iodine diet. From d 35, one group of animals was injected daily with 5 g T3 for an additional 5-d period to induce hyperthyroidism. Another group was injected with 5 g T3 and 5 g T4; on d 35, the animals in this group were killed 2 hours after the T3/T4 injection.	4000	T3 found to regulate more than 200 genes, more than 100 of which were not previously described. 60% of these genes showed dependence on the TR gene for T3 regulation.	Rapid or transient effects of T3 on lipogenic genes. Long-term effects of T3 on genes for the mitochondrial respiratory chain transcription factors and protein turnover.
Miller et al., 2004 [46]	8- to 10-week-old (TR and TR) male siblings (mice).Studied tissues: cerebellum, heart, and white adipose.	TR mice contain a cytosine insertion in exon 10 of the TR1 gene at nucleotide position 1,642 of the TR1 cDNA that leads to a frameshift of the carboxy- terminal 14 amino acids of TR1. For 7 days, T3 (5 g/mouse/day) was administered by i.p. injection.	11500	163 genes responsive to T3 treatment and 187 genes differentially expressed between TR mice and wild-type littermates.	T3 primarily acts to repress gene expression. TR has a powerful modulating effect in the heart. Novel physiologic candidates for T3 action are changes in immune-gene expression and in the induction of antiproliferative genes. The relative levels of TR isoforms lead to dramatic differential effects on gene expression.

TABLE 3: *Cont.*

Authors	Experimental model	Treatments	Number of genes	Number of genes affected by T3	Identified affected pathways and major findings
Ventura-Holman et al., 2007 [47]	Murine non-transfected hepatocyte cell line AML 12, expressing endogenous TRs.	RNA obtained from cells incubated for 3 hours or 24 hours 10nM T3, in the presence of 10% stripped fetal bovine serum. Cells also incubated in the presence of cyclohexi-mide (10 ig/ml) 10 nM T3 for 3 hours to discriminate between primary and secondary responses.	15000	12 genes upregulated and 5 genes downregulated in the presence of T3.	Novel T3 responsive genes were identified. Insights were obtained into the role of T3 in processes such as cholesterol metabolism, bile acid secretion, and oncogenesis.
Dong et al., 2007 [48]	Hypothyroid juvenile mice. Studied tissue: liver	Gene expression analyzed in livers of mice rendered hypothyroid by treating pregnant mice from gestational d 13 to postnatal d 15 with 6-propyl-2-thiouracil in drinking water.	approximately 20000	96 differentially expressed genes were identified. Of these, 72 genes encode proteins of known function, 15 of which had previously been identified as regulated by TH.	Metabolism, development, cell proliferation, apoptosis, and signal transduction. A potential thyroid response element 1218 to 1188 bp upstream of the promoter region of Nr4a1 was identified and demonstrated to bind TH receptor TR and TR.
Rat (in vivo and in vitro studies)					
Weitzel et al., 2001 [49]	Adult male Wistar rats.Studied tissue: liver.	Hypothyroidism induced by i.p. injection of Na131I (250 Ci/100 g body weight) 28 days before the experiments. Hyperthyroidism provoked by i.p. injection of T3 (50 g/100 g body weight); repeated after 24 hours. Rats killed at 0, 6, 24, and 48 hours after thyroid hormone.	4608	Sixty-two of the genes were reproducibly T3-responsive.	Beside the "classical" pathway of T3-mediated gene regulation by TRs binding to TREs, an additional pathway appears to be mediated by transcription factors like NRF-1 and PPAR and coactivators (like the PGC-1 family of coactivators).

TABLE 3: *Cont.*

Authors	Experimental model	Treatments	Number of genes	Number of genes affected by T3	Identified affected pathways and major findings
Miller et al., 2001 [50]	GC cells (rat pituitary cell-line expressing functional TRs).	Cells were incubated without or with T3 (100 nM) for 1, 3, 6, 12, 24, or 72 hours. At each time-point, cells were harvested for total RNA preparation.	4400	358 responsive genes were identified. 88% had not previously been reported to be modulated by T3. A few genes showed biphasic expression patterns. In total, 203 genes were upregulated and the remainder were downregulated by T3.	Glucose metabolism, biosynthesis, transcriptional regulation, protein degradation, and detoxification in T3-induced cell proliferation.
Human (in vivo and in vitro studies)					
Viguerie and Langin, 2003 [51]	Human adipose tissue obtained from the s.c. abdominal fat depots of Caucasian women for cDNA array and RT competitive PCR experiments.	Surgical adipose tissue samples were dissected from skin and vessels, and cultured adipocytes were obtained. Cultures were treated with T3 (100 nm) or vehicle for 24 hours. Medium-free T3 concentration was measured at 1 and 24 hours after addition of T3 (using RIA kits).	1 176	Among the statistically significant changes in mRNA levels of more than 1.3-fold, 13 and 6 genes were positively or negatively regulated, respectively.	Signal transduction, lipid metabolism, apoptosis, and inflammatory responses.

TABLE 3: *Cont.*

Authors	Experimental model	Treatments	Number of genes	Number of genes affected by T3	Identified affected pathways and major findings
Moeller et al., 2005 [52]	Skin fibroblasts of normal individuals.	Human skin was obtained by punch biopsy from three normal individuals and two patients with RTH. Fibroblasts were grown in supplemented with 10% bovine calf serum (BCS). At confluency, the medium was replaced with one containing TH-depleted BCS (TxBCS), obtained from a thyroidectomized calf. For microarrays, incubation with T3 was for 24 hours.	more than 15000	Microarray analysis identified 148 genes induced by 1.4-fold or more and five genes repressed to 0.7 or less 24 hours after treatment with 2 10-9 M T3. Taking into account duplicate genes, these represented 91 up-regulated and five downregulated genes, respectively.	Aldo-keto reductase family 1 C1-3, collagen type VI 3, member RAS oncogene family brain antigen RAB3B, platelet phosphofructokinase, hypoxia-inducible factor-1, and enolase 1 genes, previously known to be induced by TH, were identified and validated. These genes have a variety of regulatory functions in development and metabolism.
Clement et al., 2002 [53]	Healthy male Caucasian volunteers (22–33 years old). The same investigations were performed on day 0 and day 14. Studied tissue: vastus lateralis muscle by percutaneous biopsies.	Participants took one tablet of 25 g of T3 three times a day (75 g per day) for 14 days.	24000	A transcriptional profile of 383 genes regulated by T3 was obtained (381 were upregulated and only two downregulated).	Novel target genes for T3 were identified. They belong to functional classes including transcriptional control, mRNA maturation, protein turnover, signal transduction, cellular trafficking, and energy metabolism.

TABLE 3: *Cont.*

Authors	Experimental model	Treatments	Number of genes	Number of genes affected by T3	Identified affected pathways and major findings
Visser et al., 2009 [54]	Thyroidectomized patients treated for differentiated thyroid carcinoma (DTC) off and on L-thyroxine replacement. Studied tissue: skeletal muscle	Included were patients who had been diagnosed with DTC and had received initial therapy consisting of near-total thyroidectomy and radioiodine ablation therapy. Four weeks after L-thyroxine withdrawal and 8 wk after subsequent L-thyroxine replacement, patients were admitted to the clinical research unit. A catheter was inserted in a dorsal hand vein to collect blood samples for measurement of serum TSH, free T4 (fT4), and T3. Muscle biopsies were taken from the quadriceps muscle (vastus lateralis).	54674	607 differentially expressed genes on L-thyroxine treatment were identified, of which approximately 60% were positively and approximately 40% were negatively regulated.	New genes associated with thyroid state and involved in energy and fuel metabolism were overrepresented among the up-regulated genes. L-thyroxine therapy induced a large downregulation of the primary transcripts of the noncoding microRNA pair miR-206/miR-133b.

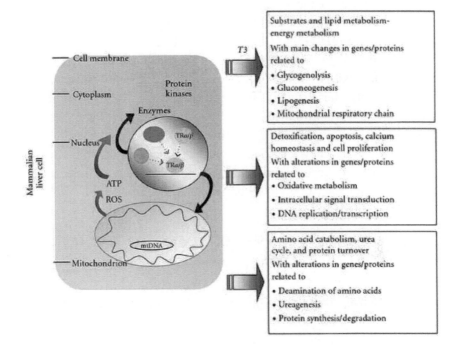

FIGURE 3: Overview of the main T3-induced transcriptomic and proteomic alterations occurring in mammalian liver. Schematic representation of the alterations in gene/protein expression underlying the response of liver to T3. Schematized are the main events and mechanisms underlying the actions of T3. Summarized are data obtained in cDNA microarray/proteomic-based studies in various mammalian models (mouse, rat, and human) (for details, see text and Tables 3 and 4).

Very recently, Visser et al. [54] examined the skeletal muscle transcriptome in thyroidectomized patients being treated for differentiated thyroid carcinoma, and compared it between those who were off or on L-thyroxine replacement. They reported for the first time that in humans as in animals, a large proportion of muscle genes (~43%) is significantly downregulated by L-thyroxine treatment. They also reported significant regulation of the primary transcripts of the noncoding RNAs miR-206 and miR-133b, which are key regulators in muscle differentiation and proliferation and may affect numerous target genes. The potential of T3 to regulate miRs may be of particular importance since this level of control would add an additional layer of complexity by which T3 may regulate cellular processes.

Collectively, these studies (summarized in Table 3) have provided a cornucopia of novel information (schematized in Figures 2 and 3) on the regulation of transcription by THs. However, the intrinsic nature of these studies means that they provide no information concerning the status of the corresponding encoded proteins, and this is particularly relevant because of the influence of thyroid hormone on protein half-life.

2.3.3 PROTEOMIC ANALYSIS PERTAINING TO THE ACTIONS OF THYROID HORMONES

As stated above, overall T3 signaling can be modulated at many levels (i.e., the thyroid hormone-receptor isoforms present in the tissue, the DNA-response element in the regulated gene, the availability of receptor-binding partners, interactions with coactivators and corepressors, ligand availability, mRNA and protein stabilities, protein translocation, and metabolic interference) [72, 90–93]. Consequently, for a deeper investigation of the biological events modulated by T3 within target organs, a systematic analysis of the T3-induced changes in protein profile would appear to be appropriate.

We recently performed, on samples taken from rats in different thyroid states, high-resolution differential proteomic analysis, combining 2D-E and subsequent MALDI-ToF MS [55, 56]. These studies (summarized in Table 4) were the first application of proteomic technology to the study of the modulations that T3 exerts in vivo over tissue proteins, and they provided the first systematic identification of T3-induced changes in the protein expression profiles of rat liver and skeletal muscle. In the liver, we unambiguously identified 14 differentially expressed proteins involved in substrate and lipid metabolism, energy metabolism, detoxification of cytotoxic products, calcium homeostasis, amino acid catabolism, and the urea cycle [55]. We found that T3 treatment affected the expressions of enzymes such as mitochondrial aldehyde dehydrogenase, α-enolase, sorbitol dehydrogenase, acyl-CoA dehydrogenase, 3-ketoacyl-CoA thiolase, and 3-hydroxyanthranilate 3,4-dioxygenase. Interestingly, the first two enzymes were upregulated, while the others were downregulated.

Our data were in accordance with the reported role played by thyroid hormone in the stimulation of the rate of ethanol elimination [110], and they provided further insight into the mechanisms actuated by T3 in that pathway. T3 is known to stimulate gluconeogenesis and glucose production in the liver, thereby opposing the action of insulin on hepatic glucose production [111]. Our results extended this knowledge by showing that T3 significantly enhances the level of α-enolase, thereby participating in glycolysis and gluconeogenesis. In addition, T3 administration induced a significant increase in the hepatic ATP synthase α-chain content (in accordance with the ability of T3 to stimulate ATP synthesis) and concomitantly reduced the expression level of electron transfer flavoprotein α-subunit (-ETF), and also that of the acyl-CoA dehydrogenases [112]. T3 treatment is associated with significant reductions in the expression levels of both peroxisomal catalase and cytoplasmic glutathione-S-transferase [55], the former being important in the protection of cells against the toxic effects of hydrogen peroxide while the latter is implicated in the cellular detoxification of a number of xenobiotics by means of their conjugation to reduced glutathione. T3 treatment of hypothyroid rats is also associated with a selective upregulation of HSP60, a molecular chaperone [113]. SMP30, also known as regucalcin, which was previously not known to be affected by T3, has now been identified as a T3 target [55]. This opens new perspectives in our understanding of the molecular pathways related to intracellular T3-dependent signaling, raising the possibility that T3 may modulate a plethora of cellular events while also acting on multifunctional proteins such as SMP30, which in turn is able to modulate the levels of second messengers such as calcium.

TABLE 4: Summary of the models used and of the major findings obtained by applying 2D-E and MS to the study of THs effects.

Authors	Experimental model	Treatments	Number of protein spots analyzed	Number of identified proteins affected by T3	Identified affected pathways and major findings
			Rat		
Silvestri et al., 2006 [55]	3-months-old male Wistar rats. Studied tissue: liver.	Hypothyroidism was induced by i.p. administration of PTU (1 mg/100 g BW) for 4 weeks together with a weekly i.p. injection of IOP (6 mg/100 g BW). T3 was chronically administered by giving seven daily i.p. injections of 15 g T3/100 g BW to hypothyroid rats, while the control euthyroid and hypothyroid rats received saline injections.	600	14	The whole cell protein content of rat liver was analyzed following T3 administration. Identified proteins were involved in substrates and lipid metabolism, energy metabolism, detoxification of cytotoxic products, calcium homeostasis, amino acid catabolism, and the urea cycle.
Silvestri et al., 2007 [56]	3-months-old male Wistar rats. Studied tissue: skeletal muscle.	Hypothyroidism and hyperthyroidism were induced as above (Silvestri et al 2006 [55]).	220	20	The whole-cell protein content of gastrocnemius muscles was analyzed. The differentially expressed proteins unambiguously identified were involved in substrates and energy metabolism, stress response, cell structure, and gene expression.

T3-treated rats exhibit significant reductions in the protein levels of both ornithine carbamoyltransferase and arginase 1 [55]. These data are in accordance with a previous report [114], and in line with the idea that in the hypothyroid state there are decreases in protein synthesis and turnover.

Concerning skeletal muscle, the whole-cell protein content of gastrocnemius muscle has been analyzed, and twenty differentially expressed proteins among euthyroid, hypothyroid, and hyperthyroid rats have been identified [56]. The largest group of affected proteins (50%) was involved in substrate and energy metabolism, another important group was represented by stress-induced proteins (HSPs) (21.4%), and the remainder were implicated in structural features or gene expression (transcription, translation), each of these two groups representing 14.3% of the identified proteins [56]. The thyroid state was found to induce structural shifts in gastrocnemius muscle, toward a slower phenotype in hypothyroidism and toward a faster phenotype in hyperthyroidism [56].

Among the proteins involved in substrate metabolism, three glycolytic enzymes have been identified, namely, beta-enolase, pyruvate kinase, and triosephosphate isomerase. Beta-enolase protein levels were increased following T3 treatment (hyperthyroidism), while pyruvate kinase and triosephosphate isomerase levels were decreased in hypothyroidism and elevated in hyperthyroidism [56]. This is in accordance with (a) a major T3-dependence on pyruvate kinase and triosephosphate isomerase and a generally decreased metabolic dependence on glycolysis in hypothyroidism, and (b) an increased reliance on glycolysis in hyperthyroidism [115]. Accordingly, hyperthyroidism was found to be associated with an increased expression of cytoplasmic malate dehydrogenase. Moreover, phosphatidylethanolamine-binding protein, a basic protein that shows preferential affinity in vitro for phosphatidylethanolamine, was significantly increased in both the hypo- and hyperthyroid gastrocnemius (versus the euthyroid controls), most likely reflecting a thyroid state-associated cell-remodeling [56].

The expression level of the ATP synthase beta subunit was increased in both hypothyroid and hyperthyroid muscle (versus euthyroid controls), with a slight decrease in hyperthyroid animals versus hypothyroid ones. Cytosolic creatine kinase, on the other hand, was decreased in hypothyroidism versus both euthyroidism and hyperthyroidism, suggesting a decreased dependence of energy metabolism on the creatine kinase shuttle in hypothyroid muscle [56].

The expression level of HSP70 was found to be significantly increased in hypothyroid muscle (versus both euthyroid and hyperthyroid muscle), paralleling the changes in MHCIb [56]. A similar expression pattern was found for HSP20 which, despite not being a heat-inducible HSP, is biologically regulated by several cellular signalling pathways. Also identified was HSP27, which has been demonstrated to play important roles in smooth muscle cells (actin polymerization, remodeling, and even cross-bridge cycling), and which can, moreover, act as a chaperone in the regulation of contractile-protein activation [116] and also combat insulin resistance [117].

Concerning cell structure, in accordance with a predominant expression of MHCIb over MHCIIb in hypothyroidism and a reversal of the ratio between the two fiber-type isoforms after T3 administration, the expression level of myosin regulatory light chain 2, typical of slow-twitch fibers, was strongly increased in hypothyroidism, with hyperthyroidism significantly reducing it (in each case, versus euthyroidism) [56].

Finally, both hypothyroidism and hyperthyroidism induced chromo-domain-helicase-DNA-binding protein 1 (CHD 1), as well as eukaryotic translation initiation factor 3 subunit 10 (IF3A), two proteins that play important roles in different steps of gene expression: (1) initiation of transcription; and (2) initiation of translation [56].

2.4 CONCLUSIONS

In conclusion, although the biochemical and cellular mechanisms that underlie sarcopenia in ageing muscle and the effects elicited by thyroid hormones are only beginning to be elucidated, array-based transcriptomic studies, together with MS-based proteomic ones, are producing new insights into the pathophysiological mechanisms behind such complex phenomena.

As can be seen from the above discussion, the approaches used in the cited studies have allowed the identification of previously unrecognized proteins, thereby increasing our awareness of the large repertoire of proteins, and the multiple cell processes and signaling pathways that are affected by T3 and by ageing (for a schematic representation, see Figures 2 and 3). However, as the majority of the cited studies were performed in vivo, the possibility remains that certain hormones and/or other factors that are affected by such metabolic situations may have been partially responsible for the observed results.

On the basis of what has been achieved so far, the authors feel justified in championing the use of combined transcriptomic and proteomic approaches in living animals for the study of complex physiological, as well as pathophysiological, systems. Such approaches should also prove valuable for drug-design, enabling the agonist and/or antagonist properties of drugs (as well as their side effects) to be characterized on the basis of the changes they induce in protein-expression patterns.

REFERENCES

1. M. Schena, D. Shalon, R. W. Davis, and P. O. Brown, "Quantitative monitoring of gene expression patterns with a complementary DNA microarray," Science, vol. 270, no. 5235, pp. 467–470, 1995.
2. D. Gerhold, T. Rushmore, and C. T. Caskey, "DNA chips: promising toys have become powerful tools," Trends in Biochemical Sciences, vol. 24, no. 5, pp. 168–173, 1999.
3. G. G. Lennon, "High-throughput gene expression analysis for drug discovery," Drug Discovery Today, vol. 5, no. 2, pp. 59–66, 2000.
4. J. E. Celis, M. Østergaard, N. A. Jensen, I. Gromova, H. H. Rasmussen, and P. Gromov, "Human and mouse proteomic databases: novel resources in the protein universe," FEBS Letters, vol. 430, no. 1-2, pp. 64–72, 1998.
5. R. D. Appel, C. Hoogland, A. Bairoch, and D. F. Hochstrasser, "Constructing a 2-D database for the World Wide Web," Methods in Molecular Biology, vol. 112, pp. 411–416, 1999.
6. M. J. Dunn, "Studying heart disease using the proteomic approach," Drug Discovery Today, vol. 5, no. 2, pp. 76–84, 2000.
7. P. G. Righetti, A. Castagna, F. Antonucci et al., "Critical survey of quantitative proteomics in two-dimensional electrophoretic approaches," Journal of Chromatography A, vol. 1051, no. 1-2, pp. 3–17, 2004.
8. A. Vlahou and M. Fountoulakis, "Proteomic approaches in the search for disease biomarkers," Journal of Chromatography B, vol. 814, no. 1, pp. 11–19, 2005.
9. A. A. Vandervoort, "Aging of the human neuromuscular system," Muscle and Nerve, vol. 25, no. 1, pp. 17–25, 2002.
10. L. J. Melton III, S. Khosla, C. S. Crowson, M. K. O'Connor, W. M. O'Fallon, and B. L. Riggs, "Epidemiology of sarcopenia," Journal of the American Geriatrics Society, vol. 48, no. 6, pp. 625–630, 2000.

11. L. J. S. Greenlund and K. S. Nair, "Sarcopenia—consequences, mechanisms, and potential therapies," Mechanisms of Ageing and Development, vol. 124, no. 3, pp. 287–299, 2003.
12. E. Carmeli, R. Coleman, and A. Z. Reznick, "The biochemistry of aging muscle," Experimental Gerontology, vol. 37, no. 4, pp. 477–489, 2002.
13. L. Larsson, "The age-related motor disability: underlying mechanisms in skeletal muscle at the motor unit, cellular and molecular level," Acta Physiologica Scandinavica, vol. 163, no. 3, pp. S27–S29, 1998.
14. D. V. Rao, G. M. Boyle, P. G. Parsons, K. Watson, and G. L. Jones, "Influence of ageing, heat shock treatment and in vivo total antioxidant status on gene-expression profile and protein synthesis in human peripheral lymphocytes," Mechanisms of Ageing and Development, vol. 124, no. 1, pp. 55–69, 2003.
15. D. H. Ly, D. J. Lockhart, R. A. Lerner, and P. G. Schultz, "Mitotic misregulation and human aging," Science, vol. 287, no. 5462, pp. 2486–2492, 2000.
16. K. J. Kyng, A. May, S. Kølvraa, and V. A. Bohr, "Gene expression profiling in Werner syndrome closely resembles that of normal aging," Proceedings of the National Academy of Sciences of the United States of America, vol. 100, no. 21, pp. 12259–12264, 2003.
17. T. Lu, Y. Pan, S.-Y. Kao et al., "Gene regulation and DNA damage in the ageing human brain," Nature, vol. 429, no. 6994, pp. 883–891, 2004.
18. L. Erraji-Benchekroun, M. D. Underwood, V. Arango et al., "Molecular aging in human prefrontal cortex is selective and continuous throughout adult life," Biological Psychiatry, vol. 57, no. 5, pp. 549–558, 2005.
19. S. Welle, A. I. Brooks, J. M. Delehanty et al., "Skeletal muscle gene expression profiles in 20-29 year old and 65-71 year old women," Experimental Gerontology, vol. 39, no. 3, pp. 369–377, 2004.
20. A. B. Csoka, S. B. English, C. P. Simkevich et al., "Genome-scale expression profiling of Hutchinson-Gilford progeria syndrome reveals widespread transcriptional misregulation leading to mesodermal/mesenchymal defects and accelarated atherosclerosis," Aging Cell, vol. 3, no. 4, pp. 235–243, 2004.
21. J. M. Zahn, R. Sonu, H. Vogel et al., "Transcriptional profiling of aging in human muscle reveals a common aging signature," PLoS genetics, vol. 2, no. 7, article e115, 2006.
22. G. E. J. Rodwell, R. Sonu, J. M. Zahn et al., "A transcriptional profile of aging in the human kidney," PLoS Biology, vol. 2, no. 12, article e427, 2004.
23. S. Welle, A. I. Brooks, J. M. Delehanty, N. Needler, and C. A. Thornton, "Gene expression profile of aging in human muscle," Physiological Genomics, vol. 14, pp. 149–159, 2003.
24. J. Tower, "Sex-specific regulation of aging and apoptosis," Mechanisms of Ageing and Development, vol. 127, no. 9, pp. 705–718, 2006.
25. C.-K. Lee, R. G. Klopp, R. Weindruch, and T. A. Prolla, "Gene expression profile of aging and its retardation by caloric restriction," Science, vol. 285, no. 5432, pp. 1390–1393, 1999.
26. A. Lombardi, E. Silvestri, F. Cioffi et al., "Defining the transcriptomic and proteomic profiles of rat ageing skeletal muscle by the use of a cDNA array, 2D- and

Blue native-PAGE approach," Journal of Proteomics, vol. 72, no. 4, pp. 708–721, 2009.

27. T. Kayo, D. B. Allison, R. Weindruch, and T. A. Prolla, "Influences of aging and caloric restriction on the transcriptional profile of skeletal muscle from rhesus monkeys," Proceedings of the National Academy of Sciences of the United States of America, vol. 98, no. 9, pp. 5093–5098, 2001.

28. R. Sreekumar, J. Unnikrishnan, A. Fu, et al., "Effects of caloric restriction on mitochondrial function and gene transcripts in rat muscle," American Journal of Physiology - Endocrinology and Metabolism, vol. 283, no. 1, pp. E38–43, 2002.

29. M. Altun, E. Edström, E. Spooner et al., "Iron load and redox stress in skeletal muscle of aged rats," Muscle and Nerve, vol. 36, no. 2, pp. 223–233, 2007.

30. J. Gannon, L. Staunton, K. O'Connell, P. Doran, and K. Ohlendieck, "Phosphoproteomic analysis of aged skeletal muscle," International Journal of Molecular Medicine, vol. 22, no. 1, pp. 33–42, 2008.

31. J. Kanski, M. A. Alterman, and C. Schöneich, "Proteomic identification of age-dependent protein nitration in rat skeletal muscle," Free Radical Biology and Medicine, vol. 35, no. 10, pp. 1229–1239, 2003.

32. J. Kanski, S. J. Hong, and C. Schöneich, "Proteomic analysis of protein nitration in aging skeletal muscle and identification of nitrotyrosine-containing sequences in vivo by nanoelectrospray ionization tandem mass spectrometry," The Journal of Biological Chemistry, vol. 280, no. 25, pp. 24261–24266, 2005.

33. K. O'Connell, P. Doran, J. Gannon, and K. Ohlendieck, "Lectin-based proteomic profiling of aged skeletal muscle: decreased pyruvate kinase isozyme M1 exhibits drastically increased levels of N-glycosylation," European Journal of Cell Biology, vol. 87, no. 10, pp. 793–805, 2008.

34. C. Gelfi, A. Viganò, M. Ripamonti et al., "The human muscle proteome in aging," Journal of Proteome Research, vol. 5, no. 6, pp. 1344–1353, 2006.

35. D. Cai, M. Li, K. Lee, K. Lee, W. Wong, and K. Chan, "Age-related changes of aqueous protein profiles in rat fast and slow twitch skeletal muscles," Electrophoresis, vol. 21, no. 2, pp. 465–472, 2000.

36. I. Piec, A. Listrat, J. Alliot, C. Chambon, R. G. Taylor, and D. Bechet, "Differential proteome analysis of aging in rat skeletal muscle," The FASEB Journal, vol. 19, no. 9, pp. 1143–1145, 2005.

37. K. O'Connell, J. Gannon, P. Doran, and K. Ohlendieck, "Proteomic profiling reveals a severely perturbed protein expression pattern in aged skeletal muscle," International Journal of Molecular Medicine, vol. 20, no. 2, pp. 145–153, 2007.

38. P. Doran, K. O'Connell, J. Gannon, M. Kavanagh, and K. Ohlendieck, "Opposite pathobiochemical fate of pyruvate kinase and adenylate kinase in aged rat skeletal muscle as revealed by proteomic DIGE analysis," Proteomics, vol. 8, no. 2, pp. 364–377, 2008.

39. J. Chang, H. Van Remmen, J. Cornell, A. Richardson, and W. F. Ward, "Comparative proteomics: characterization of a two-dimensional gel electrophoresis system to study the effect of aging on mitochondrial proteins," Mechanisms of Ageing and Development, vol. 124, no. 1, pp. 33–41, 2003.

40. D. Q. Cai, M. Li, K. K. Lee, et al., "Parvalbumin expression is downregulated in rat fast-twitch skeletal muscles during aging," Archives of Biochemistry and Biophysics, vol. 387, no. 2, pp. 202–208, 2001.

41. N. A. Dencher, S. Goto, N. H. Reifschneider, M. Sugawa, and F. Krause, "Unraveling age-dependent variation of the mitochondrial proteome," Annals of the New York Academy of Sciences, vol. 1067, no. 1, pp. 116–119, 2006.

42. J. Feng, H. Xie, D. L. Meany, L. V. Thompson, E. A. Arriaga, and T. J. Griffin, "Quantitative proteomic profiling of muscle type-dependent and age-dependent protein carbonylation in rat skeletal muscle mitochondria," Journals of Gerontology A, vol. 63, no. 11, pp. 1137–1152, 2008.

43. G. S. Cobon, N. Verrills, P. Papakostopoulos, et al., "Proteomics of ageing," Biogerontology, vol. 3, no. 1-2, pp. 133–136, 2002.

44. X. Feng, Y. Jiang, P. Meltzer, and P. M. Yen, "Thyroid hormone regulation of hepatic genes in vivo detected by complementary DNA microarray," Molecular Endocrinology, vol. 14, no. 7, pp. 947–955, 2000.

45. A. Flores-Morales, H. Gullberg, L. Fernandez et al., "Patterns of liver gene expression governed by TRβ," Molecular Endocrinology, vol. 16, no. 6, pp. 1257–1268, 2002.

46. L. D. Miller, P. McPhie, H. Suzuki, Y. Kato, E. T. Liu, and S. Y. Cheng, "Multi-tissue gene-expression analysis in a mouse model of thyroid hormone resistance," Genome Biology, vol. 5, no. 5, article R31, 2004.

47. T. Ventura-Holman, A. Mamoon, J. F. Maher, et al., "Thyroid hormone responsive genes in the murine hepatocyte cell line AML 12," Gene, vol. 396, no. 2, pp. 332–327, 2007.

48. H. Dong, C. L. Yauk, A. Williams, A. Lee, G. R. Douglas, and M. G. Wade, "Hepatic gene expression changes in hypothyroid juvenile mice: characterization of a novel negative thyroid-responsive element," Endocrinology, vol. 148, no. 8, pp. 3932–3940, 2007.

49. J. M. Weitzel, C. Radtke, and H. J. Seitz, "Two thyroid hormone-mediated gene expression patterns in vivo identified by cDNA expression arrays in rat," Nucleic Acids Research, vol. 29, no. 24, pp. 5148–5155, 2001.

50. L. D. Miller, K. S. Park, Q. M. Guo et al., "Silencing of Wnt signaling and activation of multiple metabolic pathways in response to thyroid hormone-stimulated cell proliferation," Molecular and Cellular Biology, vol. 21, no. 19, pp. 6626–6639, 2001.

51. N. Viguerie and D. Langin, "Effect of thyroid hormone on gene expression," Current Opinion in Clinical Nutrition and Metabolic Care, vol. 6, no. 4, pp. 377–381, 2003.

52. L. C. Moeller, A. M. Dumitrescu, R. L. Walker, P. S. Meltzer, and S. Refetoff, "Thyroid hormone responsive genes in cultured human fibroblasts," Journal of Clinical Endocrinology and Metabolism, vol. 90, no. 2, pp. 936–943, 2005.

53. K. Clément, N. Viguerie, M. Diehn et al., "In vivo regulation of human skeletal muscle gene expression by thyroid hormone," Genome Research, vol. 12, no. 2, pp. 281–291, 2002.

54. W. E. Visser, K. A. Heemstra, S. M. A. Swagemakers et al., "Physiological thyroid hormone levels regulate numerous skeletal muscle transcripts," Journal of Clinical Endocrinology and Metabolism, vol. 94, no. 9, pp. 3487–3496, 2009.

55. E. Silvestri, M. Moreno, L. Schiavo et al., "A proteomics approach to identify pro-tein expression changes in rat liver following administration of 3,5,3′-triiodo-L-thy-ronine," Journal of Proteome Research, vol. 5, no. 9, pp. 2317–2327, 2006.

56. E. Silvestri, L. Burrone, P. de Lange et al., "Thyroid-state influence on protein-ex-pression profile of rat skeletal muscle," Journal of Proteome Research, vol. 6, no. 8, pp. 3187–3196, 2007.

57. M. Mann and O. N. Jensen, "Proteomic analysis of post-translational modifica-tions," Nature Biotechnology, vol. 21, no. 3, pp. 255–261, 2003.

58. C. Schöneich, "Protein modification in aging: an update," Experimental Gerontol-ogy, vol. 41, no. 9, pp. 807–812, 2006.

59. J. V. Olsen, B. Blagoev, F. Gnad et al., "Global, in vivo, and site-specific phosphory-lation dynamics in signaling networks," Cell, vol. 127, no. 3, pp. 635–648, 2006.

60. K. Ohtsubo and J. D. Marth, "Glycosylation in cellular mechanisms of health and disease," Cell, vol. 126, no. 5, pp. 855–867, 2006.

61. H. H. Freeze, "Genetic defects in the human glycome," Nature Reviews Genetics, vol. 7, no. 8, pp. 660–674, 2006.

62. N. A. Dencher, M. Frenzel, N. H. Reifschneider, M. Sugawa, and F. Krause, "Pro-teome alterations in rat mitochondria caused by aging," Annals of the New York Academy of Sciences, vol. 1100, pp. 291–298, 2007.

63. S. Rexroth, J. M. W. Meyer zu Tittingdorf, F. Krause, N. A. Dencher, and H. Seelert, "Thylakoid membrane at altered metabolic state: challenging the forgotten realms of the proteome," Electrophoresis, vol. 24, no. 16, pp. 2814–2823, 2003.

64. F. Krause, "Detection and analysis of protein-protein interactions in organellar and prokaryotic proteomes by native gel electrophoresis: (Membrane) protein complex-es and supercomplexes," Electrophoresis, vol. 27, no. 13, pp. 2759–2781, 2006.

65. V. Pesce, A. Cormio, F. Fracasso et al., "Age-related mitochondrial genotypic and phenotypic alterations in human skeletal muscle," Free Radical Biology and Medi-cine, vol. 30, no. 11, pp. 1223–1233, 2001.

66. D. Dani and N. A. Dencher, "Native-DIGE: a new look at the mitochondrial mem-brane proteome," Biotechnology Journal, vol. 3, no. 6, pp. 817–822, 2008.

67. J. Chang, J. E. Cornell, H. Van Remmen, K. Hakala, W. F. Ward, and A. Richardson, "Effect of aging and caloric restriction on the mitochondrial proteome," Journals of Gerontology A, vol. 62, no. 3, pp. 223–234, 2007.

68. K. O'Connell and K. Ohlendieck, "Proteomic DIGE analysis of the mitochondria-enriched fraction from aged rat skeletal muscle," Proteomics, vol. 9, no. 24, pp. 5509–5524, 2009.

69. H. Schägger and K. Pfeiffer, "Supercomplexes in the respiratory chains of yeast and mammalian mitochondria," The EMBO Journal, vol. 19, no. 8, pp. 1777–1783, 2000.

70. D. S. Cooper, "Subclinical hypothyroidism," The New England Journal of Medi-cine, vol. 345, no. 4, pp. 260–265, 2001.

71. N. Knudsen, P. Laurberg, L. B. Rasmussen et al., "Small differences in thyroid func-tion may be important for body mass index and the occurrence of obesity in the population," Journal of Clinical Endocrinology and Metabolism, vol. 90, no. 7, pp. 4019–4024, 2005.

72. A. Oetting and P. M. Yen, "New insights into thyroid hormone action," Best Practice and Research in Clinical Endocrinology and Metabolism, vol. 21, no. 2, pp. 193–208, 2007.

73. D. Robyr, A. P. Wolffe, and W. Wahli, "Nuclear hormone receptor coregulators in action: diversity for shared tasks," Molecular Endocrinology, vol. 14, no. 3, pp. 329–347, 2000.

74. J. Zhang and M. A. Lazar, "The mechanism of action of thyroid hormones," Annual Review of Physiology, vol. 62, pp. 439–466, 2000.

75. F. Flamant, K. Gauthier, and J. Samarut, "Thyroid hormones signaling is getting more complex: STORMs are coming," Molecular Endocrinology, vol. 21, no. 2, pp. 321–333, 2007.

76. M. Busson, L. Daury, P. Seyer et al., "Avian MyoD and c-Jun coordinately induce transcriptional activity of the 3,5,3'-triiodothyronine nuclear receptor c-ErbAα1 in proliferating myoblasts," Endocrinology, vol. 147, no. 7, pp. 3408–3418, 2006.

77. C. Desbois, D. Aubert, C. Legrand, B. Pain, and J. Samarut, "A novel mechanism of action for v-ErbA: abrogation of the inactivation of transcription factor AP-1 by retinoic acid and thyroid hormone receptors," Cell, vol. 67, no. 4, pp. 731–740, 1991.

78. L. Daury, M. Busson, F. Casas, I. Cassar-Malek, C. Wrutniak-Cabello, and G. Cabello, "The triiodothyronine nuclear receptor c-ErbAα1 inhibits avian MyoD transcriptional activity in myoblasts," FEBS Letters, vol. 508, no. 2, pp. 236–240, 2001.

79. M. A. Lazar, "Thyroid hormone receptors: multiple forms, multiple possibilities," Endocrine Reviews, vol. 14, no. 2, pp. 184–193, 1993.

80. G. A. Brent, "Mechanisms of disease: the molecular basis of thyroid hormone action," The New England Journal of Medicine, vol. 331, no. 13, pp. 847–853, 1994.

81. J. Sap, A. Munoz, and K. Damm, "The c-erb-A protein is a high-affinity receptor for thyroid hormone," Nature, vol. 324, no. 6098, pp. 635–640, 1986.

82. E. D. Abel, E. G. Moura, R. S. Ahima et al., "Dominant inhibition of thyroid hormone action selectively in the pituitary of thyroid hormone-receptor-β null mice abolishes the regulation of thyrotropin by thyroid hormone," Molecular Endocrinology, vol. 17, no. 9, pp. 1767–1776, 2003.

83. J. Burnside, D. S. Sarling, F. E. Carr, and W. W. Chin, "Thyroid hormone regulation of the rat glycoprotein hormone α-subunit gene promoter activity," The Journal of Biological Chemistry, vol. 264, no. 12, pp. 6886–6891, 1989.

84. D. L. Bodenner, M. A. Mroczynski, B. D. Weintraub, S. Radovick, and F. E. Wondisford, "A detailed functional and structural analysis of a major thyroid hormone inhibitory element in the human thyrotropin β-subunit gene," The Journal of Biological Chemistry, vol. 266, no. 32, pp. 21666–21673, 1991.

85. D. Forrest and B. Vennström, "Functions of thyroid hormone receptors in mice," Thyroid, vol. 10, no. 1, pp. 41–52, 2000.

86. A. Lanni, M. Moreno, A. Lombardi, P. De Lange, and F. Goglia, "Control of energy metabolism by iodothyronines," Journal of Endocrinological Investigation, vol. 24, no. 11, pp. 897–913, 2001.

87. A. Lanni, M. Moreno, A. Lombardi, and F. Goglia, "Thyroid hormone and uncoupling proteins," FEBS Letters, vol. 543, no. 1*#8211;3, pp. 5–10, 2003.

88. E. Silvestri, L. Schiavo, A. Lombardi, and F. Goglia, "Thyroid hormones as molecular determinants of thermogenesis," Acta Physiologica Scandinavica, vol. 184, no. 4, pp. 265–283, 2005.

89. J. E. Silva, "Thermogenic mechanisms and their hormonal regulation," Physiological Reviews, vol. 86, no. 2, pp. 435–464, 2006.

90. F. Goglia, M. Moreno, and A. Lanni, "Action of thyroid hormones at the cellular level: the mitochondrial target," FEBS Letters, vol. 452, no. 3, pp. 115–120, 1999.

91. M. Gaspari, N.-G. Larsson, and C. M. Gustafsson, "The transcription machinery in mammalian mitochondria," Biochimica et Biophysica Acta, vol. 1659, no. 2-3, pp. 148–152, 2004.

92. R. C. Scarpulla, "Nuclear control of respiratory gene expression in mammalian cells," Journal of Cellular Biochemistry, vol. 97, no. 4, pp. 673–683, 2006.

93. C. Wrutniak-Cabello, F. Casas, and G. Cabello, "Thyroid hormone action in mitochondria," Journal of Molecular Endocrinology, vol. 26, no. 1, pp. 67–77, 2001.

94. A.-M. G. Psarra, S. Solakidi, and C. E. Sekeris, "The mitochondrion as a primary site of action of steroid and thyroid hormones: presence and action of steroid and thyroid hormone receptors in mitochondria of animal cells," Molecular and Cellular Endocrinology, vol. 246, no. 1-2, pp. 21–33, 2006.

95. C. Wrutniak, P. Rochard, F. Casas, A. Fraysse, J. Charrier, and G. Cabello, "Physiological importance of the T3 mitochondrial pathway," Annals of the New York Academy of Sciences, vol. 839, pp. 93–100, 1998.

96. C. H. Gouveia, J. J. Schultz, D. J. Jackson, G. R. Williams, and G. A. Brent, "Thyroid hormone gene targets in ROS 17/2.8 osteoblast like cells identified by differential display analysis," Thyroid, vol. 12, no. 8, pp. 663–671, 2002.

97. T. Iglesias, J. Caubín, A. Zaballos, J. Bernal, and A. Munoz, "Identification of the mitochondrial NADH dehydrogenase subunit 3 (ND3) as a thyroid hormone regulated gene by whole genome PCR analysis," Biochemical and Biophysical Research Communications, vol. 210, no. 3, pp. 995–1000, 1995.

98. R. J. Wiesner, T. T. Kurowski, and R. Zak, "Regulation by thyroid hormone of nuclear and mitochondrial genes encoding subunits of cytochrome-c oxidase in rat liver and skeletal muscle," Molecular Endocrinology, vol. 6, no. 9, pp. 1458–1467, 1992.

99. Y. Hiroi, H.-H. Kim, H. Ying et al., "Rapid nongenomic actions of thyroid hormone," Proceedings of the National Academy of Sciences of the United States of America, vol. 103, no. 38, pp. 14104–14109, 2006.

100. S.-Y. Cheng, J. L. Leonard, and P. J. Davis, "Molecular aspects of thyroid hormone actions," Endocrine Reviews, vol. 31, no. 2, pp. 139–170, 2010.

101. A. J. Hulbert, "Thyroid hormones and their effects: a new perspective," Biological Reviews of the Cambridge Philosophical Society, vol. 75, no. 4, pp. 519–631, 2000.

102. M. Moreno, P. de Lange, A. Lombardi, E. Silvestri, A. Lanni, and F. Goglia, "Metabolic effects of thyroid hormone derivatives," Thyroid, vol. 18, no. 2, pp. 239–253, 2008.

103. N. S. Kavok, O. A. Krasilnikova, and N. A. Babenko, "Thyroxine signal transduction in liver cells involves phospholipase C and phospholipase D activation. Genomic independent action of thyroid hormone," BMC Cell Biology, vol. 2, article 5, 2001.

104. X. Cao, F. Kambe, L. C. Moeller, S. Refetoff, and H. Seo, "Thyroid hormone induces rapid activation of Akt/protein kinase B-mammalian target of rapamycin-p70S6K cascade through phosphatidylinositol 3-kinase in human fibroblasts," Molecular Endocrinology, vol. 19, no. 1, pp. 102–112, 2005.

105. S. Seelig, C. Liaw, H. C. Towle, and J. H. Oppenheimer, "Thyroid hormone attenuates and augments hepatic gene expression at a pretranslational level," Proceedings of the National Academy of Sciences of the United States of America, vol. 78, no. 8, pp. 4733–4737, 1981.

106. H. Dong, M. Wade, A. Williams, A. Lee, G. R. Douglas, and C. Yauk, "Molecular insight into the effects of hypothyroidism on the developing cerebellum," Biochemical and Biophysical Research Communications, vol. 330, no. 4, pp. 1182–1193, 2005.

107. H.-M. Zhang, Q. Su, and M. Luo, "Thyroid hormone regulates the expression of SNAP-25 during rat brain development," Molecular and Cellular Biochemistry, vol. 307, no. 1-2, pp. 169–175, 2008.

108. D. Diez, C. Grijota-Martinez, P. Agretti et al., "Thyroid hormone action in the adult brain: gene expression profiling of the effects of single and multiple doses of triiodo-L-thyronine in the rat striatum," Endocrinology, vol. 149, no. 8, pp. 3989–4000, 2008.

109. W. E. Visser, E. C. H. Friesema, J. Jansen, and T. J. Visser, "Thyroid hormone transport by monocarboxylate transporters," Best Practice and Research in Clinical Endocrinology and Metabolism, vol. 21, no. 2, pp. 223–236, 2007.

110. J. Li, V. Nguyen, B. A. French et al., "Mechanism of the alcohol cyclic pattern: role of the hypothalamic-pituitary-thyroid axis," American Journal of Physiology, vol. 279, no. 1, pp. G118–G125, 2000.

111. T. Merkulova, A. Keller, P. Oliviero et al., "Thyroid hormones differentially modulate enolase isozymes during rat skeletal and cardiac muscle development," American Journal of Physiology, vol. 278, no. 2, pp. E330–E339, 2000.

112. M. Nagao, B. Parimoo, and K. Tanaka, "Developmental, nutritional, and hormonal regulation of tissue-specific expression of the genes encoding various acyl-CoA dehydrogenases and α-subunit of electron transfer flavoprotein in rat," The Journal of Biological Chemistry, vol. 268, no. 32, pp. 24114–24124, 1993.

113. D. A. Hood and A.-M. Joseph, "Mitochondrial assembly: protein import," Proceedings of the Nutrition Society, vol. 63, no. 2, pp. 293–300, 2004.

114. M. Sochor, P. McLean, J. Brown, and A. L. Greenbaum, "Regulation of pathways of ornithine metabolism. Effects of thyroid hormone and diabetes on the activity of enzymes at the "ornithine crossroads" in rat liver," Enzyme, vol. 26, no. 1, pp. 15–23, 1981.

115. G. D. Dimitriadis, B. Leighton, M. Parry-Billings, D. West, and E. A. Newsholme, "Effect of hypothyroidism on the sensitivity of glycolysis and glycogen synthesis to insulin in the soleus muscle of the rat," Biochemical Journal, vol. 257, no. 2, pp. 369–373, 1989.

116. S. Somara and K. N. Bitar, "Tropomyosin interacts with phosphorylated HSP27 in agonist-induced contraction of smooth muscle," American Journal of Physiology, vol. 286, no. 6, pp. C1290–C1301, 2004.

117. M. F. McCarty, "Induction of heat shock proteins may combat insulin resistance," Medical Hypotheses, vol. 66, no. 3, pp. 527–534, 2006.

CHAPTER 3

NEXT GENERATION SEQUENCING IN CANCER RESEARCH AND CLINICAL APPLICATION

DEREK SHYR AND QI LIU

3.1 INTRODUCTION

Sanger sequencing has dominated the genomic research for the past two decades and achieved a number of significant accomplishments including the completion of human genome sequence, which made the identification of single gene disorders and the detection of targeted somatic mutation for clinical molecular diagnostics possible [1,2]. Despite Sanger sequencing's accomplishments, researchers are demanding for faster and more economical sequencing, which has led to the emergence of "next-generation" sequencing technologies (NGS). NGS's ability to produce an enormous volume of data at a low price [3,4] has allowed researchers to characterize the molecular landscape of diverse cancer types and has led to dramatic advances in cancer genomic studies.

The application of NGS, mainly through whole-genome (WGS) and whole-exome technologies (WES), has produced an explosion in the con-

This chapter was originally published under the Creative Commons Attribution License. Shyr D and Liu Q. Next Generation Sequencing in Cancer Research and Clinical Application. Biological Procedures Online *15,4 (2013). doi:10.1186/1480-9222-15-4.*

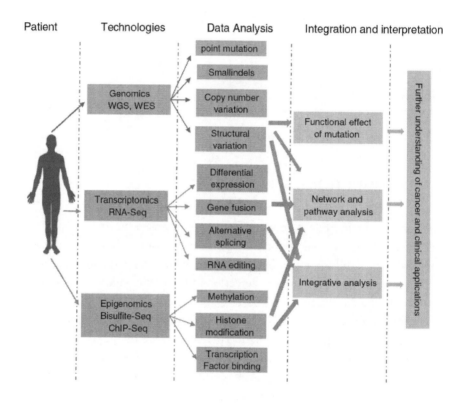

FIGURE 1: The workflow of integrating omics data in cancer research and clinical application. NGS technologies detect the genomic, transcriptomic and epigenomic alternations including mutations, copy number variations, structural variants, differentially expressed genes, fusion transcripts, DNA methylation change, etc. Various kinds of bioinformatics tools are used to analyze, integrate, and interpret the data to improve our understanding of cancer biology and develop personalized treatment strategy.

text and complexity of cancer genomic alterations, including point mutations, small insertions or deletions, copy number alternations and structural
variations. By comparing these alterations to matched normal samples, researchers have been able to distinguish two categories of variants: somatic
and germ line. The Whole transcriptome approach (RNA-Seq) can not
only quantify gene expression profiles, but also detect alternative splicing, RNA editing and fusion transcripts. In addition, epigenetic alterations,
DNA methylation change and histone modifications can be studied using
other sequencing approaches including Bisulfite-Seq and ChIP-seq. The
combination of these NGS technologies provides a high-resolution and
global view of the cancer genome. Using powerful bioinformatics tools,
researchers aim to decipher the huge amount of data to improve our understanding of cancer biology and to develop personalized treatment strategy.
Figure 1 shows the workflow of integrating omics data in cancer research
and clinical application.

3.2 CANCER RESEARCH

In the last several years, many NGS-based studies have been carried out to
provide a comprehensive molecular characterization of cancers, to identify novel genetic alterations contributing to oncogenesis, cancer progression and metastasis, and to study tumor complexity, heterogeneity and
evolution. These efforts have yielded significant achievements for breast
cancer [5-12], ovarian cancer [13], colorectal cancer [14,15], lung cancer
[16], liver cancer [17], kidney cancer [18], head and neck cancer [19],
melanoma [20], acute myeloid leukemia (AML) [21,22], etc. Table 1 sum-
marizes the recent advances in cancer genomics research applying NGS
technologies.

3.2.1 DISCOVERY OF NEW CANCER-RELATED GENES

Cancer is primarily caused by the accumulation of genetic alterations,
which may be inherited in the germ line or acquired somatically during
a cell's life cycle. The effects of these alterations in oncogenes, tumor

suppressor genes or DNA repair genes, allows cells to escape growth and regulatory control mechanisms, leading to the development of a tumor [23]. The progeny of the cancer cell may also undergo further mutations, resulting in clonal expansion [24]. As clonal expansion continues, clones eventually become invasive to its surrounding tissue and metastasize to distant areas from the primary tumor [25].

TABLE 1: Recent NGS-based studies in cancer

Cancer	Experiment Design	Description	ref
Colon cancer	72 WES, 68 RNA-seq, 2 WGS	Identify multiple gene fusions such as RSPO2 and RSPO3 from RNA-seq that may function in tumorigenesis	[15]
Breast cancer	65 WGS/WES, 80 RNA-seq	36% of the mutations found in the study were expressed. Identify the abundance of clonal frequencies in an epithelial tumor subtype	[11]
Hepatocellular carcinoma	1 WGS, 1 WES	Identify TSC1 nonsense substitution in sub-population of tumor cells, intra-tumor hetero-geneity, several chromosomal rearrange-ments, and patterns in somatic substitutions	[17]
Breast cancer	510 WES	Identify two novel protein-expression-defined subgroups and novel subtype-associ-ated mutations	[5]
Colon and rectal cancer	224 WES, 97 WGS	24 genes were found to be significantly mutated in both cancers. Similar patterns in genomic alterations were found in colon and rectum cancers	[14]
squamous cell lung cancer	178 WES, 19 WGS, 178 RNA-seq, 158 miRNA-seq	Identify significantly altered pathways including NFE2L2 and KEAP1 and potential therapeutic targets	[16]
Ovarian carci-noma	316 WES	Discover that most high-grade serous ovarian cancer contain TP53 mutations and recurrent somatic mutations in 9 genes	[13]
Melanoma	25 WGS	Identify a significantly mutated gene, PREX2 and obtain a comprehensive genomic view of melanoma	[20]
Acute myeloid leukemia	8 WGS	Identify mutations in relapsed genome and compare it to primary tumor. Discover two major clonal evolution patterns	[21]
Breast cancer	24 WGS	Highlights the diversity of somatic rearrange-ments and analyzes rearrangement patterns related to DNA maintenance	[8]

TABLE 1: *Cont.*

Cancer	Experiment Design	Description	ref
Breast cancer	31 WES, 46 WGS	Identify eighteen significant mutated genes and correlate clinical features of oestrogen-receptor-positive breast cancer with somatic alterations	[7]
Breast cancer	103 WES, 17 WGS	Identify recurrent mutation in CBFB transcription factor gene and deletion of RUNX1. Also found recurrent MAGI3-AKT3 fusion in triple-negative breast cancer	[6]
Breast cancer	100 WES	Identify somatic copy number changes and mutations in the coding exons. Found new driver mutations in a few cancer genes	[9]
Acute myeloid leukemia	24 WGS	Discover that most mutations in AML genomes are caused by random events in hematopoietic stem/progenitor cells and not by an initiating mutation	[22]
Breast cancer	21 WGS	Depict the life history of breast cancer using algorithms and sequencing technologies to analyze subclonal diversification	[12]
Head and neck squamous cell carcinoma	32 WES	Identify mutation in NOTCH1 that may function as an oncogene	[19]
Renal carcinoma	30 WES	Examine intra-tumor heterogeneity reveal branch evolutionary tumor growth	[18]

The sequencing of cancer genomes has revealed a number of novel cancer-related genes, especially in breast cancer. Recently, six papers reported their findings on large breast cancer dataset: TCGA performed exome sequencing on 510 samples from 507 patients [5], Banerji et al. conducted exome sequencing on 103 samples and whole genome sequencing on 17 samples, Ellis et al. did exome sequencing on 31 samples and whole genome sequencing on 46 samples [7], Stephens et al. applied exome sequencing on 100 samples, Shah et al. performed whole genome/exome and RNA sequencing on 65 and 80 samples of triple-negative breast cancers [11], and Nik-Zainal et al. performed whole genome sequencing on 21 tumor/normal pairs [12]. Besides confirming recurrent somatic mutations in TP53, GATA3 and PIK3CA, these studies discovered novel cancer-related mutations. Although novel mutations occur at low frequency (less than

10%), mutations of specific genes are enriched in the subtype of breast cancers and could be grouped into cancer-related pathways. For example, mutations of MAP3K1 frequently occur in luminal A subtype [5,7]. Pathways involving p53, chromatin remodeling and ERBB signaling are over-represented in mutated genes [11]. Furthermore, some mutations indicate therapeutic opportunities such as the mutant GATA3, which might be a positive predictive marker for aromatase inhibitor response [7].

Genomic sequencing has also helped characterize the mutation profile of colorectal cancer. For example, exome sequencing performed on 72 tumor-normal pairs identified 36,303 protein-altering somatic mutations. Further analysis for significantly mutated genes led to 23 candidates that included expected cancer genes such as KRAS, TP53 and PIK3CA and novel genes such as ATM, which regulates the cell cycle checkpoint. RNA sequencing identified recurrent R-spondin fusions, which might potentiate Wnt signaling and induce tumorigenesis [15]. Another example includes exome sequencing performed on 224 tumor and normal pairs. This study identified 15 highly mutated genes in the hypermutated cancers and 17 in the non-hypermutated cancers. Among the non-hypermutated cancers, novel frequent mutations in SOX9, ARID1A, ATM and FAM123B were detected besides the known APC, TP53 and KRAS mutations. The analysis of the mutations and functional roles of SOX9, ARID1A, ATM and FAM123B suggested they are highly potential colorectal cancer-related genes. Non-hypermutated colon and rectum cancers were found to have similar patterns in genomic alternation. Whole genome sequencing of 97 tumors with matched normal samples identified the recurrent NAV2-TCF7L1 fusion [14].

3.2.2 TUMOR HETEROGENEITY AND EVOLUTION

What makes cancer a difficult disease to conquer has much to do with the evolution of cancer that results from the selection and genetic instability occurring in each clone, leading to heterogeneity in tumors [26]. This idea was first proposed by Peter Nowell in 1976 as the clonal evolution model of cancer, which attempted to explain the increase in tumor aggressiveness over a period of time. Further work by other researchers in the 1980s supported this theory with studies of metastatic subclones from a mouse sarcoma cell line [26].

The wide application of NGS has revealed substantial insights into tumor heterogeneity and tumor evolution. Variations between tumors are referred to as intertumor heterogeneity, while variations within a single tumor are intratumor heterogeneity. Intertumor heterogeneity is recognized by different morphological phenotype, expression profiles and mutation and copy number variation patterns, categorizing tumors into different subtypes [27-31]. The mRNA-expression subtype was found to be associated with somatic mutation landscapes in the recent TCGA and Eillis et al.'s studies. [5,7]. As a huge amount of somatic mutations generated by NGS, the picture emerges like that individual tumor is unique, each containing distinct mutation patterns. For instance, Stephens et al. found that there were 73 different combination possibilities of mutated cancer genes among the 100 breast cancers [9].

Intratumor heterogeneity can be recognized as non-identical cellular clones or subclones within a single tumor, indicating different histology, gene expression, and metastatic and proliferative potential. The ability to generate high-resolution data makes NGS a particularly useful tool for studying intratumor heterogeneity. A recent NGS-based study on renal cell carcinoma from four patients has successfully illuminated intratumor heterogeneity [18]. For patient 1, the pre-treatment samples of the primary tumor and chest-wall metastasis went through exon-capture multi-region sequencing on DNA. Of the 128 validated mutations found in 9 regions of the primary tumor, 40 were ubiquitous, 59 were shared by some regions, and 29 were unique to specific regions, showing that genetic heterogeneity exists within a tumor and an "ongoing regional clonal evolution" [18]. Most importantly, the study showed that a single biopsy of a tumor only reveals a small part of a tumor's mutational landscape; from a single biopsy, about 55% of all mutations were detected in this tumor and 34% were shared by most regions of the tumor.

The ongoing and parallel evolution of cancer cells may establish and maintain intratumor heterogeneity. For example, phylogenetic relationships of the tumor regions in patient 1 and 2 by the renal cell carcinoma study revealed a branching rather than linear evolution of the tumor [18]. Studies have also shown branching structures of evolution in breast cancer

[26]. According to the "Trunk-Branch Model of Tumor Growth" [26], there are somatic events that promote tumor growth, which represents the trunk of the tree in the early stage of tumor development. These somatic aberrations would most likely be ubiquitous at this stage. Over time, other somatic events, known as drivers, cause tumor heterogeneity to occur, which causes branching to take place in tumors as well as in metastatic sites. Later, these branches will evolve and become more isolated, resulting in a 'Bottleneck Effect' that can result in chromosomal instability, allowing further expansion of tumor heterogeneity [26]. This leads to the tumor's ability to adapt and survive in changing environments, which affects the success of drug treatment [18]. Therefore, it is important to examine tumor clonal structure and identify common mutations located in the trunk of the phylogenetic tree, which may help understand target therapy resistance and discover more robust therapeutic approaches.

3.3 CLINICAL APPLICATION

Besides allowing researchers to understand mutations in cancer, NGS has already been applied to the clinic in many areas including prenatal diagnostics, pathogen detection, genetic mutations, and more [32]. Although genetic mutations have been identified with Sanger sequencing, PCR, and microarrays in clinical application, these three have limitations that don't apply to NGS. For example, although microarrays can detect single nucleotide variants (SNVs), they have trouble identifying larger DNA aberrations, e.g., large indels and structural rearrangements, which are common in cancer. In contrast, whole exome and whole-genome sequencing can provide the clinician a comprehensive view of the DNA aberrations, genetic recombination, and other mutations [28,32]. Therefore, NGS platforms serve as a good diagnostic and prognostic tool and help clinicians identify specific characteristics in each patient, paving the road towards personalized medicine.

NGS has already been applied in the clinic for cancer diagnosis and prognosis. For example, whole genome sequencing identified a novel insertional fusion that created a classic bcr3 PML-RARA fusion gene for a patient with acute myeloid leukemia and the findings altered the treatment plan for the patient [33]. By sequencing the tumor genome of a patient, clinicians are able to design patient-specific probes that uses DNA in the patient's blood serum to monitor the progress of a patient's treatment and detect for any signs of relapse [27-31]. The discovery of more biomarkers and the development of target-therapies will be essential in helping a clinician choose the best personalized treatment for his or her patients.

There has also been a dramatic increase in the number of clinical trials using NGS technologies since 2010 (Table 2). Ranging from WGS and WES to RNA-seq and targeted sequencing, clinical trials are using NGS to find genetic alterations that are the drivers of certain diseases in patients and apply that knowledge into the practice of clinical medicine. The information gained from these studies may help with drug development and explain the resistance of certain treatments.

3.4 METHODS AND RESOURCES

3.4.1 PIPELINE AND TOOLS FOR NGS DATA ANALYSIS

To analyze and interpret the increasing amount of sequencing data, a number of statistical methods and bioinformatics tools have been developed. For WGS and WES, the analysis generally includes read alignment, variant detection (point mutation, small indels, copy number variation and structural rearrangement) and variant functional prediction (Table 3). Reads are mapped back to the human reference genomes using MAQ [34], BWA [35,36], Bowtie2 [37], BFAST [38], SOAP2 [39], Novoalign/ NovoalignCS, SSAHA2 [40], SHRiMP [41], etc. These methods differ in their computational efficiency, sensitivity and ability to accurately map noisy reads, to deal with long or short reads and pair-end reads. Having aligned the reads to the genome, mutation calling identifies the sites in which at least one of the bases differs from a reference sequence by GATK [42], SAMtools [43], SOAPsnp [44], SNVMix [45], Varscan [46], etc. Differing in the underlying statistical models, the performances of these methods are comparable and vary on sequencing depths [47-49]. Detecting somatic mutation involves mutation calling in paired tumor-normal DNA, coupled with comparison to the reference. A naïve somatic mutation caller applies standard calling tools on the normal and tumor samples separately and then selects mutations detected in tumor but not in normal. Alternatively, a complicated caller jointly analyzes tumor-normal pair data such as Varscan2 [50], Somaticsniper [51] and JointSNVMix [52]. SIFT [53], PolyPhen [54], CHASM [55] and ANNOVAR [56] have been developed to understand the impact of the mutations on gene function and to distinguish between driver and passenger mutations. For WGS, various kinds of structural variations can be discovered using BreakDancer [57], VariationHunter [58], PEMer [59] and SVDetect [60]. RNA-seq data analysis generally includes reads alignment, gene expression quantification, differentially expressed genes/isoforms or alternative splicing detection and novel transcripts discovery (Table 4). There are two major approaches to map RNA-seq reads. One is to align reads to the reference transcriptome using standard DNA-seq reads aligner. The alternative is to map reads to the reference genome allowing for the identification of novel splice junctions using a RNA-seq specific aligner, such as TopHat [61], MapSplice [62], SpliceMap [63], GSNAP [64], and STAR [65]. Having aligned reads, expression values are quantified by aggregating reads into counts and differential expression analysis is performed based on counts (DEseq [66],edgeR [67]) or FPKM/RPKM values (CuffLinks [68,69]). Estimating isoform-level expression is very difficult since many genes have multiple isoforms and most reads are shared by different isoforms. To deal with read assignment uncertainty, Alexa-seq [70] counts only the reads that map uniquely to a single isoform, while Cufflinks [68,69] and MISO [71] construct a likelihood model that best explains all the reads obtained in the experiment. In addition, fusion transcripts can be detected using SOAPfusion, TopHat-Fusion [72], BreakFusion [73], FusionHunter [74], deFuse [75], FusionAnalyser [76], etc. To obtain a more complete view of cancer genome, an integrative approach to study diverse mutations, transcriptomes and epigenomes simultaneously on the pathways or networks is much more informative and promising. A growing number of pathway-oriented tools is now becoming available, including PARADIGM [77], NetBox [78], MEMo [79], CONEXIC [80], etc.

TABLE 2: Active cancer studies using NGS as the primary outcome measure

Study Title/Sponsor	NCT#/# Enrolled/ Start Date	Condition	Description	Sequencing Technologies
Tumor Specific Plasma DNA in Breast Cancer/Dartmouth-Hitchcock Medical Center	NCT01617915/6/ October 2012	Breast Cancer	Analyze chromosomal rearrangements and genomic alterations	Whole genome sequencing
Whole Exon Sequencing of Down Syndrome Acute Myeloid Leukemia/ Children's Oncology Group	NCT01507441/10/ February 2012	Leukemia	Examine DNA samples of patients with Leukemia and Down Syndrome and identify DNA alterations	Whole exome Sequencing
Studying Genes in Samples From Younger Patients with Adrenocortical Tumor/Children's Oncology Group	NCT01528956/10/ February 2012	Adrenocortical Carcinoma	Study genes from patients with adrenocortical tumor	Whole genome Sequencing
Feasibility Clinical Study of Targeted and Genome-Wide Sequencing/University Health Network, Toronto	NCT01345513/150/ March 2011	Solid Tumors	Identify gene mutations in cancer patients	Whole genome sequencing
An Ancillary Pilot Trial Using Whole Genome Sequencing in Patients with Advance Refractor Cancer/Scottsdale Healthcare	NCT01443390/10/ September 2011	Advanced Cancer	Investigate patients with cancer that are using Phase I drugs and its effect on the patient	Whole genome Sequencing
Cancer Genome Analysis/Seoul National University Hospital	NCT01458604/100/ August 2011	Malignant Tumor	Identify and analyze genetic alterations in tumors for therapeutic agents	Targeted Sequencing, whole exome sequencing and RNA-seq
RNA Biomarkers in Tissue Samples From Infants with Acute Meyloid Leukemia/Children's Oncology Group	NCT01229124/20/ October 2010	Leukemia	Analyze tissue samples and identify biomarkers from RNA	RNA-seq
Molecular Analysis of Solid Tumors/St. Jude Children's Research Hospital	NCT01050296/360/ January 2010	Pediatric Solid Tumors	Analyze gene expression profiles of tumor and examine genetic alterations	Whole genome Sequencing
Deep Sequencing of the Breast Cancer Transcriptome/University of Arkansas	NCT01141530/30/ Sept 2009	Breast Cancer	Examine transcriptional regulation and triple negative breast cancer	RNA-seq

TABLE 3: Computational tools for cancer genomics

Category	Program	URL	Ref
Alignment	MAQ	http://maq.sourceforge.net/	[34]
	BWA	http://bio-bwa.sourceforge.net/	[35,36]
	Bowtie2	http://bowtie-bio.sourceforge.net/bowtie2/	[37]
	BFAST	http://bfast.sourceforge.net	[38]
	SOAP2	http://soap.genomics.org.cn/soapaligner.html	[39]
	Novoalign/NovoalignCS	http://www.novocraft.com/	
	SSAHA2	http://www.sanger.ac.uk/resources/software/ssaha2/	[40]
	SHRiMP	http://compbio.cs.toronto.edu/shrimp/	[41]
Mutation calling	GATK	http://www.broadinstitute.org/gatk/	[42]
	Samtools	http://samtools.sourceforge.net/	[43]
	SOAPsnp	http://soap.genomics.org.cn/soapsnp.html	[44]
	SNVmix	http://compbio.bccrc.ca/software/snv-mix/	[45]
	VarScan	http://varscan.sourceforge.net/	[46,50]
	Somaticsniper	http://gmt.genome.wustl.edu/somatic-sniper/	[51]
	JointSNVMix	http://compbio.bccrc.ca/software/jointsnvmix/	[52]
SV detection	BreakDancer	http://breakdancer.sourceforge.net/	[57]
	VariationHunter	http://variationhunter.sourceforge.net/	[58]
	PEMer	http://sv.gersteinlab.org/pemer/	[59]
	SVDetect	http://svdetect.sourceforge.net/	[60]
Function effect of mutation	SIFT	http://sift.jcvi.org/	[53]
	CHASM	http://wiki.chasmsoftware.org	[55]
	PolyPhen-2	http://genetics.bwh.harvard.edu/pph2/	[54]
	ANNOVAR	http://www.openbioinformatics.org/annovar/	[56]

Source: http://www.clinicaltrials.gov .

TABLE 4: Computational tools for cancer transcriptomics

Category	Program	URL	ref
Spliced alignment	TopHat	http://tophat.cbcb.umd.edu/	[61,69]
	MapSplice	http://www.netlab.uky.edu/p/bioinfo/MapSplice	[62]
	SpliceMap	http://www.stanford.edu/group/wonglab/Splice-Map/	[63]
	GSNAP	http://research-pub.gene.com/gmap/	[64]
	STAR	http://gingeraslab.cshl.edu/STAR/	[65]
Differential expression	CuffDiff	http://cufflinks.cbcb.umd.edu/	[68,69]
	EdgeR	http://www.bioconductor.org/packages/2.11/bioc/html/edgeR.html	[67]
	DESeq	http://www-huber.embl.de/users/anders/DESeq/	[66]
	Myrna	http://bowtie-bio.sourceforge.net/myrna/index.shtml	[81]
Alternative splicing	CuffDiff	http://cufflinks.cbcb.umd.edu/	[68,69]
	MISO	http://genes.mit.edu/burgelab/miso/	[71]
	DEXseq	http://watson.nci.nih.gov/bioc_mirror/packages/2.9/bioc/html/DEXSeq.html	[82]
	Alexa-seq	http://www.alexaplatform.org/alexa_seq/	[70]
Gene fusion	SOAPfusion	http://soap.genomics.org.cn/SOAPfusion.html	
	TopHat-Fusion	http://tophat.cbcb.umd.edu/fusion_index.html	[72]
	BreakFusion	http://bioinformatics.mdanderson.org/main/BreakFusion	[73]
	FusionHunter	http://bioen-compbio.bioen.illinois.edu/Fusion-Hunter/	[74]
	deFuse	http://sourceforge.net/apps/mediawiki/defuse/	[75]
	FusionAnalyser	http://www.ilte-cml.org/FusionAnalyser/	[76]

3.4.2 COMPREHENSIVE CANCER PROJECTS AND RESOURCES

The vast amount of oncogenomics data are generated from large scale collaborative cancer projects (Table 5). The Cancer Genome Atlas (TCGA) and International Cancer Genome Consortium (ICGC) are the two largest representatives of such coordinated efforts. Beginning as a three-year pilot in 2006, TCGA aims to comprehensively map the important genomic changes that occur in the major types and subtypes of cancer. TCGA will examine over 11,000 samples for 20 cancer types (http://cancergenome. nih.gov/). ICGC launched in 2008 and its goal is 'to obtain a comprehensive description of genomic, transcriptomic and epigenomic changes in 50 different tumor types and/or subtypes which are of clinical and societal importance across the globe'(http://icgc.org/icgc). The Cancer Genome Project (CGP) has many efforts at the Sanger Institute and aims to identify sequence variants/mutations critical in the development of human cancers (http://www.sanger.ac.uk/genetics/CGP/). The NCI's Cancer Genome Anatomy Project (CGAP) seeks to determine the gene expression profiles of normal, precancer and cancer cells, leading eventually to improved detection, diagnosis and treatment for the patient (http://cgap.nci.nih.gov/). Recently, the Clinical Proteomic Tumor Analysis Consortium (CPTAC) has launched to systematically identify proteins that derive from alterations in cancer genomes using proteomic technologies (http://proteomics. cancer.gov/). The combination of genomic and proteomic initiatives is anticipated to produce a more comprehensive inventory of the detectable proteins in a tumor and advance our understanding of cancer biology.

The data and the results from these projects are freely available to the research community (Table 5). A number of databases and frameworks have been developed to make the data and the results easily and directly accessible. For example, the results from CGP are collated and stored in http://COSMIC [83]. The cBio Cancer Genomics Portal, containing dataset from TCGA and published papers, is specifically designed to interactively explore multidimensional cancer genomics data, including mutation, copy number variations, expression changes (microarray and RNA-seq), DNA methylation values, and protein and phosphoprotein levels [84]. Intogen is also a framework that facilitates the analysis and integration of multimensional data for the identification of genes and biological modules critical in cancer development [85]. The Broad GDAC Firehose, designed to coordinate the various tools utilized by TCGA, provides level 3 and level 4 analyses and enables researchers to easily incorporate TCGA data into their projects. Table 5 also includes resources useful for cancer research but not built on NGS data, e.g., Progenetix [86].

TABLE 5: Comprehensive cancer projects and resources

Name	Description	URL
Comprehensive cancer projects		
The Cancer Genome Atlas	A joint effort to accelerate our understanding of the molecular basis of cancer through the application of genome analysis technologies	http://cancergenome.nih.gov/
International Cancer Genome Consortium	International consortium with the goal of obtaining comprehensive description of genomic, transcriptomic, and epigenomic changes in 50 different cancer types and/or subtypes of clinical and societal importance across the globe	http://icgc.org/icgc
Cancer Genome Anatomy Project	Interdisciplinary program to determine the gene expression profiles of normal, precancer, and cancer cells, leading eventually to improved detection, diagnosis, and treatment for the patient	http://cgap.nci.nih.gov/
Cancer Genome Project	To identify somatically acquired sequence variants/mutations and hence identify genes critical in the development of human cancers	http://www.sanger.ac.uk/genetics/CGP/
The Clinical Proteomic Tumor Analysis Consortium	A comprehensive and coordinated effort to accelerate the understanding of the molecular basis of cancer through the application of proteomic technologies	http://proteomics.cancer.gov/
Resources		
COSMIC	Catalogue of Somatic Mutations in Cancer	http://www.sanger.ac.uk/genetics/CGP/cosmic/
Progenetix	Copy number abnormalities in human cancer from CGH experiments	http://www.progenetix.org/cgi-bin/pgHome.cgi
MethyCancer	An information resource and analysis platform for study interplay of DNA methylation, gene expression and cancer	http://methycancer.psych.ac.cn/
IntOGen	Integrates multidimensional OncoGenomics Data for the identification of genes and groups of genes involved in cancer development	http://www.intogen.org/

TABLE 5: *Cont.*

Name	Description	URL
Oncomine	A cancer microarray database and integrated data-mining platform	http://www.oncomine.org/
cBio	Provides visualization, analysis and download of large-scale cancer genomics data sets	http://www.cbioportal.org/
Firehose	Provides L3 data and L4 analyses packaged in a form amenable to immediate algorithmic analysis	https://confluence.broadinstitute.org/display/GDAC/Home
UCSC Cancer Genomics Browser	A suite of web-based tools to visualize, integrate and analyze cancer genomics and its associated clinical data	https://genome-cancer.soe.ucsc.edu/
Cancer Genome Workbench	Hosts mutation, copy number, expression, and methylation data from a number of projects, including TCGA, TARGET, COSMIC, GSK, NCI60. It has tools for visualizing sample-level genomic and transcription alterations in various cancers.	https://cgwb.nci.nih.gov/

3.5 CHALLENGES AND PERSPECTIVE

Although NGS has already helped researchers discover a plethora of information in the field of cancer, challenges in translating the large amounts of oncogenomics data into information that can be easily interpretable and accessible for cancer care still lie ahead. From a computational point of view, many technical and statistical issues remain unsolved. For example, repetitive DNA represents a major obstacle for the accuracy of read alignment and assembly, as well as structure variation detection [87]. Furthermore, it is difficult to distinguish rare mutations in tumor from sequencing and alignment artifacts, especially when a tumor has low purity. Despite new methods to comprehensively catalogue genomic variants, the prediction of their functional effect and the identification of disease-causal variants are still in an early phase [88]. Current algorithms for quantifying isoform expression are not computationally trivial and are incredibly difficult to explain. Although the concept of integrative analysis is not new, predictive networks or pathway models that combine various omics data are still underway. Most importantly, since sequencing technologies and methodologies are both evolving rapidly, it is a difficult challenge to store, analyze and present the data in a method that is transparent and reproducible [89]. On the other hand, tumor complexity and heterogeneity make the analysis and the interpretation of sequencing data even harder. Heterogeneity is dynamic and evolves over time. This challenges the simple notion of binning mutations as tumorigenesis 'driver' and neutral 'passenger,' since some passengers are also drivers just waiting for the right context [90].

From a clinical point of view, a major challenge is to assess genomic variants as potential therapeutic targets. Although many diverse variants are demonstrated to converge on similar deregulated pathways, there is still a lack of pathway-targeted therapies. With the discovery of intra-tumor heterogeneity, questions have been raised about how well a glimpse of a tumor's genomic landscape can steer the treatment. Currently, many clinicians decide a treatment based on the genetic markers from a few biopsies. Whether these markers are over- or under-represented in the tumor is unknown, causing the selection of treatment to be difficult [29]. In addition to heterogeneity, the tumor's ability to evolve allows it to have more opportunities to adapt and survive to various treatments. Some researchers hope that with current target therapies, intratumor heterogeneity will decrease to a certain point [29] so that clinicians can then target the non-responsive clones before a tumor re-growth and more mutations can occur; however, choosing an appropriate target therapy will be a challenge. A few researchers have already shown certain treatments, such as the cytotoxic therapies, that have increased genome instability and diversity, resulting in a faster tumor evolution rate and, thus, heterogeneity. The fact is that this area of cancer is understudied [26]; however, one of the key challenges researchers must solve is identifying branched subclones are resistant to which target therapies. More knowledge of network medicine and the interaction between the trunk and branch mutations may lead to appropriate target therapies and personalized therapeutic strategies that can prevent drug resistance and effectively eradicate cancer [26,91].

To accelerate the rate of translating genomic data into clinical practice, a sustained collaboration among multiple centers and effective communication among bioinformaticians, statistical geneticists, molecular biologists and physician are required. Bioinformaticians and statistical geneticists are responsible for providing reproducible and accurate analysis, identifying 'drivers' in the unstable and evolving cancer genome and building powerful and flexible integrative model to consider interactions among genomic, transcriptomic, metabolomics, proteomics and epigenomic alterations in the context of tumor microenvironment. Biologists interpret and confirm the functional relevance of variants to cancer. Physicians assess relationships of variants to cancer prognosis and response to therapy. Appropriate infrastructure within each research institution that integrates the clinic for patient samples, wet lab for sequencing, and Bioinformatics for data analysis should allow the sequenced data to be processed efficiently, producing results that can create effective personalized therapies applicable to the clinic. In addition, easily accessible and understandable databases that connect genomic findings with clinical outcome are also required. With these efforts and developments, NGS will greatly potentiate genome-based cancer diagnosis and personalized treatment strategies.

REFERENCES

1. Taylor BS, Ladanyi M: Clinical cancer genomics: how soon is now? J Pathol 2011, 223:318-326.
2. Sosman JA, Kim KB, Schuchter L, Gonzalez R, Pavlick AC, Weber JS, McArthur GA, Hutson TE, Moschos SJ, Flaherty KT, Hersey P, Kefford R, Lawrence D, Puzanov I, Lewis KD, Amaravadi RK, Chmielowski B, Lawrence HJ, Shyr Y, Ye F, Li J, Nolop KB, Lee RJ, Joe AK, Ribas A: Survival in BRAF V600-mutant advanced melanoma treated with vemurafenib. N Engl J Med 2012, 366:707-714.
3. Metzker ML: Sequencing technologies - the next generation. Nat Rev Genet 2010, 11:31-46.
4. Wold B, Myers RM: Sequence census methods for functional genomics. Nat Methods 2008, 5:19-21.
5. Cancer Genome Atlas Research Network: Comprehensive molecular portraits of human breast tumours. Nature 2012, 490:61-70.
6. Banerji S, Cibulskis K, Rangel-Escareno C, Brown KK, Carter SL, Frederick AM, Lawrence MS, Sivachenko AY, Sougnez C, Zou L, Cortes ML, Fernandez-Lopez JC, Peng S, Ardlie KG, Auclair D, Bautista-Pina V, Duke F, Francis J, Jung J, Maffuz-Aziz A, Onofrio RC, Parkin M, Pho NH, Quintanar-Jurado V, Ramos AH, Rebollar-Vega R, Rodriguez-Cuevas S, Romero-Cordoba SL, Schumacher SE, Stransky N, Thompson KM, Uribe-Figueroa L, Baselga J, Beroukhim R, Polyak K, Sgroi DC, Richardson AL, Jimenez-Sanchez G, Lander ES, Gabriel SB, Garraway LA, Golub TR, Melendez-Zajgla J, Toker A, Getz G, Hidalgo-Miranda A, Meyerson M: Sequence analysis of mutations and translocations across breast cancer subtypes. Nature 2012, 486:405-409.
7. Ellis MJ: Whole-genome analysis informs breast cancer response to aromatase inhibition. Nature 2012, 486:353-360.
8. Stephens PJ: Complex landscapes of somatic rearrangement in human breast cancer genomes. Nature 2009, 462:1005-1010.
9. Stephens PJ: The landscape of cancer genes and mutational processes in breast cancer. Nature 2012, 486:400-404.
10. Nik-Zainal S: The life history of 21 breast cancers. Cell 2012, 149:994-1007.
11. Shah SP: The clonal and mutational evolution spectrum of primary triple-negative breast cancers. Nature 2012, 486:395-399.
12. Nik-Zainal S: Mutational processes molding the genomes of 21 breast cancers. Cell 2012, 149:979-993.
13. Cancer Genome Atlas Research Network: Integrated genomic analyses of ovarian carcinoma. Nature 2011, 474:609-615.
14. Cancer Genome Atlas Research Network: Comprehensive molecular characterization of human colon and rectal cancer. Nature 2012, 487:330-337.
15. Seshagiri S, Stawiski EW, Durinck S, Modrusan Z, Storm EE, Conboy CB, Chaudhuri S, Guan Y, Janakiraman V, Jaiswal BS, Guillory J, Ha C, Dijkgraaf GJ, Stinson J, Gnad F, Huntley MA, Degenhardt JD, Haverty PM, Bourgon R, Wang W, Koep-

pen H, Gentleman R, Starr TK, Zhang Z, Largaespada DA, Wu TD, de Sauvage FJ: Recurrent R-spondin fusions in colon cancer. Nature 2012, 488:660-664.

16. Hammerman PS, Hayes DN, Wilkerson MD, Schultz N, Bose R, Chu A, Collisson EA, Cope L, Creighton CJ, Getz G, Herman JG, Johnson BE, Kucherlapati R, Ladanyi M, Maher CA, Robertson G, Sander C, Shen R, Sinha R, Sivachenko A, Thomas RK, Travis WD, Tsao MS, Weinstein JN, Wigle DA, Baylin SB, Govindan R, Meyerson M: Comprehensive genomic characterization of squamous cell lung cancers. Nature 2012, 489:519-525.

17. Totoki Y, Tatsuno K, Yamamoto S, Arai Y, Hosoda F, Ishikawa S, Tsutsumi S, Sonoda K, Totsuka H, Shirakihara T, Sakamoto H, Wang L, Ojima H, Shimada K, Kosuge T, Okusaka T, Kato K, Kusuda J, Yoshida T, Aburatani H, Shibata T: High-resolution characterization of a hepatocellular carcinoma genome. Nat Genet 2011, 43:464-469.

18. Gerlinger M, Rowan AJ, Horswell S, Larkin J, Endesfelder D, Gronroos E, Martinez P, Matthews N, Stewart A, Tarpey P, Varela I, Phillimore B, Begum S, McDonald NQ, Butler A, Jones D, Raine K, Latimer C, Santos CR, Nohadani M, Eklund AC, Spencer-Dene B, Clark G, Pickering L, Stamp G, Gore M, Szallasi Z, Downward J, Futreal PA, Swanton C: Intratumor heterogeneity and branched evolution revealed by multiregion sequencing. N Engl J Med 2012, 366:883-892.

19. Agrawal N, Frederick MJ, Pickering CR, Bettegowda C, Chang K, Li RJ, Fakhry C, Xie TX, Zhang J, Wang J, Zhang N, El-Naggar AK, Jasser SA, Weinstein JN, Trevino L, Drummond JA, Muzny DM, Wu Y, Wood LD, Hruban RH, Westra WH, Koch WM, Califano JA, Gibbs RA, Sidransky D, Vogelstein B, Velculescu VE, Papadopoulos N, Wheeler DA, Kinzler KW, Myers JN: Exome sequencing of head and neck squamous cell carcinoma reveals inactivating mutations in NOTCH1. Science 2011, 333:1154-1157.

20. Berger MF: Melanoma genome sequencing reveals frequent PREX2 mutations. Nature 2012, 485:502-506.

21. Ding L: Clonal evolution in relapsed acute myeloid leukaemia revealed by whole-genome sequencing. Nature 2012, 481:506-510.

22. Welch JS: The origin and evolution of mutations in acute myeloid leukemia. Cell 2012, 150:264-278.

23. Wong KM, Hudson TJ, McPherson JD: Unraveling the genetics of cancer: genome sequencing and beyond. Annu Rev Genomics Hum Genet 2011, 12:407-430.

24. Cahill DP, Kinzler KW, Vogelstein B, Lengauer C: Genetic instability and darwinian selection in tumours. Trends Cell Biol 1999, 9:M57-M60.

25. Brosnan JA, Iacobuzio-Donahue CA: A new branch on the tree: next-generation sequencing in the study of cancer evolution. Semin Cell Dev Biol 2012, 23:237-242.

26. Swanton C: Intratumor heterogeneity: evolution through space and time. Cancer Res 2012, 72:4875-4882.

27. Russnes HG, Navin N, Hicks J, Borresen-Dale AL: Insight into the heterogeneity of breast cancer through next-generation sequencing. J Clin Invest 2011, 121:3810-3818.

28. Samuel N, Hudson TJ: Translating Genomics to the Clinic. Clinical chemistry: Implications of Cancer Heterogeneity; 2012.

29. Almendro V, Fuster G: Heterogeneity of breast cancer: etiology and clinical relevance. Clinical & translational oncology: official publication of the Federation of Spanish Oncology Societies and of the National Cancer Institute of Mexico 2011, 13:767-773.

30. Yancovitz M, Litterman A, Yoon J, Ng E, Shapiro RL, Berman RS, Pavlick AC, Darvishian F, Christos P, Mazumdar M, Osman I, Polsky D: Intra- and inter-tumor heterogeneity of BRAF(V600E))mutations in primary and metastatic melanoma. PLoS One 2012, 7:e29336.

31. Curtis C, Shah SP, Chin SF, Turashvili G, Rueda OM, Dunning MJ, Speed D, Lynch AG, Samarajiwa S, Yuan Y, Graf S, Ha G, Haffari G, Bashashati A, Russell R, McKinney S, Langerod A, Green A, Provenzano E, Wishart G, Pinder S, Watson P, Markowetz F, Murphy L, Ellis I, Purushotham A, Borresen-Dale AL, Brenton JD, Tavare S, Caldas C, Aparicio S: The genomic and transcriptomic architecture of 2,000 breast tumours reveals novel subgroups. Nature 2012, 486:346-352.

32. Desai AN, Jere A: Next-generation sequencing: ready for the clinics? Clin Genet 2012, 81:503-510.

33. Welch JS, Westervelt P, Ding L, Larson DE, Klco JM, Kulkarni S, Wallis J, Chen K, Payton JE, Fulton RS, Veizer J, Schmidt H, Vickery TL, Heath S, Watson MA, Tomasson MH, Link DC, Graubert TA, DiPersio JF, Mardis ER, Ley TJ, Wilson RK: Use of whole-genome sequencing to diagnose a cryptic fusion oncogene. JAMA 2011, 305:1577-1584.

34. Li H, Ruan J, Durbin R: Mapping short DNA sequencing reads and calling variants using mapping quality scores. Genome Res 2008, 18:1851-1858.

35. Li H, Durbin R: Fast and accurate short read alignment with Burrows-Wheeler transform. Bioinformatics 2009, 25:1754-1760.

36. Li H, Durbin R: Fast and accurate long-read alignment with Burrows-Wheeler transform. Bioinformatics 2010, 26:589-595.

37. Langmead B, Salzberg SL: Fast gapped-read alignment with Bowtie 2. Nat Methods 2012, 9:357-359.

38. Homer N, Merriman B, Nelson SF: BFAST: an alignment tool for large scale genome resequencing. PLoS One 2009, 4:e7767.

39. Li R, Yu C, Li Y, Lam TW, Yiu SM, Kristiansen K, Wang J: SOAP2: an improved ultrafast tool for short read alignment. Bioinformatics 2009, 25:1966-1967.

40. Ning Z, Cox AJ, Mullikin JC: SSAHA: a fast search method for large DNA databases. Genome Res 2001, 11:1725-1729.

41. Rumble SM, Lacroute P, Dalca AV, Fiume M, Sidow A, Brudno M: SHRiMP: accurate mapping of short color-space reads. PLoS Comput Biol 2009, 5:e1000386.

42. DePristo MA, Banks E, Poplin R, Garimella KV, Maguire JR, Hartl C, Philippakis AA, del Angel G, Rivas MA, Hanna M, McKenna A, Fennell TJ, Kernytsky AM, Sivachenko AY, Cibulskis K, Gabriel SB, Altshuler D, Daly MJ: A framework for variation discovery and genotyping using next-generation DNA sequencing data. Nat Genet 2011, 43:491-498.

43. Li H, Handsaker B, Wysoker A, Fennell T, Ruan J, Homer N, Marth G, Abecasis G, Durbin R: The Sequence Alignment/Map format and SAMtools. Bioinformatics 2009, 25:2078-2079.

44. Li R, Li Y, Fang X, Yang H, Wang J, Kristiansen K: SNP detection for massively parallel whole-genome resequencing. Genome Res 2009, 19:1124-1132.

45. Goya R, Sun MG, Morin RD, Leung G, Ha G, Wiegand KC, Senz J, Crisan A, Marra MA, Hirst M, Huntsman D, Murphy KP, Aparicio S, Shah SP: SNVMix: predicting single nucleotide variants from next-generation sequencing of tumors. Bioinformatics 2010, 26:730-736.

46. Koboldt DC, Chen K, Wylie T, Larson DE, McLellan MD, Mardis ER, Weinstock GM, Wilson RK, Ding L: VarScan: variant detection in massively parallel sequencing of individual and pooled samples. Bioinformatics 2009, 25:2283-2285.

47. Lam HY, Pan C, Clark MJ, Lacroute P, Chen R, Haraksingh R, O'Huallachain M, Gerstein MB, Kidd JM, Bustamante CD, Snyder M: Detecting and annotating genetic variations using the HugeSeq pipeline. Nat Biotechnol 2012, 30:226-229.

48. Liu Q, Guo Y, Li J, Long J, Zhang B, Shyr Y: Steps to ensure accuracy in genotype and SNP calling from Illumina sequencing data. BMC Genomics 2012, 13:S8.

49. Wang W, Wei Z, Lam TW, Wang J: Next generation sequencing has lower sequence coverage and poorer SNP-detection capability in the regulatory regions. Sci Rep 2011, 1:55.

50. Koboldt DC, Zhang Q, Larson DE, Shen D, McLellan MD, Lin L, Miller CA, Mardis ER, Ding L, Wilson RK: VarScan 2: somatic mutation and copy number alteration discovery in cancer by exome sequencing. Genome Res 2012, 22:568-576.

51. Larson DE, Harris CC, Chen K, Koboldt DC, Abbott TE, Dooling DJ, Ley TJ, Mardis ER, Wilson RK, Ding L: SomaticSniper: identification of somatic point mutations in whole genome sequencing data. Bioinformatics 2012, 28:311-317.

52. Roth A, Ding J, Morin R, Crisan A, Ha G, Giuliany R, Bashashati A, Hirst M, Turashvili G, Oloumi A, Marra MA, Aparicio S, Shah SP: JointSNVMix: a probabilistic model for accurate detection of somatic mutations in normal/tumour paired next-generation sequencing data. Bioinformatics 2012, 28:907-913.

53. Kumar P, Henikoff S, Ng PC: Predicting the effects of coding non-synonymous variants on protein function using the SIFT algorithm. Nat Protoc 2009, 4:1073-1081.

54. Adzhubei IA, Schmidt S, Peshkin L, Ramensky VE, Gerasimova A, Bork P, Kondrashov AS, Sunyaev SR: A method and server for predicting damaging missense mutations. Nat Methods 2010, 7:248-249.

55. Wong WC, Kim D, Carter H, Diekhans M, Ryan MC, Karchin R: CHASM and SNVBox: toolkit for detecting biologically important single nucleotide mutations in cancer. Bioinformatics 2011, 27:2147-2148.

56. Wang K, Li M, Hakonarson H: ANNOVAR: functional annotation of genetic variants from high-throughput sequencing data. Nucleic Acids Res 2010, 38:e164.

57. Chen K, Wallis JW, McLellan MD, Larson DE, Kalicki JM, Pohl CS, McGrath SD, Wendl MC, Zhang Q, Locke DP, Shi X, Fulton RS, Ley TJ, Wilson RK, Ding L, Mardis ER: BreakDancer: an algorithm for high-resolution mapping of genomic structural variation. Nat Methods 2009, 6:677-681.

58. Hormozdiari F, Hajirasouliha I, Dao P, Hach F, Yorukoglu D, Alkan C, Eichler EE, Sahinalp SC: Next-generation VariationHunter: combinatorial algorithms for transposon insertion discovery. Bioinformatics 2010, 26:i350-i357.

59. Korbel JO, Abyzov A, Mu XJ, Carriero N, Cayting P, Zhang Z, Snyder M, Gerstein MB: PEMer: a computational framework with simulation-based error models for inferring genomic structural variants from massive paired-end sequencing data.

60. Genome Biol 2009, 10:R23.

61. Zeitouni B, Boeva V, Janoueix-Lerosey I, Loeillet S, Legoix-ne P, Nicolas A, Delattre O, Barillot E: SVDetect: a tool to identify genomic structural variations from paired-end and mate-pair sequencing data. Bioinformatics 2010, 26:1895-1896.

62. Trapnell C, Pachter L, Salzberg SL: TopHat: discovering splice junctions with RNA-Seq. Bioinformatics 2009, 25:1105-1111.

63. Wang K, Singh D, Zeng Z, Coleman SJ, Huang Y, Savich GL, He X, Mieczkowski P, Grimm SA, Perou CM, MacLeod JN, Chiang DY, Prins JF, Liu J: MapSplice: accurate mapping of RNA-seq reads for splice junction discovery. Nucleic Acids Res 2010, 38:e178.

64. Au KF, Jiang H, Lin L, Xing Y, Wong WH: Detection of splice junctions from paired-end RNA-seq data by SpliceMap. Nucleic Acids Res 2010, 38:4570-4578.

65. Wu TD, Nacu S: Fast and SNP-tolerant detection of complex variants and splicing in short reads. Bioinformatics 2010, 26:873-881.

66. Dobin A, Davis CA, Schlesinger F, Drenkow J, Zaleski C, Jha S, Batut P, Chaisson M, Gingeras TR: STAR: ultrafast universal RNA-seq aligner. Bioinformatics 2013, 29:15-21.

67. Anders S, Huber W: Differential expression analysis for sequence count data. Genome Biol 2010, 11:R106.

68. Robinson MD, McCarthy DJ, Smyth GK: edgeR: a Bioconductor package for differential expression analysis of digital gene expression data. Bioinformatics 2010, 26:139-140.

69. Trapnell C, Hendrickson DG, Sauvageau M, Goff L, Rinn JL, Pachter L: Differential analysis of gene regulation at transcript resolution with RNA-seq. Nat Biotechnol 2012, 31:46-53.

70. Trapnell C, Roberts A, Goff L, Pertea G, Kim D, Kelley DR, Pimentel H, Salzberg SL, Rinn JL, Pachter L: Differential gene and transcript expression analysis of RNA-seq experiments with TopHat and Cufflinks. Nat Protoc 2012, 7:562-578.

71. Griffith M, Griffith OL, Mwenifumbo J, Goya R, Morrissy AS, Morin RD, Corbett R, Tang MJ, Hou YC, Pugh TJ, Robertson G, Chittaranjan S, Ally A, Asano JK, Chan SY, Li HI, McDonald H, Teague K, Zhao Y, Zeng T, Delaney A, Hirst M, Morin GB, Jones SJ, Tai IT, Marra MA: Alternative expression analysis by RNA sequencing. Nat Methods 2010, 7:843-847.

72. Katz Y, Wang ET, Airoldi EM, Burge CB: Analysis and design of RNA sequencing experiments for identifying isoform regulation. Nat Methods 2010, 7:1009-1015.

73. Kim D, Salzberg SL: TopHat-Fusion: an algorithm for discovery of novel fusion transcripts. Genome Biol 2011, 12:R72.

74. Chen K, Wallis JW, Kandoth C, Kalicki-Veizer JM, Mungall KL, Mungall AJ, Jones SJ, Marra MA, Ley TJ, Mardis ER, Wilson RK, Weinstein JN, Ding L: BreakFusion: targeted assembly-based identification of gene fusions in whole transcriptome paired-end sequencing data. Bioinformatics 2012, 28:1923-1924.

75. Li Y, Chien J, Smith DI, Ma J: FusionHunter: identifying fusion transcripts in cancer using paired-end RNA-seq. Bioinformatics 2011, 27:1708-1710.

76. McPherson A, Hormozdiari F, Zayed A, Giuliany R, Ha G, Sun MG, Griffith M, Heravi Moussavi A, Senz J, Melnyk N, Pacheco M, Marra MA, Hirst M, Nielsen TO, Sahinalp SC, Huntsman D, Shah SP: deFuse: an algorithm for gene fusion discovery in tumor RNA-Seq data. PLoS Comput Biol 2011, 7:e1001138.

77. Piazza R, Pirola A, Spinelli R, Valletta S, Redaelli S, Magistroni V, Gambacorti-Passerini C: FusionAnalyser: a new graphical, event-driven tool for fusion rearrangements discovery. Nucleic Acids Res 2012, 40:e123.

78. Vaske CJ, Benz SC, Sanborn JZ, Earl D, Szeto C, Zhu J, Haussler D, Stuart JM: Inference of patient-specific pathway activities from multi-dimensional cancer genomics data using PARADIGM. Bioinformatics 2010, 26:i237-i245.

79. Cerami E, Demir E, Schultz N, Taylor BS, Sander C: Automated network analysis identifies core pathways in glioblastoma. PLoS One 2010, 5:e8918.

80. Ciriello G, Cerami E, Sander C, Schultz N: Mutual exclusivity analysis identifies oncogenic network modules. Genome Res 2012, 22:398-406.

81. Akavia UD, Litvin O, Kim J, Sanchez-Garcia F, Kotliar D, Causton HC, Pochanard P, Mozes E, Garraway LA, Pe'er D: An integrated approach to uncover drivers of cancer. Cell 2010, 143:1005-1017.

82. Langmead B, Hansen KD, Leek JT: Cloud-scale RNA-sequencing differential expression analysis with Myrna. Genome Biol 2010, 11:R83.

83. Anders S, Reyes A, Huber W: Detecting differential usage of exons from RNA-seq data. Genome Res 2012, 22:2008-2017.

84. Forbes SA, Bindal N, Bamford S, Cole C, Kok CY, Beare D, Jia M, Shepherd R, Leung K, Menzies A, Teague JW, Campbell PJ, Stratton MR, Futreal PA: COSMIC: mining complete cancer genomes in the Catalogue of Somatic Mutations in Cancer. Nucleic Acids Res 2011, 39:D945-D950.

85. Cerami E, Gao J, Dogrusoz U, Gross BE, Sumer SO, Aksoy BA, Jacobsen A, Byrne CJ, Heuer ML, Larsson E, Antipin Y, Reva B, Goldberg AP, Sander C, Schultz N: The cBio cancer genomics portal: an open platform for exploring multidimensional cancer genomics data. Cancer Discov 2012, 2:401-404.

86. Gundem G, Perez-Llamas C, Jene-Sanz A, Kedzierska A, Islam A, Deu-Pons J, Furney SJ, Lopez-Bigas N: IntOGen: integration and data mining of multidimensional oncogenomic data. Nat Methods 2010, 7:92-93.

87. Baudis M, Cleary ML: Progenetix.net: an online repository for molecular cytogenetic aberration data. Bioinformatics 2001, 17:1228-1229.

88. Treangen TJ, Salzberg SL: Repetitive DNA and next-generation sequencing: computational challenges and solutions. Nat Rev Genet 2012, 13:36-46.

89. Cooper GM, Shendure J: Needles in stacks of needles: finding disease-causal variants in a wealth of genomic data. Nat Rev Genet 2011, 12:628-640.

90. Nekrutenko A, Taylor J: Next-generation sequencing data interpretation: enhancing reproducibility and accessibility. Nat Rev Genet 2012, 13:667-672.

91. Eisenstein M: Reading cancer's blueprint. Nat Biotechnol 2012, 30:581-584.

92. Katsios C, Papaloukas C, Tzaphlidou M, Roukos DH: Next-generation sequencing-based testing for cancer mutational landscape diversity: clinical implications? Expert Rev Mol Diagn 2012, 12:667-670.

CHAPTER 4

SCIENTIFIC CHALLENGES AND IMPLEMENTATION BARRIERS TO TRANSLATION OF PHARMACOGENOMICS IN CLINICAL PRACTICE

Y. W. FRANCIS LAM

4.1 INTRODUCTION

Variability in clinical response to standard therapeutic dosage regimen was reported in the 1950s by many pioneers in the field. Since then, the association between monogenic polymorphisms and variations of drugs' metabolism, transport, or target had been identified and the vision of personalized drug therapy in health care envisioned [1, 2]. Pharmacogenomic-guided drug therapy for patient is based on the premise that a large portion of interindividual variability in drug response (efficacy and/or toxicity) is genetically determined. Despite the widespread recognition of the scientific rationale and the clinical implementation of pharmacogenomic tests at several major academic medical institutions [3–7], most clinicians and researchers engaged in the discipline would agree that the early vision of achieving personalized therapy in the form of therapeutic regimens tailored to an individual's genetic profile remains some years away.

This chapter was originally published under the Creative Commons Attribution License. Lam YWF. Scientific Challenges and Implementation Barriers to Translation of Pharmacogenomics in Clinical Practice. ISRN Pharmacology **2013** *(2013). http://dx.doi.org/10.1155/2013/641089.*

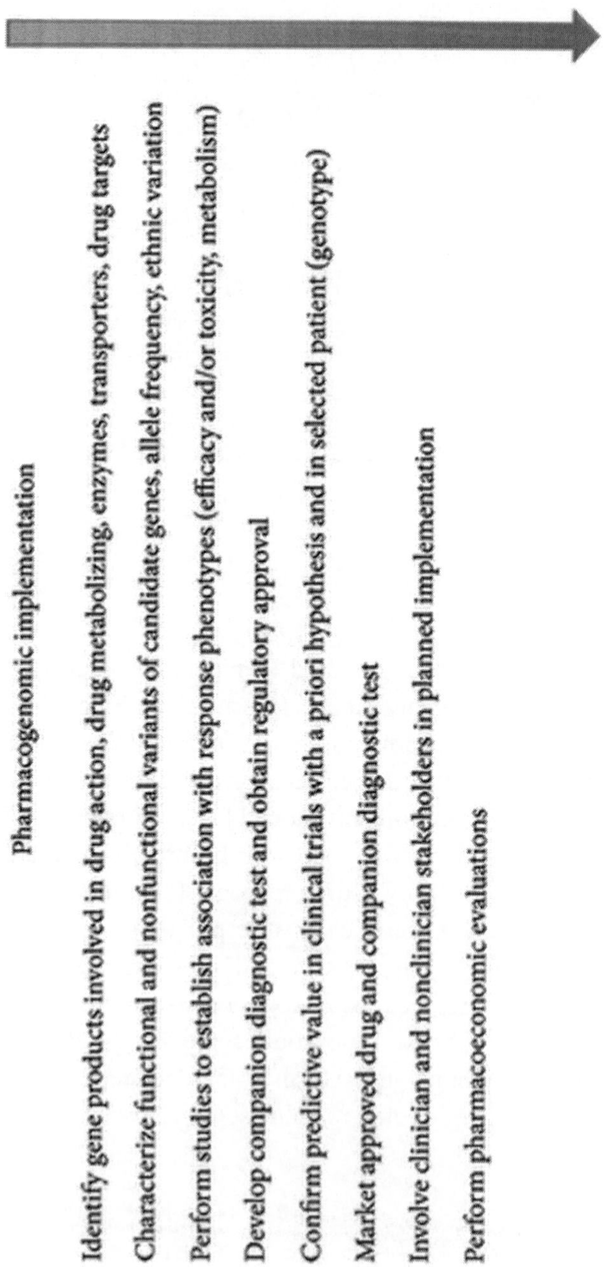

Pharmacogenomic implementation

Identify gene products involved in drug action, drug metabolizing, enzymes, transporters, drug targets

Characterize functional and nonfunctional variants of candidate genes, allele frequency, ethnic variation

Perform studies to establish association with response phenotypes (efficacy and/or toxicity, metabolism)

Develop companion diagnostic test and obtain regulatory approval

Confirm predictive value in clinical trials with a priori hypothesis and in selected patient (genotype)

Market approved drug and companion diagnostic test

Involve clinician and nonclinician stakeholders in planned implementation

Perform pharmacoeconomic evaluations

FIGURE 1: Sequence of scientific developments and implementation steps for pharmacogenomics testing in clinical practice.

Broadly speaking, the development and implementation pathways for pharmacogenomic tests consist of several stages (Figure 1): first, discovery of pharmacogenomic biomarkers and validation in well-controlled studies with independent populations; second, replication of drug-gene(s) association and demonstration of utility in at-risk patients; third, development and regulatory approval of companion-diagnostic test; fourth, assessing the clinical impact and cost-effectiveness of the pharmacogenomic biomarkers; fifth, involvement of all stakeholders in clinical implementation. Lessons learned in making pharmacogenomic-guided therapy useful to clinicians have identified multiple scientific challenges and implementation barriers existing within these stages, each of which is fueled by multitude of stakeholders with varied goals and interests [8]. This paper will provide a perspective on these existing challenges and barriers in the complex process of implementing pharmacogenomics in clinical practice, as well as incorporating pharmacogenomics into the drug development process.

4.2 SCIENTIFIC CHALLENGES AND COMPLEXITY

4.2.1 GENETIC VARIABILITIES AND NONGENETIC INFLUENCES ON GENOTYPE-PHENOTYPE ASSOCIATION

Many pharmacogenomic biomarkers have been identified over the last decade, but only few of them have been utilized to different extents in clinical setting (Table 1) [9]. One of the major challenges for translating most discovered biomarkers to their clinical implementation as genomic tests has been the inconsistent replication result of genetic associations, whether alone or in combination. Traditionally, the candidate gene approach incorporating a panel of genes that encode known drug targets, metabolizing enzymes, and membrane transporters is used in pharmacogenomic studies to test the hypothesis of an association between single nucleotide polymorphisms (SNPs) and a pharmacological or therapeutic endpoint. A good example of inconsistent replication result of genetic associations is the atypical antipsychotic clozapine with its complex pharmacological effects via the dopaminergic, serotonergic, adrenergic, and histaminergic receptors within the central nervous system. Over the years, conflicting study results exist in the literature for association between clozapine response with either SNPs of each known pharmacological receptor subtype [10–13], combinations of polymorphisms [14], and metabolizing enzymes and transporters [15]. It is also of note that the original association regarding combination of polymorphisms was not replicated in a subsequent study [16]. The recent identification of yet another new candidate gene for clozapine treatment response [15] illustrates the limitation of candidate gene approach in that there is always the possibility of involvement of other yet-to-be-identified genes, including those that have not been known to be linked to the pharmacology of the drug, that could account for additional variability in patient's therapeutic response. More importantly, the effect size of most genetic variants is small to modest. When evaluated or used alone, most of these markers are likely of insufficient sensitivity and specificity to provide clinically useful prediction, especially of efficacy.

The recognition of multiple gene variants, rather than SNPs, each accounting for part of the disposition and response phenotypes, has led to the increased use of whole genome approach for discovery of new biological pathways and identification of associations between pharmacogenomic biomarkers and response phenotypes. Genome-wide association study (GWAS) approach screens large number of SNPs (up to 2.3 million per array) across the whole genome in order to determine the most significant SNPs associated with response phenotypes. In contrast to the hypothesis-driven candidate gene approach, there is no a priori knowledge of specific gene for the discovery-driven GWAS approach. Rather, the large numbers of SNP analyses test multiple hypotheses and necessitate large sample size, sophisticated computing and platforms (e.g., Affymetrix GeneChips), and high cost. In addition, the level of significance associated with each test needs to be corrected for multiple hypothesis testings. Refinement of the GWAS approach takes a two-step design, using high-density array to discover the SNP associations in a population cohort followed by replicating the initial findings above the genome-wide significance with additional patient sets in a more hypothesis-driven study of sufficient sample size. While this approach has been successfully applied in the pharmacogenomics of clopidogrel, flucloxacillin, simvastatin, and warfarin [17–22], the implications of the results are less clear for other drugs such as the psychotropics [23–30].

A middle-of-the-road approach would be to limit the number of SNPs that warrant analysis. Based on the phenomenon of linkage disequilibrium among SNPs, whereby two or more SNPs are inherited together in haplotype blocks more frequently than would be expected based on chance alone [31], a single representative SNP within a haplotype block could serve as a "tag SNP" (tSNP) for the haplotype. By genotyping a smaller number of carefully chosen tSNPs to identify haplotype blocks of DNA sequences that are inherited together, researchers can capture other commonly associated SNPs within the same region. The HapMap database created by the International HapMap Project (http://www.hapmap.org/) is freely available for selection of these tSNPs. Based on the HapMap database, many GWASs of drug responses have been completed [18, 19, 32, 33]. It is hoped that some of the scientific challenges for study replication related to SNP genotyping may be alleviated through this approach [34].

Regardless of the choice of approach to identify the genotype-phenotype association, population variations in prevalence and relative importance of different allele variants, for example, *CYP2D6, HLA-B, UGT1A1*, and *SL-C6A4*, remind investigators of the importance of ethnicity and population stratification [35, 36], which could magnify the sample size requirement for statistical power in most pharmacogenomic studies. For example, although the algorithms based on the work of Gage et al. [37] and the International Warfarin Pharmacogenetics Consortium (IWPC) [38, 39] are clinically useful, they do not include detection of the *CYP2C9*8*, an allele commonly occurring in African Americans. The lower success with algorithm-based dose prediction in African Americans [40] is likely related to exclusion of this allele in most dosing algorithms. Another example is *HLA-B*1502* being a strong predictor of carbamazepine-induced severe cutaneous drug reactions in Han Chinese and most Southeast Asians but not in Caucasians, who do not carry the allele variant [41–43]. If not accounted for, these ethnicity-or population-related variables will confound the results of most pharmacogenomic association studies and could complicate the result interpretation. In addition, there is no universal agreement among different test platforms as to which allele variant should be tested routinely for some genetic polymorphisms, for example, *CYP2D6* and *UGT1A1*.

TABLE 1: Selected examples of drugs with relevant pharmacogenomic biomarkers and context of use.

Drugs	Pharmacogenomic biomarker or variant allele	Response phenotype	Regulatory decision and/or clinical recommendation
Abacavir	HLA-B*5701	Hypersensitivity reactions	FDA and EMA warn of increased risk in patients with HLA-B*5701. Genetic screening recommended before starting therapy. Patients tested positive should not receive abacavir.
Azathioprine and 6-mercaptopurine (e.g. TMPT*2)	Defective TPMT alleles	Myelosuppression	Increased risk for myelotoxicity in homozygotes treated with conventional doses. FDA recommends genetic testing prior to treatment.
Carbamazepine	HLA-B*1502	Stevens-Johnson syndrome (SJS) and toxic epidermal necrolysis (TEN)	FDA warns of increased risk for increased risk of SJS and TEN in patients with HLA-B*1502. Patients from high-risk regions (e.g., Southeast Asia) should be screened for HLA-B*1502 before starting carbamazepine.
Cetuximab and panitumumab	EGRF, KRAS	Efficacy	With clinical benefit limited to patients with EGRF-positive tumors, both chemotherapeutic drugs are indicated for EGRF-expressing colorectal cancer with wild-type KRAS. They may be ineffective in patients with tumors expressing KRAS mutation. Mandatory testing required.
Codeine	Duplicated or amplified CYP2D6 alleles	CNS depression	FDA warning regarding patients who are ultrarapid metabolizers secondary to the CYP2D6*2XN genotype would have much higher morphine concentration, and at increased risk for CNS symptoms related to overdose, even when treated with standard doses.

TABLE 1: *Cont.*

Drugs	Pharmacogenomic biomarker or variant allele	Response phenotype	Regulatory decision and/or clinical recommendation
Clopidogrel	Defective *CYP2C19* alleles (e.g. *CYP2C19*2*, *CYP2C19*3*)	Efficacy	FDA warns of possible reduced effectiveness in *CYP2C19* homozygotes.
Crizotinib	*ALK*	Efficacy	Mandatory testing required by the FDA to confirm the presence of lymphoma kinase (ALK) mutation prior to drug use.
Gefitinib	*EGRF*	Efficacy	Approved by EMA for treatment of EGRF-expressing tumors.
Imatinib	*BCR-ABL translocation*	Efficacy	Mandatory testing required by the FDA for confirmation of disease and selection of patients for which the drug is indicated.
Irinotecan	*UGT1A1*28*	Neutropenia	FDA recommends dosage reduction by one level in homozygotes.
Maraviroc	*CCR-5*	Efficacy	FDA and EMA approved indication is only for HIV infection with CCR-5-tropic-HIV-1.
Trastuzumab	*HER2*	Efficacy	FDA and EMA require mandatory testing for HER2-overexpressing cancers prior to treatment.
Vemurafenib	*BRAF V600E* mutation	Efficacy	FDA requires mandatory testing for the mutation prior to drug use.
Warfarin	*CYP2C9* *VKORC1*	Efficacy and toxicity (bleeding)	FDA provides dose recommendations according to *CYP2C9* and *VKORC1* genotypes.

In addition to the aforementioned ethnicity-related considerations, the drug disposition and response phenotypes can be affected by patient-specific variables. Phenocopying with a change in metabolic phenotype secondary to concurrent enzyme inhibitor [44, 45] could create genotype-phenotype discordance and affect the ability to predict possible drug response based on genotype-guided dosing and achievable drug concentration. Inflammatory responses elicited by extrahepatic tumors have been shown to release cytokines such as interleukin-6 (IL-6) and resulted in transcriptional downregulation of the human *CYP3A4* gene [46]. Therefore, lower docetaxel clearance reported in cancer patients could be related to tumor-associated inflammation and subsequent transcriptional repression of *CYP3A4*, potentially leading to unanticipated toxicity despite normal enzymatic activity in the patient. IL-6-mediated downregulation of cytochrome P-450 enzyme activities also likely contributed to a recent report of significant increase in clozapine concentration in a patient with infection and inflammation [47]. An additional challenge for applying pharmacogenomic biomarkers in targeted cancer therapeutics is sampling of tumor tissue that carries the somatic mutations (e.g., testing for the epidermal growth factor receptor 1 (HER1) mutation in patients treated with gefitinib for nonsmall cell lung cancer and testing for overexpression of the human epidermal growth factor receptor 2 (HER2) protein in patients receiving trastuzumab for breast cancer). The presence of tumor cell heterogeneity might result in intra- and interindividual variabilities in tumor tissue content and, hence, measurable level of the biomarker. In spite of this limitation, there have been multiple successful clinical applications of pharmacogenomics biomarkers in selecting chemotherapeutic drugs [48].

Furthermore, there is an increasing appreciation that genetic heterogeneity alone cannot explain interindividual variations in drug responses. Yet currently, much less is known about the influence of environmental variables and gene-environment interactions on drug disposition and response phenotypes such as mutations and polymorphisms [49–51]. Epigenetics refers to changes in gene expression without nucleotide sequence alteration. Environmental factors, through their participation in epigenetic mechanisms, could result in many different phenotypes within a population. In the not too distant future, pharmacoepigenetic investigations focusing on studying the interaction among drugs, environment, and genes could provide additional insight of drug response variations beyond the level of genetic polymorphisms [52].

4.2.2 ANALYTICAL VALIDITY, CLINICAL VALIDITY, AND CLINICAL UTILITY OF PHARMACOGENOMIC BIOMARKERS

After demonstration of a genetic association with response phenotype, there is the need of validating the biomarker, regardless of whether it is to be developed as a companion diagnostic test. For the purpose of personalized therapy, a companion diagnostic for a drug can be defined as a biomarker that is critical to the safe and effective use of the drug. The ACCE (analytical validity, clinical validity, clinical utility and associated ethical, legal, and social implications (ELSI)) Model Project [53] sponsored by the Office of Public Health Genomics, Centers for Disease Control and Prevention (CDC), has been recently advocated by some investigators to be the basis for evaluation of pharmacogenomic biomarker tests. Analytical validity determines how well a diagnostic test measures what it is intended to measure, regardless of whether it is an expression pattern, a mutation, or a protein. Clinical validity measures the ability of the test to differentiate between responders and nonresponders, or to identify patients who are at risk for adverse drug reactions. The clinical utility measures the ability of the test result to predict outcome in a clinical environment and the additional value over nontesting, that is, standard empirical treatment.

In 2004, the CDC launched the Evaluation of Genomic Applications in Practice and Prevention (EGAPP) initiative, which aims to establish an evidence-based process, including assessments of analytical validity, clinical validity, and clinical utility, for evaluating genetic tests and genomic technology that are being translated from research to clinical practice. For the pharmacogenomics discipline, one often-cited publication was the 2007 EGAPP Working Group evidence-based review of the literature on the use of CYP genotyping for clinical management of depressed patients with the selective serotonin reuptake inhibitors (SSRIs). Based on strong evidence of analytical validity, possible demonstration of clinical validity, and lack of study data to support evaluation of potential clinical utility, the working group does not recommend the application of *CYP2D6* genotyping for SSRI pharmacotherapy [54].

Since approval of most CYP genotyping tests by the Food and Drug Administration (FDA) is dependent on their technical performance in detecting CYP450 gene variants, the strong evidence of analytical validity is to be expected. The weak evidence of association between genotype and phenotypes (different metabolic phenotypes, responders versus non-responders) is also not unexpected, since most SSRIs rely on multiple but not necessarily polymorphic enzymes for metabolism and have a flat dose-response relationship with wide therapeutic index. The clinical validity of the CYP genotyping tests to differentiate response phenotypes is further limited by the CYP genotype-metabolic phenotype discordance that can occur as a result of drug-drug interactions [44, 45] or environmental influences. Given these limitations as well as the lack of cost-effectiveness data, it is not surprising that the SSRIs are not good candidates for genotype-based pharmacogenomic therapy and, hence, the recommendation of the EGAPP Working Group. Other pharmacogenomic biomarkers could be better candidates for testing association between specific genotype and clinical phenotype [55–63], as indicated by published guidelines. Pharmacogenetic dosing algorithms [37, 39] based on the patient's *CYP2C9* and *VKORC1* genotypes and other nongenetic factors (e.g., age, body size, and concurrent interacting drug) have been used to determine warfarin dosage regimens. As shown for clopidogrel, simvastatin, and warfarin, replication of the association in multiple cohorts or inclusion of replication data would provide further evidence of clinical validity [17, 18, 64, 65].

4.2.3 THE COMPLEXITY OF DEFINING WHAT CONSTITUTES CLINICAL UTILITY

Establishing the clinical utility of pharmacogenomic biomarkers has been advocated to ensure that their use is appropriate, cost-effective, and ultimately improves clinical outcome in patients. Yet within the clinical and scientific communities, there are constant debates with little agreement regarding the required levels of evidence for proof of clinical utility of diagnostic tests that are scientifically appropriate but at the same time realistically achievable [66–71]. The gold standard for demonstration of clinical utility of a drug is the use of randomized controlled trials (RCTs). Given the current evidence-based driven clinical environment, many investigators advocate that hypothesis-driven, prospective, double-blind RCTs would provide the ideal approach to validate the clinical utility of pharmacogenomic biomarkers. However, within the context of personalized medicine, the biomarker as a companion diagnostic test is intended for use with a drug to produce the optimal efficacy and safety. This makes it difficult to distinguish the clinical utility of the test that is different from that of the drug or the drug-test combination.

In addition, the traditional assessment of evidence of drug efficacy and safety with the use of RCTs may not necessarily portray the benefit of pharmacogenomic biomarkers. Complex disease etiologies, heterogeneous patient population, placebo effects, and drug response variabilities per se all contribute to statistical power issues that necessitate large patient cohort for RCT. All too often, the end result is achievement of small average benefit in the entire heterogeneous patient cohort, despite the trial being costly in terms of time and sample size. In contrast to evidence-based practice, the emphasis and value of pharmacogenomics are more geared towards incremental advantages in efficacy and safety for the outliers (the poor metabolizers, the ultra-rapid metabolizers, the nonresponders, or those susceptible to develop adverse drug reactions) over traditional therapy or standard dosing regimen. For example, the IWPC showed that a pharmacogenetic dosing algorithm was most predictive of therapeutic anticoagulation in 46% of the patients cohort who required <25 mg/week or >49 mg/week [39].

Therefore, a balance between the scientific demands of RCTs and the practical value of genotyping for patient care seems appropriate. Given the low prevalence of genetic variants associated with drug response and the desire to generate more robust evidence, many investigators and sponsors have advocated the use of prospective enrichment design clinical trials [72] to include patients who are more likely to respond or at least be stratified according to disease subtypes [73] and/or exclude patients who are highly susceptible to adverse drug reactions. However, even with the assumption of (and sometimes proven) association between genetic variabilities and drug response, both advantages and disadvantages exist for this study design [8]. A recent simulation study of trial designs suggested that conducting more trials with smaller sample sizes and lessened evidence-based criteria might contribute substantially to cancer survival, and assessment relying solely on the current traditional, risk-averse trial design might slow long-term progress [74]. In this regard, it is of note that the FDA recently approved crizotinib and vemurafenib with their respective pharmacogenomic biomarker tests solely on data from two single-arm studies. Finally, ethical concerns might preclude conducting RCT in patients with specific genetic polymorphisms [75]. Examples would be prescribing of abacavir in patients tested positive for HLA-$B*5701$ and 6-mercaptopurine or azathioprine in homozygous carriers of $TMPT$ mutations. Likewise, conducting a pharmacogenomic add-on as part of a head-to-head efficacy comparison of two antipsychotics in patients who are carriers of the Del allele of the $-141C$ Ins/Del polymorphism in the dopamine D_2 receptor gene would be difficult. The Del allele is associated with poor antipsychotic response [76]; yet, all currently marketed antipsychotics are D_2 blocker, albeit with different extent of blockade.

Not surprisingly, pharmaceutical companies have very little financial incentive to conduct time- and cost-intensive RCTs, especially for out-of-patent marketed drugs. To move the discipline forward to eventual implementation, we have to rethink the types of study design and/or the quality of study data for evidence of clinical validity and utility. The concept of conducting practical clinical trials in real-world setting had been previously proposed for regulatory decision-making [77, 78]. The recent study by Anderson et al. provided evidence of comparative effectiveness between pharmacogenetic-guided warfarin therapy in 504 patients versus standard care in 1,866 patients and a strong validation to the clinical benefit associated with the use of pharmacogenomic biomarkers in a real world setting [79]. At the "grassroot level," the concept of practical clinical trial can even be modified and adopted on a much smaller scale in clinics or physician offices. As an example, elimination of tolbutamide is known to be 50% and 84% slower in carriers of $CYP2C9*2$ and $CYP2C9*3$ variants, respectively, than in homozygous carriers of $CYP2C9*1$ [80]. Yet, to-date, there is no prospective RCT to evaluate the appropriateness of 50% to 90% dose reductions for patients who are carriers of the two allelic variants. In contrast, evaluating tolbutamide efficacy can be easily done after implementation of these dosage reductions. Therefore, such effort in clinical practice, instead of expensive and time-consuming RCT, could constitute the first step of obtaining evidence of clinical utility of *CYP2C9* genotyping in optimizing tolbutamide therapy.

For patient care, a good example for the need of balance between evidence-based medicine and personalized medicine is clopidogrel. Despite the extensive evidence of clopidogrel efficacy linked to *CYP2C19* genetic polymorphism [81, 82], debates continue over the routine use of *CYP2C19* genotyping to guide clopidogrel therapy [83–85]. This prevents more widespread use of the biomarker in individualized therapy, despite the significantly higher rates of stent thrombosis and the associated mortality rates in carriers of the reduced-function *CYP2C19*2* allele. Based on lack of outcomes data, the joint clinical alert issued in 2010 by the American College of Cardiology and the American Heart Association did not recommend routine genotyping and suggested the need of large, prospective, controlled trials. One such trial is the Pharmacogenomics of Antiplatelet Intervention-2 (PAPI-2) trial that evaluates the effect of genotype-guided antiplatelet therapy versus standard care on cardiovascular events among 7,200 patients undergoing percutaneous coronary intervention (PCI) (clinicaltrials.gov NCT01452152). However, the results will likely not be available until 2015. The questions then become are we in the meantime sacrificing patient care on the insistence of waiting for proof of value via the evidence-based approach? If no study results are available in the near future, should we focus on steps that can facilitate the genotyping implementation in clinical setting and examine the cost-effectiveness of genotypes-guided antiplatelet therapy with a variety of different approaches?

4.2.4 EVALUATION OF COST-EFFECTIVENESS

For many healthcare facilities and systems, it is also critical to assess whether a test offers a good return on investment. Therefore, in addition to clinical validity and clinical utility, another potential barrier to test implementation is demonstration of cost-effectiveness of the companion diagnostic test. Ideally, the pharmacogenomic biomarker will result in cost-effective improved clinical care in patients who will benefit from individualized therapy with the drug and avoidance of cost-ineffective treatment for patients who likely will not benefit from the drug, either as a result of lack of response or increased adverse drug reactions [86, 87].

Traditional cost-effectiveness analysis compares the relative costs and outcomes of two different approaches, typically visualized on a cost-effectiveness plane divided into four quadrants [88]. As mentioned in the last paragraph, avoidance of cost-ineffective treatment is one component of cost-effective improved clinical care. Along this line, the antipsychotic drugs offer an alternative approach to cost-effectiveness evaluation for pharmacogenomics biomarkers. With an annual cost that is at least ten times higher, the atypical antipsychotic agents are more expensive yet no more efficacious and, hence, likely to be less cost-effective, than the typical antipsychotic agents [89, 90]. Rather than focusing on using biomarkers to predict efficacy of the more expensive atypical antipsychotic agents [10–15], genotyping for the *Glycine9* allele of the *Ser9Gly* polymorphism in the dopamine 3 receptor gene [91, 92] might be used to identify patients susceptible to tardive dyskinesia, a highly prevalent adverse drug reaction associated with the use of the less expensive typical antipsychotic agents. The genetic testing might enable appropriate dose reduction for the typical antipsychotic agents and lessen the incidence of adverse drug reaction.

Additional approaches of demonstrating cost-effectiveness of pharmacogenomic-based therapy can range from clinical trial comparing per-patient cost for specific clinical outcome between genotype-based regimen and standard regimen [93] to decision model-based study using simulated patient cohort [94–96]. Alternative approach exists even within the context of cost-effectiveness comparison between genotype-based regimen and standard regimen with no genetic testing. With generic availability of clopidogrel, a cost-effectiveness study of the value of pharmacogenomic biomarker should compare clopidogrel use in *CYP2C19* EMs and UMs versus the use of prasugrel or ticagrelor for PMs.

Regardless of the specific approach, it should be understood that the economic impact and cost-effectiveness of screening could be affected by different variables. Two separate studies utilized modeling techniques with simulated patient cohorts to evaluate the potential clinical and economic outcomes for pharmacogenomic-guided warfarin dosing. While the relatively high cost of *CYP2C9* and *VKORC1* bundled test ($326 to $570) resulted in only modest improvements (quality-adjusted life years, survival rates, and total adverse rates), the investigators also suggested that improvements in the cost-effectiveness can be achieved in several ways, specifically further cost reduction of the genotyping test and utilizing genotype-guided warfarin dosing algorithm in outliers (patients with out-of-range INRs and/or those who are at high risk for hemorrhage [97, 98]). The benefits of pharmacogenomics-guided therapy for patient subpopulations have been discussed earlier. Other variables such as different population prevalence of a specific variant and cost of alternative treatment approaches would also impact the economic impact analysis.

In summary, clinical utility and cost-effectiveness cannot be the only measures in determining the relative value of pharmacogenomics for drug therapy optimization in individual patients. Rather, they should be used to supplement the best practice strategies currently in place to achieve optimal drug therapy.

4.2.5 REGULATORY APPROVAL OF PHARMACOGENOMIC DIAGNOSTIC TESTS

Over the last decade, the FDA has progressively acknowledged the importance of biomarkers and provided new recommendations on pharmacogenomic diagnostic tests and data submission. These efforts included the publication of FDA Guidance for Pharmacogenomic Data Submission, Guidance on Pharmacogenetic Tests and Genetic Tests for Heritable Markers, and draft guidance for "In Vitro Diagnostic Multivariate Index Assays" (IVDMIAs), the introduction of the Voluntary Data Submission Program, and formation of an Interdisciplinary Pharmacogenomic Review Group (IPRG) to evaluate the voluntary submissions, as well as the approval and classification of different biomarkers [99]. Obviously, any biomarker with FDA approval will generate more confidence for clinicians, healthcare facility administrators, and payers, and could enhance test implementation and utilization in the clinical settings. Additional regulatory efforts also provide an impetus of pharmacogenomic data submission for drug approval and additional research to address the debate over the utility of the information incorporated in the revised labels, for example, for clopidogrel [83–85].

Within the United States, there are separate regulatory oversights for a pharmacogenomic biomarker developed as an in-house test by a clinical laboratory versus that for an in vitro diagnostic device developed by a medical device manufacturer. Quality standards for clinical laboratory tests are governed by the Clinical Laboratory Improvement Amendments (CLIA). In addition, the laboratories are accredited either by the College of American Pathologists, the Joint Commission on Accreditation of Healthcare Organizations, or Health Department of each individual state, that take into consideration of CLIA compliance and laboratory standard practices that are in line with Good Laboratory Practice (GLP) regulations enforced by the FDA. Although there is internal validation within the laboratory, there is no external regulatory review process for the test itself.

On the other hand, the GLP regulations govern the testing of in vitro medical diagnostic device. Although currently there is no formal regulatory process for submission of companion diagnostic tests, the FDA previously ruled that evaluation and approval of the AmpliChip CYP450 Test as an in vitro diagnostic device was required. In addition, the regulatory agency had fast track approved trastuzumab with the companion diagnostic Hercep Test in 2001 for detecting overexpression of HER2 protein in breast cancer tissue by immunohistochemistry and more recently for tests that utilize fluorescence in situ hybridization to amplify the *HER2* gene. Further examples of FDA assuming a greater role were the respective companion diagnostic tests approved for crizotinib and vemurafenib. With the formation of a personalized medicine group within the Office of In Vitro Diagnostic Device, Center for Device Evaluation and radiological Health, it is likely that more FDA-approved tests would be available in the future [100]. Although no similar frameworks for premarketing regulatory review and approval of pharmacogenomic biomarkers exist in the European Union and the United Kingdom, there are regulations applicable for postmarketing approval. Gefitinib was approved by the European Medicines Agency (EMA) in June 2009, followed by subsequent approval of a companion diagnostic test for *HER1* mutations.

4.3 INTEGRATION OF PHARMACOGENOMIC BIOMARKER WITHIN THE HEALTHCARE SYSTEM

There are several challenges and practical aspects related to clinical decision support infrastructure and training of healthcare professionals (Table 2) that need to be addressed before pharmacogenomic biomarkers can be successfully utilized in any healthcare setting. These are further discussed in the following sections.

TABLE 2: Practical issues involved in clinical implementation of pharmacogenomic testing in healthcare system.

Issue	Challenge
Test performance	Reasonable turnaround time for delivery of test result
Interpretation of result	Not a straightforward normal versus abnormal interpretation
	Education of clinicians is crucial to proper use
Education of health professionals	Variable time and content devoted to educating future clinicians within health professional schools
	Overwhelming information for most current practicing clinicians
Cost reimbursement by payers	Almost exclusively based on proof of cost-effectiveness
Acceptance by clinicians	Potential additional workload
	Potential legal liability
	Health disparity concern for patient
Acceptance by patients	Privacy and discrimination concern
	Health disparity concern
	Ownership of genetic information

4.3.1 THE MULTIFACET PROCESS OF CLINICAL IMPLEMENTATION

Even with a decrease in genotyping cost over time, a relatively low demand for specific biomarker test at institutional clinical laboratories may not justify the cost of equipment and technical upkeep associated with in-house testing. This not only precludes the ideal point-of-care consultation at the bedside or within the clinic, but also results in long turnaround time for obtaining test results from external clinical laboratories or research institutions. The impact of the time delay would depend on the "urgen-cy" of the test, for example, HER2 expression or *CYP2C19* genotyping prior to scheduled PCI versus on-the-spot warfarin dosing adjustment or in the setting of emergency PCI. Nevertheless, progress has been made in this aspect. A recent commentary of pharmacogenomics in primary care reported acceptable turnaround time of 24 hours for a feasibility study of warfarin pharmacogenetic testing in a family practice clinic [101]. In addition, a point-of-care *CYP2C19* genotyping device with a turnaround time of about an hour has been developed and recently used to explore the feasibility of incorporating *CYP2C19*2* testing into clinical protocol for antiplatelet dosing [102]. In addition to technology advances, the concept and adoption of preemptive (preprescription) genotyping [5, 103–105] with result stored in electronic medical record for subsequent use would also help minimize the inconvenience of time delay in test reporting. The issue of health informatics technology will be discussed in a later section.

Not unexpectedly, patients expect healthcare professionals to be able to explain the pharmacogenomic diagnostic test results and answer their questions regarding treatment access and choices. While interpretation of genotype result for deciding the appropriateness of a specific drug for a pa-tient is usually not difficult, for example, the presence of the *HLA-B*5701* variant for excluding abacavir therapy in patients with HIV-1 infection, the contrary would be true when the genotype result is used for dosing adjustment. The challenges for genotype-based doing guidelines [106] are related to the multitude of genetic and nongenetic variables that can affect drug disposition and response, the significant interindividual variabilities in activities of most of the metabolizing enzymes, and the possibility of phenocopying with metabolic phenotype change in the presence of drug-drug interaction [44, 45]. This difference in interpretation complexity re-lated to the intended use of the test is likely one of the reasons for the FDA to previously separate pharmacogenomic biomarkers into three catego-ries. Despite these challenges, warfarin dosing recommendations based on *CYP2C9* and *VKORC1* genotypes have been incorporated by the FDA into the updated product label in 2010. The dose table provided in the product label was reported to provide better dose prediction than empiric dosing [99, 107].

However, the inclusion of most of the pharmacogenomics biomarkers as informational pharmacogenetic tests by the FDA on the revised labels of many drugs, without clear guidance on dosing recommendation and/or therapeutic alternatives, usually results in a "knowledge vacuum" for the clinicians. All stakeholders would agree that lack of sufficient pharmacogenomics education for health professionals remains a major barrier for practical implementation of pharmacogenomics within the healthcare system [8]. The need of adequate training was echoed in a recent USA survey of more than 10,000 physicians. Although 98% of all respondents agreed that the genetic profile of a patient could influence drug therapy decision, only 29% had received some pharmacogenomics education during their medical training, and only 10% felt they were adequately trained to apply the knowledge in clinical practice [108]. Although the International Society for Pharmacogenomics recommended incorporating pharmacogenomics education in medical, pharmacy, and health science curricula [109], pharmacogenomics courses or materials have only been included to a variable extent at most pharmacy schools [110, 111]. The gap in knowledge can currently be addressed through clinical guidelines available from professional organizations (Clinical Pharmacogenetics Implementation Consortium, the International AIDS Society-USA panel, the European Science Foundation, the British Association of Dermatology, and the Pharmacogenomics Working Group of the Royal Dutch Association for the Advancement of Pharmacy) [55–63, 112], availability of simple dosing algorithm such as that for warfarin [79], and further effort to include specific dosing recommendation in product label [99, 107].

The most logical setting for initial implementation of pharmacogenomics would be healthcare facilities affiliated with academic institutions. The concept of pharmacogenomics-guided drug therapy is similar to that of clinical pharmacokinetics consultation service (CPCS) or therapeutic drug monitoring (TDM) program. In this regard, the familiarity of the CPCS or TDM program should be emphasized to clinicians who view the adoption of pharmacogenomics with some skepticism. Likewise, hospitals with established CPCS or TDM program might find the task of introducing pharmacogenetic testing less formidable simply by expanding or modifying their existing clinical services. The availability of consultation service, in any format, should be complemented by educational training of clinicians to achieve specific competences. Crews et al. reported significant increase in ordering of the CYP2D6 genotyping test one year after its availability via the CPCS [3]. In a similar manner, once more clinicians are educated about the utility of pharmacogenomic approach to drug therapy, especially how to use the information, they would over time integrate pharmacogenomic findings and technologies into their practice.

The importance of healthcare informatics for implementation of pharmacogenomics in clinical practice could not be overemphasized. At the level of patient care, integration of genotyping order template and/or genotype result into a robust system of electronic medical record (EMR) with pop-up action alert and order templates for actionable pharmacogenomic tests to be used by physicians will be necessary [113, 114]. At the level of research, the health information technology would enable organizational management of all research data and accessibility by the EMR [115–118]. Both the patient care- and research-level informatics should incorporate updated information when available and be linked to other health informatics such as billing, clinical laboratory, and clinical trials within the healthcare facility. Although adoption of EMR is not universal [119], health information technology is a critical area for investment by healthcare system administrators, perhaps through collaborative efforts with the technology industry and the government. Successful examples incorporating a coordinated team approach (physicians, pharmacists, information technology and laboratory personnels) with appropriate infrastructure support (informatics) to facilitate clinical implementation of pharmacogenomics have been reported at several institutions [3–5].

To fully integrate the multifacet process of the pharmacogenomics service, other organizational aspects of clinical decision support should include fostering effective communication and collaboration between laboratory staff and clinicians, creating flexible workflow with minimal disruption to the daily activities of the practitioners, delineating policies and reward systems that allow equitable schedule to minimize the additional "time burdens" perceived by some healthcare providers, and standardizing procedures to incorporate up-to-date pharmacogenomics-related information into formulary review and decision by the pharmacy and therapeutics committee. All these steps would facilitate implementation with minimal effect on work efficiency and cost for the healthcare system.

4.3.2 REIMBURSEMENT ISSUES

Successful implementation of pharmacogenomic biomarkers in clinical practice not only involves multidisciplinary coordination among physicians, pharmacists, clinical laboratories, health information specialists, and healthcare system administrators, but also requires collaborative efforts and willingness from the payer, a significant stakeholder in this endeavor. With the current healthcare landscape and the high cost of providing healthcare, the reimbursability of any particular test plays a significant role in deciding its implementation status in most healthcare facilities. While the cost of testing for several oncologic biomarkers and thiopurine S-methyltransferase in the United States is reimbursed in some hospitals, that is not the case for most pharmacogenomic biomarker tests. Both federal and private payers are reluctant to reimburse the cost of the tests on the basis of either (1) lack of evidence of clinical utility (which is usually associated with endorsement by professional organizations), (2) tests being not medically necessary (because it has never been classified by the FDA as required test), or (3) lack of cost-effectiveness analysis and/or comprehensive comparative effectiveness analysis. Even with the product labeling information regarding the impact of CYP variants for warfarin, the Centers for Medicare and Medicaid Services recently denied coverage for genetic testing except when the test is provided for the purpose of clinical trials. This reluctance stance is consistent with the findings by Cohen et al. [120] who reported that most payers do not consider test accuracy in identifying subpopulations of interest, test cost, medication adherence, and off-label use as relevant factors in their consideration for reimbursement. In their survey of 12 payers, the most consistent determining factor is conclusive evidence linking the use of the diagnostic test with health outcome.

Even though most payers understand the implications of pharmacogenomics in healthcare and the potential return on investment, their reluctance to pay for diagnostic tests costing much less (most costing ≤ $500) than what they actually pay for the more expensive drugs (for which the diagnostic tests could be useful) primarily reflects their expectation of demonstration of clinical utility and comparative effectiveness [120, 121]. Accordingly, inconsistent assessment of clinical utility and benefit could only result in confusion regarding the appropriate use and interpretation of biomarker-based pharmacogenomic diagnostic tests. Hopefully, more realistic clinical practice guidelines from diverse groups of organizations and expert panels that take into consideration of the issues discussed earlier, would pave the way to greater extent of implementation. To that end, it is of note that regulatory guidance [122] has been published to support the recommendation of the clinical practice guidelines. In addition, additional clarification from regulatory agencies regarding definition of clinical utility, especially in the context of distinguishing the difference between utility of a diagnostic test versus test/drug combination versus the drug itself, would be very helpful in dealing with issues of implementation decision and test reimbursement.

It should also be noted that even for trastuzumab, which is reimbursed by most insurers, there have been few cost-effectiveness analysis of HER2 protein expression and treatment with trastuzumab [123]. For most pharmacogenomic biomarkers, the ideal analyses might not be available until years after the diagnostic test is marketed. With limited comprehensive pharmacoeconomic data for cost-effectiveness evaluations [124, 125], other evaluation approaches ranging from comparing per-patient cost for specific clinical outcome within in-patient setting [93] to decision model-based study that utilizes simulated patient cohort [94–96] should be considered. In addition, all stakeholders should recognize that a "negative" cost-effectiveness conclusion based primarily on high cost of genotyping needs to be interpreted with the high likelihood of lower cost of genotyping in the foreseeable future.

Since revenue generation from a pharmacogenomic diagnostic companion test would likely be significantly less than that for a drug, there is not much incentive for pharmaceutical companies to include a thorough cost-effectiveness analysis as part of drug development. With much less financial resources than pharmaceutical companies, the lack of incentive for conducting similar evaluations also applies to diagnostic companies developing the biomarkers. In a way similar to the mutually beneficial codevelopment of proprietary drug and diagnostic test [126, 127], one possible solution is for diagnostic companies to collaborate with other stakeholders, such as pharmacy benefit manager (PBM), to generate the evidence deemed necessary for reimbursement by both private payers and regulatory agencies. Medco is the first PBM to use claims data in demonstrating a 28% reduction in bleeding or thromboembolic events in patients whose physicians were provided with *CYP2C9* and *VKORC1* genotypes results, when compared to patients without genetic testing. Concurrent with the clinical effectiveness data is a $910 cost saving over a 6-month study period in the genotyped group [128]. This type of economic impact data for pharmacogenomic testing could be used as evidence of cost-effectiveness to insurance payers and administrators of healthcare systems for consideration of potential implementation.

Given the dilemma of insistence of evidence-based data for reimbursement and the limited financial resource of most diagnostic companies in developing the biomarker, some paradigm shifts in thinking about approaches to reimbursement decision could be offered to the payers. Instead of a universal reimbursement for all patients tested for a pharmacogenomic biomarker, an action-based reimbursement could be instituted. Using clopidogrel as an example, the differential reimbursement could take the form of no payment for the *CYP2C19* genotype test, if no PCI is performed and clopidogrel is not prescribed, or even different amount of payment based on the risk of PCI. This differential pay concept is currently in place for most prescription drugs in the form of copayment, as well as in coverage amount between within-network versus out-of-network physician visits [110]. Adopting such approach would lessen the financial burden for payer since the cost of the one-time test could be easily covered through cost saving associated with not using the drug when it is ineffective or harmful in specific patient populations, and it could provide a work-around to some payers' insisting on conclusive evidence of linking diagnostic tests to health outcomes [120].

4.3.3 ETHICAL, LEGAL, AND SOCIAL ISSUES

Implementation of pharmacogenomic testing could result in situations where an individual's disease or medical condition is revealed to other parties, however unintended, as well as potential for discrimination and ineligibility for employment and insurance. Therefore, even though the public is in general receptive to genetic-based prescribing [129, 130], effort should be directed towards alleviating their concern regarding privacy and confidentiality for the purposes of employment and insurance coverage decisions. They should be informed that there are ways to both protect patients' privacy whilst at the same time promote the pharmacogenomic implementation in clinical practice [131, 132]. In addition, provisions from the 2008 Genetic Information Nondiscrimination Act were designed to protect individuals from genetic discrimination. Addressing these concerns also encourages informed patients to participate in necessary research [115, 133], for example, comparative effectiveness requested by other stakeholders, as well as facilitate healthcare professionals' willingness to fully integrate genomic services into clinical practice. Despite this, other existing concerns include ownership of genetic materials, availability and access to the information (both locally and across different health system facilities similar to that of the Veterans Affairs EMR), and patient's awareness of the consequences of storing genetic materials and phenotypic data. These concerns would need to be addressed to the satisfaction of all stakeholders, especially the patients.

Most discussions and debates on the ethical, legal, and social implications of genetic tests usually make few distinctions between pharmacogenomic biomarkers designed for drug therapy individualization and genetic tests predicting disease susceptibility that usually carry a much greater potential for abuse. For the purpose of implementation, it would seem appropriate that consent for pharmacogenomic biomarker tests designed to individualize their drug therapy (choice and/or dosage regimen) not be treated the same extent of scrutiny and requirement as genetic testing for disease susceptibility. A lessening in regulation and consent requirements for pharmacogenomic markers might make it easier for their implementation. However, this issue of is very much open for further discussion before consensus can be made.

Social concerns also arise from clinical implementation of pharmacogenomic biomarkers within the healthcare systems. In the United States, patients are required to pay for some of the cost of the medical service, either in the form of copayment or coinsurance. Therefore, an individual patient's socioeconomic status could preclude any potential beneficial pharmacogenomic test information and exacerbate health-care disparities among different patients. In addition, for patients who are identified by pharmacogenomic test either as nonresponders or at high risk of adverse drug reaction to a specific drug, the use of pharmacogenomic test as a "gatekeeper" of accessibility to drug treatment might pose a problem if there is no suitable alternative drug available. As discussed earlier in this paper, carriers of the Del allele of the 141C Ind/Del polymorphism of the dopamine D_2 receptor gene are predicted to have poor response to antipsychotic treatment; yet, all currently marketed antipsychotic treatments possess blockade. How then should those patients be advised and treated? Is it ethical or appropriate if the patient and/or the physician decide to use a drug regardless of the unfavorable response and/or risk associated with a specific genotype? These are relevant questions since the clinical validity and clinical utility of most pharmacogenomic tests have not been universally accepted in clinical practice. Another potential concern is liability for the healthcare provider. If a pharmacogenetic test (e.g., *CYP2C29*) is used to guide therapy with one drug (e.g., warfarin) and the patient is later prescribed another drug that is also affected by the gene previously tested (e.g., phenytoin), should the clinician be responsible to act on the genotype results when dosing the second drug? If the answer is affirmative, then some point-of-care mechanism must be in place, for example, in an EMR with pop-up action alert containing the pharmacogenomic information, so that the clinician is aware of genetic test results relevant to the prescribed drug. The immediate implication with availability of pharmacogenomic information within the EMR is that the information should not be ignored for clinical, ethical, and legal reasons.

Pharmacogenomic biomarker tests are a subset of the increasing universe of genetic tests advertised over the internet directly to the consumer. Most of these direct-to-consumer (DTC) genetic tests are "home brew" and not subject to regulatory oversight by the FDA and/or CLIA compliance for test quality standards and proficiency. In addition, companies selling DTC genetic tests can develop and market them without establishing clinical utility, which contrasts significantly to that demanded for pharmacogenomic biomarkers discussed earlier in this paper. The lack of regulatory oversight and concern of test validity likely contribute to the conclusion that most DTC genetic tests are not useful in predicting disease risk [134, 135]. Current knowledge suggests that genomic profiling based on a single SNP, a common feature to most DTC genetic tests, is not necessarily clinically accurate or useful. In this regard, the recent report of a DTC genome-wide platform [136] could provide a useful example of the impact of pharmacogenomic profiling on patient care. Despite the increased consumer desire for health-related information and personalized medicine, most patients would need the help of clinicians to differentiate the relevance of different pharmacogenomics tests. This underscores the importance of educating clinicians and preparing them to provide the appropriate test interpretation for clinical decision-making.

4.4 INCORPORATING PHARMACOGENOMICS INTO DRUG DEVELOPMENT

Incorporating pharmacogenomics into the entire drug development process holds significant potentials for more efficient and effective clinical trials as well as financial implications for the industry. However, the issues of sufficient sample size, the cost and time associated with conducting a RCT to address a specific study hypothesis, and the logistics of ensuring privacy concerns of institutional review board with possible delay in study approval and subject enrollment have posted a significant challenge and deterrent for the industry to fully incorporate pharmacogenomics in different phases of drug development [137]. In addition, the blockbuster drug concept and its financial impact on revenue have historically played a major role in pharmaceutical drug development. As such, the concept of pharmacogenomics and the resultant segmented (and smaller) market tailored to a subpopulation with specific genotype have been viewed unfavorably because of lower revenue and decreased profit. However, trastuzumab provides a good example of the benefit of paradigm shift in thinking about market share and revenue. The manufacturer's development of trastuzumab along with the diagnostic device results in capturing the market share associated with breast cancer drug treatment in all, albeit at a smaller number, of the women overexpressing the HER2 protein.

There are additional drug development advantages associated with this "mental shift" in business model from the traditional approach of product differentiation to the new commerce of market segmentation, sometimes even with little or no competition. Identifying patients likely to respond to participate in clinical trials could enable benefits to be shown in a smaller number of patients, resulting in more efficient phases II and III studies conducted in shorter time frame and reducing the overall cost of drug development. It could also screen out patients likely to have unfavorable side effects that only appear in phase IV postmarketing surveillance studies, and such undesirable events sometime could lead to the inevitable and unfavorable outcomes of postmarketing product recall and litigation. The litigation and financial burden could be further minimized if the pharmaceutical company works with regulatory agencies to incorporate the pharmacogenomic information into a drug label that more accurately describes contraindications, precautions, and warnings [138]. Finally, as indicated earlier in this paper, beneficial partnership to develop and market a companion diagnostic test can also lead to additional revenue stream [127].

With more than 50% of new chemical entities failing in expensive phase III clinical trials, high attrition rate in drug development is a well-known fact for the pharmaceutical industry, and a much less discussed and explored role of pharmacogenomics is the potential of "rescuing" drugs that fail clinical trials during drug development. The prime example for this benefit is gefitinib, which originally was destined to failure because only a small number of patients with small cell lung cancer responded to the drug. However, in 2004, published results showed that tumor response to the drug was linked to mutations in *HER1*. Subsequently, development of pharmacogenomic biomarker tests for *HER1* mutations in patients enables identification of responders for gefitinib [139–143]. This example showed that investigational drugs found to be ineffective or unsafe during phase II or III clinical trials might deserve a second look from the perspective of pharmacogenomics. Another example is lumiracoxib, a selective cyclooxygenase-2 inhibitor that was withdrawn in 2005 from most global pharmaceutical markets because of hepatotoxicity. Recently, Singer et al. reported a strong association between patients with *HLA-DQ* variant alleles, especially *HLA-DQA1*0102*, and elevated transferase levels secondary to lumiracoxib-related liver injury [144, 145]. As a result, the manufacturer of lumiracoxib has submitted an application to the EMA for its use in targeted subpopulations.

Therefore, as demonstrated by gefitinib and possibly lumiracoxib, "failing" drugs can be further developed with a smaller target population with the genetic profile predictive of improved efficacy and/or reduced toxicity. This result can then be used for approval with appropriate product label containing the pharmacogenomic information. In reality, a go-ahead decision by the pharmaceutical company for such "drug rescue" with potential drug approval is dependent not only on the cost and time associated with developing a companion diagnostic test but also measurable better efficacy than competitor drugs in a smaller number of patients. To facilitate this aspect of drug development, regulatory "decision incentives" in the form of conditional approval with subsequent requirement of phase IV trial or approval similar to those developed and submitted under the Orphan Drug Act could go a long way to provide sufficient incentive for the pharmaceutical industry.

Regulatory agencies worldwide, primarily the FDA, the EMA, and the Japanese Pharmaceuticals and Medical Devices Agency, have recognized the opportunity to utilize pharmacogenomics in predicting drug response and incorporated pharmacogenomic information into revised labels of approved drugs as well as regulatory review, for example, by the IPRG of the FDA, that is independent of the drug review itself. Nevertheless, relevant drug efficacy and safety data and issues that are important for regulatory decision-making were developed long before the era of pharmacogenomics, and it is unclear how traditional regulatory review would approach the inclusion of any pharmacogenomic data in a new drug application (NDA) package. As described earlier, the FDA has developed multidisciplinary workshop [146] as well as regulatory initiatives such as the Voluntary Exploratory Data Submission in the USA, and the Pharmacogenomics Briefing Meetings in Europe and Japan have attempted to encourage the use and submission of pharmacogenomic data by the pharmaceutical industry. However, concerns and questions remain regarding what type of pharmacogenomic data is necessary and when they should be incorporated in the NDA process [8].

4.5 CONCLUSION

Although significant scientific and technological advances enable identification of variants in (or haplotypes linked to) genes that regulate the disposition and target pathways of drugs, translating the pharmacogenomic findings into clinical practice has been met with continued scientific debates, as well as commercial, economical, educational, ethical, legal, and societal barriers. Despite the well-known potentials of improving drug efficacy and safety, as well as the efficiency of the drug development process, the logistical issues and challenges identified for incorporating pharmacogenomics into clinical practice and drug development could only be addressed with all stakeholders in the field working together and occasionally accepting a paradigm change in their current approach.

REFERENCES

1. W. E. Evans and M. V. Relling, "Pharmacogenomics: translating functional genomics into rational therapeutics," Science, vol. 286, no. 5439, pp. 487–491, 1999.
2. E. S. Vesell, "New directions in pharmacogenetics: introduction," Federation Proceedings, vol. 43, no. 8, pp. 2319–2325, 1984.
3. K. R. Crews, S. J. Cross, J. N. Mccormick et al., "Development and implementation of a pharmacist-managed clinical pharmacogenetics service," American Journal of Health-System Pharmacy, vol. 68, no. 2, pp. 143–150, 2011.
4. D. R. Nelson, M. Conlon, C. Baralt, J. A. Johnson, M. J. Clare-Salzler, and M. Rawley-Payne, "University of florida clinical and translational science institute: transformation and translation in personalized medicine," Clinical and Translational Science, vol. 4, no. 6, pp. 400–402, 2011.
5. J. M. Pulley, J. C. Denny, J. F. Peterson et al., "Operational implementation of prospective genotyping for personalized medicine: the design of the vanderbilt PREDICT project," Clinical Pharmacology and Therapeutics, vol. 92, pp. 87–95, 2012.
6. P. H. O'Donnell, A. Bush, J. Spitz et al., "The 1200 patients project: creating a new medical model system for clinical implementation of pharmacogenomics," Clinical Pharmacology and Therapeutics, vol. 92, pp. 446–449, 2012.
7. L. Cavallari and E. A. Nutescu, "A team approach to warfarin pharmacogenetics," in Pharmacy Practice Nerws, 2012.
8. Y. W. F. Lam, "Translating pharmacogenomic research to therapeutic potentials," in Pharmacogenomics: Challenges and Opportunities in Therapeutic Implementation, Y. W. F. Lam and L. H. Cavallari, Eds., Elsevier, 2013.
9. I. Zineh, G. D. Pebanco, C. L. Aquilante et al., "Discordance between availability of pharmacogenetics studies and pharmacogenetics-based prescribing information

for the top 200 drugs," The Annals of Pharmacotherapy, vol. 40, no. 4, pp. 639–644, 2006.

10. A. K. Malhotra, D. Goldman, R. W. Buchanan et al., "The dopamine D3 receptor (DRD3) Ser9Gly polymorphism and schizophrenia: a haplotype relative risk study and association with clozapine response," Molecular Psychiatry, vol. 3, no. 1, pp. 72–75, 1998.

11. A. K. Malhotra, D. Goldman, C. Mazzanti, A. Clifton, A. Breier, and D. Pickar, "A functional serotonin transporter (5-HTT) polymorphism is associated with psychosis in neuroleptic-free schizophrenics," Molecular Psychiatry, vol. 3, no. 4, pp. 328–332, 1998.

12. A. K. Malhotra, D. Goldman, N. Ozaki, A. Breier, R. Buchanan, and D. Pickar, "Lack of association between polymorphisms in the 5-HT(2A) receptor gene and the antipsychotic response to clozapine," The American Journal of Psychiatry, vol. 153, no. 8, pp. 1092–1094, 1996.

13. M. Masellis, V. Basile, H. Y. Meltzer et al., "Serotonin subtype 2 receptor genes and clinical response to clozapine in schizophrenia patients," Neuropsychopharmacology, vol. 19, no. 2, pp. 123–132, 1998.

14. M. J. Arranz, J. Munro, J. Birkett et al., "Pharmacogenetic prediction of clozapine response," The Lancet, vol. 355, no. 9215, pp. 1615–1616, 2000.

15. S. T. Lee, S. Ryu, S. R. Kim et al., "Association study of 27 annotated genes for clozapine pharmacogenetics: validation of preexisting studies and identification of a new candidate gene, ABCB1, for treatment response," Journal of Clinical Psychopharmacology, vol. 32, no. 4, pp. 441–448, 2012.

16. J. Schumacher, T. G. Schulze, T. F. Wienker, M. Rietschel, and M. M. Nothen, "Pharmacogenetics of clozapine response," The Lancet, vol. 356, no. 9228, pp. 506–507, 2000.

17. A. R. Shuldiner, J. R. O'Connell, K. P. Bliden et al., "Association of cytochrome P450 2C19 genotype with the antiplatelet effect and clinical efficacy of clopidogrel therapy," Journal of the American Medical Association, vol. 302, no. 8, pp. 849–858, 2009.

18. E. Link, S. Parish, J. Armitage et al., "SLCO1B1 variants and statin-induced myopathy—a genomewide study," The New England Journal of Medicine, vol. 359, no. 8, pp. 789–799, 2008.

19. G. M. Cooper, J. A. Johnson, T. Y. Langaee et al., "A genome-wide scan for common genetic variants with a large influence on warfarin maintenance dose," Blood, vol. 112, no. 4, pp. 1022–1027, 2008.

20. P. C. Cha, T. Mushiroda, A. Takahashi et al., "Genome-wide association study identifies genetic determinants of warfarin responsiveness for Japanese," Human Molecular Genetics, vol. 19, no. 23, pp. 4735–4744, 2010.

21. F. Takeuchi, R. McGinnis, S. Bourgeois et al., "A genome-wide association study confirms VKORC1, CYP2C9, and CYP4F2 as principal genetic determinants of warfarin dose," PLoS Genetics, vol. 5, no. 3, 2009.

22. A. K. Daly, P. T. Donaldson, P. Bhatnagar et al., "HLA-B*5701 genotype is a major determinant of drug-induced liver injury due to flucloxacillin," Nature Genetics, vol. 41, no. 7, pp. 816–819, 2009.

23. K. Åberg, D. E. Adkins, J. Bukszár et al., "Genomewide association study of movement-related adverse antipsychotic effects," Biological Psychiatry, vol. 67, no. 3, pp. 279–282, 2010.
24. D. E. Adkins, K. Åberg, J. L. McClay et al., "Genomewide pharmacogenomic study of metabolic side effects to antipsychotic drugs," Molecular Psychiatry, vol. 16, no. 3, pp. 321–332, 2011.
25. D. E. Adkins, K. Berg, J. L. McClay et al., "A genomewide association study of citalopram response in major depressive disorder-a psychometric approach," Biological Psychiatry, vol. 68, no. 6, pp. e25–e27, 2010.
26. A. Alkelai, L. Greenbaum, A. Rigbi, K. Kanyas, and B. Lerer, "Genome-wide association study of antipsychotic-induced parkinsonism severity among schizophrenia patients," Psychopharmacology, vol. 206, no. 3, pp. 491–499, 2009.
27. C. Lavedan, L. Licamele, S. Volpi et al., "Association of the NPAS3 gene and five other loci with response to the antipsychotic iloperidone identified in a whole genome association study," Molecular Psychiatry, vol. 14, no. 8, pp. 804–819, 2009.
28. J. L. McClay, D. E. Adkins, K. Åberg et al., "Genome-wide pharmacogenomic analysis of response to treatment with antipsychotics," Molecular Psychiatry, vol. 16, no. 1, pp. 76–85, 2011.
29. S. Volpi, C. Heaton, K. MacK et al., "Whole genome association study identifies polymorphisms associated with QT prolongation during iloperidone treatment of schizophrenia," Molecular Psychiatry, vol. 14, no. 11, pp. 1024–1031, 2009.
30. S. Volpi, S. G. Potkin, A. K. Malhotra, L. Licamele, and C. Lavedan, "Applicability of a genetic signature for enhanced iloperidone efficacy in the treatment of schizophrenia," Journal of Clinical Psychiatry, vol. 70, no. 6, pp. 801–809, 2009.
31. J. W. Belmont, P. Hardenbol, T. D. Willis et al., "The international HapMap project," Nature, vol. 426, no. 6968, pp. 789–796, 2003.
32. M. I. Lucena, M. Molokhia, Y. Shen et al., "Susceptibility to amoxicillin-clavulanate-induced liver injury is influenced by multiple HLA class I and II alleles," Gastroenterology, vol. 141, no. 1, pp. 338–347, 2011.
33. K. A. Jablonski, J. B. McAteer, P. I. W. De Bakker et al., "Common variants in 40 genes assessed for diabetes incidence and response to metformin and lifestyle intervention in the diabetes prevention program," Diabetes, vol. 59, no. 10, pp. 2672–2681, 2010.
34. R. Lubomirov, J. D. Lulio, A. Fayet et al., "ADME pharmacogenetics: investigation of the pharmacokinetics of the antiretroviral agent lopinavir coformuiated with ritonavir," Pharmacogenetics and Genomics, vol. 20, no. 4, pp. 217–230, 2010.
35. G. Suarez-Kurtz, D. D. Vargens, V. A. Sortica, and M. H. Hutz, "Accuracy of NAT2 SNP genotyping panels to infer acetylator phenotypes in African, Asian, Amerindian and admixed populations," Pharmacogenomics, vol. 13, no. 8, pp. 851–854, 2012.
36. G. Suarez-Kurtz, S. D. Pena, and M. H. Hutz, "Application of the FST statistics to explore pharmacogenomic diversity in the Brazilian population," Pharmacogenomics, vol. 13, no. 7, pp. 771–777, 2012.
37. B. F. Gage, C. Eby, J. A. Johnson et al., "Use of pharmacogenetic and clinical factors to predict the therapeutic dose of warfarin," Clinical Pharmacology and Therapeutics, vol. 84, no. 3, pp. 326–331, 2008.

38. R. P. Owen, R. B. Altman, and T. E. Klein, "PharmGKB and the international warfarin pharmacogenetics consortium: the changing role for pharmacogenomic databases and single-drug pharmacogenetics," Human Mutation, vol. 29, no. 4, pp. 456–460, 2008.

39. T. E. Klein, R. B. Altman, N. Eriksson et al., "Estimation of the warfarin dose with clinical and pharmacogenetic data," The New England Journal of Medicine, vol. 360, pp. 753–764, 2009.

40. J. Shin and D. Cao, "Comparison of warfarin pharmacogenetic dosing algorithms in a racially diverse large cohort," Pharmacogenomics, vol. 12, no. 1, pp. 125–134, 2011.

41. M. McCormack, A. Alfirevic, S. Bourgeois et al., "HLA-A*3101 and carbamazepine-induced hypersensitivity reactions in Europeans," The New England Journal of Medicine, vol. 364, pp. 1134–1143, 2011.

42. P. Chen, J. J. Lin, C. S. Lu et al., "Carbamazepine-induced toxic effects and HLA-B*1502 screening in Taiwan," The New England Journal of Medicine, vol. 364, no. 12, pp. 1126–1133, 2011.

43. V. L. Yip, A. G. Marson, A. L. Jorgensen, M. Pirmohamed, and A. Alfirevic, "HLA genotype and carbamazepine-induced cutaneous adverse drug reactions: a systematic review," Clinical Pharmacology and Therapeutics, vol. 92, pp. 757–765, 2012.

44. C. L. Alfaro, Y. W. F. Lam, J. Simpson, and L. Ereshefsky, "CYP2D6 status of extensive metabolizers after multiple-dose fluoxetine, fluvoxamine, paroxetine, or sertraline," Journal of Clinical Psychopharmacology, vol. 19, no. 2, pp. 155–163, 1999.

45. C. L. Alfaro, Y. W. F. Lam, J. Simpson, and L. Ereshefsky, "CYP2D6 inhibition by fluoxetine, paroxetine, sertraline, and venlafaxine in a crossover study: intraindividual variability and plasma concentration correlations," Journal of Clinical Pharmacology, vol. 40, no. 1, pp. 58–66, 2000.

46. G. R. Robertson, C. Liddle, and S. J. Clarke, "Inflammation and altered drug clearance in cancer: transcriptional repression of a human CYP3A4 transgene in tumor-bearing mice," Clinical Pharmacology and Therapeutics, vol. 83, no. 6, pp. 894–897, 2008.

47. K. A. Espnes, K. O. Heimdal, and O. Spigset, "A puzzling case of increased serum clozapine levels in a patient with inflammation and infection," Therapeutic Drug Monitoring, vol. 34, no. 5, pp. 489–492, 2012.

48. F. Stegmeier, M. Warmuth, W. R. Sellers, and M. Dorsch, "Targeted cancer therapies in the twenty-first century: lessons from imatinib," Clinical Pharmacology and Therapeutics, vol. 87, no. 5, pp. 543–552, 2010.

49. A. Gomez and M. Ingelman-Sundberg, "Pharmacoepigenetics: its role in interindividual differences in drug response," Clinical Pharmacology and Therapeutics, vol. 85, no. 4, pp. 426–430, 2009.

50. W. Baer-Dubowska, A. Majchrzak-Celińska, and M. Cichocki, "Pharmocoepigenetics: a new approach to predicting individual drug responses and targeting new drugs," Pharmacological Reports, vol. 63, no. 2, pp. 293–304, 2011.

51. V. Perera, A. S. Gross, and A. J. McLachlan, "Influence of environmental and genetic factors on CYP1A2 activity in individuals of South Asian and European ancestry," Clinical Pharmacology and Therapeutics, vol. 92, no. 4, pp. 511–519, 2012.

52. M. Kacevska, M. Ivanov, and M. Ingelman-Sundberg, "Epigenomics and interindividual differences in drug response," Clinical Pharmacology and Therapeutics, vol. 92, no. 6, pp. 727–736, 2012.

53. Centers for Disease Control and Prevention, Genetic testing: ACCE model system for collecting, analyzing and disseminating information on genetic tests ,2010.

54. A. O. Berg, M. Piper, K. Armstrong et al., "Recommendations from the EGAPP working group: testing for cytochrome P450 polymorphisms in adults with nonpsychotic depression treated with selective serotonin reuptake inhibitors," Genetics in Medicine, vol. 9, no. 12, pp. 819–825, 2007.

55. L. Becquemont, A. Alfirevic, U. Amstutz et al., "Practical recommendations for pharmacogenomics-based prescription: 2010 ESF-UB conference on pharmacogenetics and pharmacogenomics," Pharmacogenomics, vol. 12, no. 1, pp. 113–124, 2011.

56. K. R. Crews, A. Gaedigk, H. M. Dunnenberger et al., "Clinical Pharmacogenetics Implementation consortium (CPIC) guidelines for codeine therapy in the context of cytochrome P450 2D6 (CYP2D6) genotype," Clinical Pharmacology and Therapeutics, vol. 91, pp. 321–326, 2012.

57. E. A. Fargher, K. Tricker, W. Newman et al., "Current use of pharmacogenetic testing: a national survey of thiopurine methyltransferase testing prior to azathioprine prescription," Journal of Clinical Pharmacy and Therapeutics, vol. 32, no. 2, pp. 187–195, 2007.

58. J. A. Johnson, L. Gong, M. Whirl-Carrillo et al., "Clinical Pharmacogenetics Implementation Consortium Guidelines for CYP2C9 and VKORC1 genotypes and warfarin dosing," Clinical Pharmacology and Therapeutics, vol. 90, pp. 625–629, 2011.

59. M. A. Martin, T. E. Klein, B. J. Dong, M. Pirmohamed, D. W. Haas, and D. L. Kroetz, "Clinical pharmacogenetics implementation consortium guidelines for hla-B genotype and abacavir dosing," Clinical Pharmacology and Therapeutics, vol. 91, pp. 734–738, 2012.

60. M. V. Relling, E. E. Gardner, W. J. Sandborn et al., "Clinical pharmacogenetics implementation consortium guidelines for thiopurine methyltransferase genotype and thiopurine dosing," Clinical Pharmacology and Therapeutics, vol. 89, pp. 387–391, 2011.

61. S. A. Scott, K. Sangkuhl, E. E. Gardner et al., "Clinical pharmacogenetics implementation consortium guidelines for cytochrome P450-2C19 (CYP2C19) genotype and clopidogrel therapy," Clinical Pharmacology and Therapeutics, vol. 90, no. 2, pp. 328–332, 2011.

62. M. A. Thompson, J. A. Aberg, P. Cahn et al., "Antiretroviral treatment of adult HIV infection: 2010 recommendations of the International AIDS Society-USA panel," Journal of the American Medical Association, vol. 304, no. 3, pp. 321–333, 2010.

63. R. A. Wilke, L. B. Ramsey, S. G. Johnson et al., "The clinical pharmacogenomics implementation consortium: CPIC guideline for SLCO1B1 and simvastatin-induced myopathy," Clinical Pharmacology and Therapeutics, vol. 92, pp. 112–117, 2012.

64. S. L. Chan, C. Suo, S. C. Lee, B. C. Goh, K. S. Chia, and Y. Y. Teo, "Translational aspects of genetic factors in the prediction of drug response variability: a case study of warfarin pharmacogenomics in a multi-ethnic cohort from Asia," The Pharmacogenomics Journal, vol. 12, no. 4, pp. 312–318, 2012.

65. N. A. Limdi, M. Wadelius, L. Cavallari et al., "Warfarin pharmacogenetics: a single VKORC1 polymorphism is predictive of dose across 3 racial groups," Blood, vol. 115, no. 18, pp. 3827–3834, 2010.

66. R. B. Altman, "Pharmacogenomics: "Noninferiority" is sufficient for initial imple-
     mentation," Clinical Pharmacology and Therapeutics, vol. 89, no. 3, pp. 348–350,
     2011.

67. C. M. Kelly and K. I. Pritchard, "CYP2D6 genotype as a marker for benefit of adju-
     vant tamoxifen in postmenopausal women: lessons learned," Journal of the National
     Cancer Institute, vol. 104, no. 6, pp. 427–428, 2012.

68. M. J. Khoury, M. Gwinn, W. D. Dotson, and M. S. Bowen, "Is there a need for PGx-
     ceptionalism?" Genetics in Medicine, vol. 13, pp. 866–867, 2011.

69. Y. Nakamura, M. J. Ratain, N. J. Cox, H. L. McLeod, D. L. Kroetz, and D. A. Flock-
     hart, "Re: CYP2D6 genotype and tamoxifen response in postmenopausal women
     with endocrine-responsive breast cancer: the breast International Group 1-98 trial,"
     Journal of the National Cancer Institute, vol. 104, no. 16, pp. 1266–1268, 2012.

70. J. Woodcock, "Assessing the clinical utility of diagnostics used in drug therapy,"
     Clinical Pharmacology and Therapeutics, vol. 88, no. 6, pp. 765–773, 2010.

71. Y. W. F. Lam, "How much evidence is necessary for pharmacogenomic testing im-
     plementation?" Clinical and Experimental Pharmacology, vol. 2, p. e107, 2012.

72. J. C. Stingl and J. Brockmöller, "Why, when, and how should pharmacogenetics be
     applied in clinical studies: current and future approaches to study designs," Clinical
     Pharmacology and Therapeutics, vol. 89, no. 2, pp. 198–209, 2011.

73. C. S. Karapetis, S. Khambata-Ford, D. J. Jonker et al., "K-ras mutations and benefit
     from cetuximab in advanced colorectal cancer," The New England Journal of Medi-
     cine, vol. 359, no. 17, pp. 1757–1765, 2008.

74. M. C. Deley, K. V. Ballman, J. Marandet, and D. Sargent, "Taking the long view:
     how to design a series of phase III trials to maximize cumulative therapeutic ben-
     efit," Clinical Trials, vol. 9, pp. 283–292, 2012.

75. S. D. Patterson, N. Cohen, M. Karnoub et al., "Prospective-retrospective biomarker
     analysis for regulatory consideration: white paper from the industry pharmacoge-
     nomics working group," Pharmacogenomics, vol. 12, no. 7, pp. 939–951, 2011.

76. J. P. Zhang, T. Lencz, and A. K. Malhotra, "D2 receptor genetic variation and clinical
     response to antipsychotic drug treatment: a meta-analysis," The American Journal of
     Psychiatry, vol. 167, no. 7, pp. 763–772, 2010.

77. E. P. Brass, "The gap between clinical trials and clinical practice: the use of prag-
     matic clinical trials to inform regulatory decision making," Clinical Pharmacology
     and Therapeutics, vol. 87, no. 3, pp. 351–355, 2010.

78. S. R. Tunis, D. B. Stryer, and C. M. Clancy, "Practical clinical trials: increasing the
     value of clinical research for decision making in clinical and health policy," Journal
     of the American Medical Association, vol. 290, no. 12, pp. 1624–1632, 2003.

79. J. L. Anderson, B. D. Horne, S. M. Stevens et al., "A randomized and clinical effec-
     tiveness trial comparing two pharmacogenetic algorithms and standard care for indi-
     vidualizing warfarin dosing (CoumaGen-II)," Circulation, vol. 125, pp. 1997–2005,
     2012.

80. J. Kirchheiner, S. Bauer, I. Meineke et al., "Impact of CYP2C9 and CYP2C19 poly-
     morphisms on tolbutamide kinetics and the insulin and glucose response in healthy
     volunteers," Pharmacogenetics, vol. 12, no. 2, pp. 101–109, 2002.

81. J. L. Mega, T. Simon, J. P. Collet et al., "Reduced-function CYP2C19 genotype and
     risk of adverse clinical outcomes among patients treated with clopidogrel predomi-

nantly for PCI: a meta-analysis," Journal of the American Medical Association, vol. 304, no. 16, pp. 1821–1830, 2010.

82. A. M. Harmsze, J. W. van Werkum, J. M. Ten Berg et al., "CYP2C19*2 and CYP2C9*3 alleles are associated with stent thrombosis: a case-control study," European Heart Journal, vol. 31, no. 24, pp. 3046–3053, 2010.

83. G. N. Levine, E. R. Bates, J. C. Blankenship et al., "ACCF/AHA/SCAI guideline for percutaneous coronary intervention: executive summary: a report of the American College of Cardiology Foundation/American heart association task force on practice guidelines and the society for cardiovascular angiography and interventions," Circulation, vol. 124, pp. 2574–2609, 2011.

84. J. A. Johnson, D. M. Roden, L. J. Lesko, E. Ashley, T. E. Klein, and A. R. Shuldiner, "Clopidogrel: a case for indication-specific pharmacogenetics," Clinical Pharmacology and Therapeutics, vol. 91, pp. 774–776, 2012.

85. D. R. Holmes, G. J. Dehmer, S. Kaul, D. Leifer, P. T. O'Gara, and C. M. Stein, "ACCF/AHA clinical alert: ACCF/AHA clopidogrel clinical alert: approaches to the FDA "boxed warning" a report of the American college of cardiology foundation task force on clinical expert consensus documents and the American heart association," Circulation, vol. 122, no. 5, pp. 537–557, 2010.

86. R. Lubomirov, S. Colombo, J. Di Iulio et al., "Association of pharmacogenetic markers with premature discontinuation of first-line anti-HIV therapy: an observational cohort study," Journal of Infectious Diseases, vol. 203, no. 2, pp. 246–257, 2011.

87. T. L. Kauf, R. A. Farkouh, S. R. Earnshaw, M. E. Watson, P. Maroudas, and M. G. Chambers, "Economic efficiency of genetic screening to inform the use of abacavir sulfate in the treatment of HIV," PharmacoEconomics, vol. 28, no. 11, pp. 1025–1039, 2010.

88. W. C. Black, "The CE plane: a graphic representation of cost-effectiveness," Medical Decision Making, vol. 10, no. 3, pp. 212–214, 1990.

89. R. A. Rosenheck, D. L. Leslie, and J. A. Doshi, "Second-generation antipsychotics: cost-effectiveness, policy options, and political decision making," Psychiatric Services, vol. 59, no. 5, pp. 515–520, 2008.

90. L. Sikich, J. A. Frazier, J. McClellan et al., "Double-blind comparison of first- and second-generation antipsychotics in early-onset schizophrenia and schizoaffective disorder: findings from the treatment of early-onset schizophrenia spectrum disorders (TEOSS) study," American Journal of Psychiatry, vol. 165, no. 11, pp. 1420–1431, 2008.

91. B. Lerer, R. H. Segman, H. Fangerau et al., "Pharmacogenetics of tardive dyskinesia: combined analysis of 780 patients supports association with dopamine D3 receptor gene Ser9Gly polymorphism," Neuropsychopharmacology, vol. 27, no. 1, pp. 105–119, 2002.

92. P. R. Bakker, P. N. van Harten, and J. van Os, "Antipsychotic-induced tardive dyskinesia and the Ser9Gly polymorphism in the DRD3 gene: a meta analysis," Schizophrenia Research, vol. 83, no. 2-3, pp. 185–192, 2006.

93. T. Furuta, N. Shirai, M. Kodaira et al., "Pharmacogenomics-based tailored versus standard therapeutic regimen for eradication of H. pylori," Clinical Pharmacology and Therapeutics, vol. 81, no. 4, pp. 521–528, 2007.

94. R. H. Perlis, D. A. Ganz, J. Avorn et al., "Pharmacogenetic testing in the clinical management of schizophrenia: a decision-analytic model," Journal of Clinical Psychopharmacology, vol. 25, no. 5, pp. 427–434, 2005.

95. G. F. Guzauskas, D. A. Hughes, S. M. Bradley, and D. L. Veenstra, "A risk-benefit assessment of prasugrel, clopidogrel, and genotype-guided therapy in patients undergoing percutaneous coronary intervention," Clinical Pharmacology and Therapeutics, vol. 91, pp. 829–837, 2012.

96. E. S. Reese, C. Daniel Mullins, A. L. Beitelshees, and E. Onukwugha, "Cost-effectiveness of cytochrome P450 2C19 genotype screening for selection of antiplatelet therapy with clopidogrel or prasugrel," Pharmacotherapy, vol. 32, no. 4, pp. 323–332, 2012.

97. M. H. Eckman, J. Rosand, S. M. Greenberg, and B. F. Gage, "Cost-effectiveness of using pharmacogenetic information in warfarin dosing for patients with nonvalvular atrial fibrillation," Annals of Internal Medicine, vol. 150, no. 2, pp. 73–83, 2009.

98. J. H. S. You, K. K. N. Tsui, R. S. M. Wong, and G. Cheng, "Potential clinical and economic outcomes of CYP2C9 and VKORC1 genotype-guided dosing in patients starting warfarin therapy," Clinical Pharmacology and Therapeutics, vol. 86, no. 5, pp. 540–547, 2009.

99. F. W. Frueh, S. Amur, P. Mummaneni et al., "Pharmacogenomic biomarker information in drug labels approved by the United States Food and Drug Administration: prevalence of related drug use," Pharmacotherapy, vol. 28, no. 8, pp. 992–998, 2008.

100. M. A. Hamburg and F. S. Collins, "The path to personalized medicine," The New England Journal of Medicine, vol. 363, no. 4, pp. 301–304, 2010.

101. G. Bartlett, N. Zgheib, A. Manamperi et al., "Pharmacogenomics in primary care: a crucial entry point for global personalized medicine?" Current Pharmacogenomics and Personalized Medicine, vol. 10, no. 2, pp. 101–105, 2012.

102. J. D. Roberts, G. A. Wells, M. R. Le May et al., "Point-of-care genetic testing for personalisation of antiplatelet treatment (RAPID GENE): a prospective, randomised, proof-of-concept trial," The Lancet, vol. 379, no. 9827, pp. 1705–1711, 2012.

103. J. T. Delaney, A. H. Ramirez, E. Bowton et al., "Predicting clopidogrel response using DNA samples linked to an electronic health record," Clinical Pharmacology and Therapeutics, vol. 91, pp. 257–263, 2012.

104. J. S. Schildcrout, J. C. Denny, E. Bowton et al., "Optimizing drug outcomes through pharmacogenetics: a case for preemptive genotyping," Clinical Pharmacology and Therapeutics, vol. 92, pp. 235–242, 2012.

105. C. A. Fernandez, C. Smith, W. Yang et al., "Concordance of DMET plus genotyping results with those of orthogonal genotyping methods," Clinical Pharmacology and Therapeutics, vol. 92, pp. 360–365, 2012.

106. J. Kirchheiner, K. Brøsen, M. L. Dahl et al., "CYP2D6 and CYP2C19 genotype-based dose recommendations for antidepressants: a first step towards subpopulation-specific dosages," Acta Psychiatrica Scandinavica, vol. 104, no. 3, pp. 173–192, 2001.

107. B. S. Finkelman, B. F. Gage, J. A. Johnson, C. M. Brensinger, and S. E. Kimmel, "Genetic warfarin dosing: tables versus algorithms," Journal of the American College of Cardiology, vol. 57, no. 5, pp. 612–618, 2011.

108. E. J. Stanek, C. L. Sanders, K. A. Taber et al., "Adoption of pharmacogenomic testing by US physicians: results of a nationwide survey," Clinical Pharmacology and Therapeutics, vol. 91, pp. 450–458, 2012.

109. D. Gurwitz, J. E. Lunshof, G. Dedoussis et al., "Pharmacogenomics education: International Society of pharmacogenomics recommendations for medical, pharmaceutical, and health schools deans of education," The Pharmacogenomics Journal, vol. 5, pp. 221–225, 2005.

110. Y. W. F. Lam, "Rethinking pharmacogenomics education beyond health professionals: addressing the "Know-Do" gap across the personalized medicine innovation ecosystem," Current Pharmacogenomics and Personalized Medicine, vol. 10, no. 4, pp. 277–287, 2012.

111. K. B. Mccullough, C. M. Formea, K. D. Berg et al., "Assessment of the pharmacogenomics educational needs of pharmacists," American Journal of Pharmaceutical Education, vol. 75, no. 3, p. 51, 2011.

112. J. J. Swen, M. Nijenhuis, A. De Boer et al., "Pharmacogenetics: from bench to byte an update of guidelines," Clinical Pharmacology and Therapeutics, vol. 89, no. 5, pp. 662–673, 2011.

113. C. L. Overby, P. Tarczy-Hornoch, J. I. Hoath, I. J. Kalet, and D. L. Veenstra, "Feasibility of incorporating genomic knowledge into electronic medical records for pharmacogenomic clinical decision support," BMC Bioinformatics, vol. 11, supplement 9, p. S10, 2010.

114. J. K. Hicks, K. R. Crews, J. M. Hoffman et al., "A clinician-driven automated system for integration of pharmacogenetic interpretations into an electronic medical record," Clinical Pharmacology and Therapeutics, vol. 92, pp. 563–566, 2012.

115. C. A. McCarty, A. Nair, D. M. Austin, and P. F. Giampietro, "Informed consent and subject motivation to participate in a large, population-based genomics study: the marshfield clinic personalized medicine research project," Community Genetics, vol. 10, no. 1, pp. 2–9, 2006.

116. H. Xu, S. P. Stenner, S. Doan, K. B. Johnson, L. R. Waitman, and J. C. Denny, "MedEx: a medication information extraction system for clinical narratives," Journal of the American Medical Informatics Association, vol. 17, no. 1, pp. 19–24, 2010.

117. A. N. Kho, J. A. Pacheco, P. L. Peissig et al., "Electronic medical records for genetic research: results of the eMERGE consortium," Science Translational Medicine, vol. 3, no. 79, Article ID 79re1, 2011.

118. A. H. Ramirez, Y. Shi, J. S. Schildcrout et al., "Predicting warfarin dosage in European-Americans and African-Americans using DNA samples linked to an electronic health record," Pharmacogenomics, vol. 13, pp. 407–418, 2012.

119. A. K. Jha, C. M. Desroches, E. G. Campbell et al., "Use of electronic health records in U.S. hospitals," The New England Journal of Medicine, vol. 360, no. 16, pp. 1628–1638, 2009.

120. J. Cohen, A. Wilson, and K. Manzolillo, "Clinical and economic challenges facing pharmacogenomics," The Pharmacogenomics Journal, 2012.

121. J. R. Trosman, S. L. Van Bebber, and K. A. Phillips, "Health technology assessment and private payers's coverage of personalized medicine," The American Journal of Managed Care, vol. 17, supplement 5, pp. SP53–SP60, 2011.

122. Food Drug Administration, "Guidance on pharmacogenetic tests and genetic tests for heritable markers".

123. E. B. Elkin, M. C. Weinstein, E. P. Winer, K. M. Kuntz, S. J. Schnitt, and J. C. Weeks, "HER-2 testing and trastuzumab therapy for metastatic breast cancer: a cost-effectiveness analysis," Journal of Clinical Oncology, vol. 22, no. 5, pp. 854–863, 2004.

124. M. E. Van den Akker-Van, D. Gurwitz, S. B. Detmar et al., "Cost-effectiveness of pharmacogenomics in clinical practice: a case study of thiopurine methyltransferase genotyping in acute lymphoblastic leukemia in Europe," Pharmacogenomics, vol. 7, no. 5, pp. 783–792, 2006.

125. A. R. Hughes, W. R. Spreen, M. Mosteller et al., "Pharmacogenetics of hypersensitivity to abacavir: from PGx hypothesis to confirmation to clinical utility," The Pharmacogenomics Journal, vol. 8, no. 6, pp. 365–374, 2008.

126. U. S. Food and Drug Administration, "FDA approves Xalkori with companion diagnostic for a type of late-stage lung cancer," 2011.

127. E. D. Blair, "Assessing the value-adding impact of diagnostic-type tests on drug development and marketing," Molecular Diagnosis and Therapy, vol. 12, no. 5, pp. 331–337, 2008.

128. R. S. Epstein, T. P. Moyer, R. E. Aubert et al., "Warfarin genotyping reduces hospitalization rates results from the MM-WES (Medco-Mayo Warfarin Effectiveness study)," Journal of the American College of Cardiology, vol. 55, no. 25, pp. 2804–2812, 2010.

129. J. L. Bevan, J. A. Lynch, T. N. Dubriwny et al., "Informed lay preferences for delivery of racially varied pharmacogenomics," Genetics in Medicine, vol. 5, no. 5, pp. 393–399, 2003.

130. S. B. Haga, J. M. O'Daniel, G. M. Tindall, I. R. Lipkus, and R. Agans, "Survey of US public attitudes toward pharmacogenetic testing," The Pharmacogenomics Journal, pp. 197–204, 2011.

131. G. Loukides, A. Gkoulalas-Divanis, and B. Malin, "Anonymization of electronic medical records for validating genome-wide association studies," Proceedings of the National Academy of Sciences of the United States of America, vol. 107, no. 17, pp. 7898–7903, 2010.

132. J. Aberdeen, S. Bayer, R. Yeniterzi et al., "The MITRE identification scrubber toolkit: design, training, and assessment," International Journal of Medical Informatics, vol. 79, no. 12, pp. 849–859, 2010.

133. L. G. Dressler and S. F. Terry, "How will GINA influence participation in pharmacogenomics research and clinical testing?" Clinical Pharmacology and Therapeutics, vol. 86, no. 5, pp. 472–475, 2009.

134. A. C. J. W. Janssens, M. Gwinn, L. A. Bradley, B. A. Oostra, C. M. van Duijn, and M. J. Khoury, "A critical appraisal of the scientific basis of commercial genomic profiles used to assess health risks and personalize health interventions," American Journal of Human Genetics, vol. 82, no. 3, pp. 593–599, 2008.

135. D. H. Spencer, C. Lockwood, E. Topol et al., "Direct-to-consumer genetic testing: reliable or risky?" Clinical Chemistry, vol. 57, no. 12, pp. 1641–1644, 2011.

136. E. A. Ashley, A. J. Butte, M. T. Wheeler et al., "Clinical assessment incorporating a personal genome," The Lancet, vol. 375, no. 9725, pp. 1525–1535, 2010.

137. N. Grecco, N. Cohen, A. W. Warner et al., "PhRMA survey of pharmacogenomic and pharmacodynamic evaluations: what next?" Clinical Pharmacology and Therapeutics, vol. 91, pp. 1035–1043, 2012.

138. Strattera, Eli Lilly and Company, IN, USA, 2011.

139. A. Inoue, T. Suzuki, T. Fukuhara et al., "Prospective phase II study of gefitinib for chemotherapy-naïve patients with advanced non-small-cell lung cancer with epidermal growth factor receptor gene mutations," Journal of Clinical Oncology, vol. 24, no. 21, pp. 3340–3346, 2006.

140. L. V. Sequist, R. G. Martins, D. Spigel et al., "First-line gefitinib in patients with advanced non-small-cell lung cancer harboring somatic EGFR mutations," Journal of Clinical Oncology, vol. 26, no. 15, pp. 2442–2449, 2008.

141. K. Tamura, I. Okamoto, T. Kashii et al., "Multicentre prospective phase II trial of gefitinib for advanced non-small cell lung cancer with epidermal growth factor receptor mutations: results of the west Japan thoracic oncology group trial (WJ-TOG0403)," British Journal of Cancer, vol. 98, no. 5, pp. 907–914, 2008.

142. C. H. Yang, C. J. Yu, J. Y. Shih et al., "Specific EGFR mutations predict treatment outcome of stage IIIB/IV patients with chemotherapy-naive non-small-cell lung cancer receiving first-line gefitinib monotherapy," Journal of Clinical Oncology, vol. 26, no. 16, pp. 2745–2753, 2008.

143. K. Kobayashi, A. Inoue, K. Usui et al., "First-line gefitinib for patients with advanced non-small-cell lung cancer harboring epidermal growth factor receptor mutations without indication for chemotherapy," Journal of Clinical Oncology, vol. 27, no. 9, pp. 1394–1400, 2009.

144. J. B. Singer, S. Lewitzky, E. Leroy et al., "A genome-wide study identifies HLA alleles associated with lumiracoxib-related liver injury," Nature Genetics, vol. 42, no. 8, pp. 711–714, 2010.

145. C. F. Spraggs, L. R. Parham, C. M. Hunt, and C. T. Dollery, "Lapatinib-induced liver injury characterized by class II HLA and Gilbert's syndrome genotypes," Clinical Pharmacology and Therapeutics, vol. 91, pp. 647–652, 2012.

146. L. C. Surh, M. A. Pacanowski, S. B. Haga et al., "Learning from product labels and label changes: how to build pharmacogenomics into drug-development programs," Pharmacogenomics, vol. 11, no. 12, pp. 1637–1647, 2010.

PART II

TRANSLATIONAL AND PERSONALIZED MEDICINE

CHAPTER 5

CLINICAL PROTEOMICS AND OMICS CLUES USEFUL IN TRANSLATIONAL MEDICINE RESEARCH

ELENA LÓPEZ, LUIS MADERO, JUAN LÓPEZ-PASCUAL, AND MARTIN LATTERICH

5.1 INTRODUCTION

5.1.1 THE POST-GENOME ERA: ADVANCES IN CLINICAL PROTEOMIC RESEARCH

Improved biomarkers are of vital importance for cancer detection, diagnosis and prognosis. While significant advances in understanding the molecular basis of disease are underway in genomics, proteomics will ultimately delineate the functional units of a cell: proteins and their intricate interactive networks and signalling pathways in health and disease.

Much progress has been made to characterize thousands of proteins qualitatively and quantitatively in complex biological systems by the use of multi-dimensional sample fractionation strategies, mass spectrometry (MS) and protein micro-arrays. Comparative/quantitative analysis of high-quality clinical biospecimens (e.g., tissue and biofluids) of the human cancer

This chapter was originally published under the Creative Commons Attribution License. López E, Madero L, López-Pascual J, and Latterich M. Clinical Proteomics and OMICS Clues Useful in Translational Medicine Research. Proteome Science **10**,*35, (2012). doi:10.1186/1477-5956-10-35.*

proteome landscape can potentially reveal protein/peptide biomarkers responsible for this disease by means of their altered levels of expression, post-translational modifications (PTMs), as well as different forms of protein variants. Despite technological advances in proteomics, major hurdles still exist at every step of the biomarker development pipeline [1-12].

In the post-genome era, the field of proteomics has incited great interest in the pursuit of protein/peptide biomarker discovery especially since MS has been shown to be capable of characterizing a large number of proteins and their PTMs

In complex biological systems, in some instances even quantitatively. Technological advances such as protein/antibody chips, depletion of multiple high abundance proteins by affinity columns, and affinity enrichment of targeted protein analytes as well as multidimensional chromatographic fractionation, have all expanded the dynamic range of detection for low abundance proteins by several orders of magnitude in serum or plasma, making it possible to detect the more abundant disease-relevant proteins in these complex biological matrices [13-21]. However, plasma and cell-extract based discovery research studies aimed at identifying low abundance proteins (e.g. some kinases) are extremely difficult. Therefore, it is necessary to develop significant technological improvements related to identifying this low abundance, although high biological impact molecules. Moreover, if these protein kinases to be studied contain PTMs, it is important to know that spatial and temporal factors can decrease the efficiency of our study (e.g. many kinases are regulated by phosphorylation of the activation loop, which then directly reflects cellular kinase activity).

Furthermore, proteomics has been widely applied in several areas of science, ranging from deciphering molecular pathogenesis of diseases, the characterization of novel drug targets, to the discovery of potential diagnostic and prognostic biomarkers, where technology is capable of identifying and quantifying proteins associated with a particular disease by means of their altered levels of expression [22-24] and/or PTMs [25-27] between the control and disease states (e.g., biomarker candidates). This type of comparative (semi-quantitative) analysis enables correlations to be drawn between the range of proteins, their variations and modifications produced by a cell, tissue and biofluids and the initiation, progression, therapeutic monitoring or remission of a disease state.

PTMs including phosphorylation, glycosylation, acetylation and oxidation, in particular, have been of great interest in this field as they have been demonstrated to being linked to disease pathology and are useful targets for therapeutics.

In addition to MS-based large-scale protein and peptide sequencing, other innovative approaches including self-assembling protein microarrays [28] and bead-based flow cytometry [29] to identify and quantify proteins and protein- protein interaction in a high throughput manner have furthered our understanding of the molecular mechanisms involved in diseases.

In summary, clinical proteomics has come a long way in the past decade in terms of technology/platform development and protein chemistry, to identify molecular signatures of diseases based on protein pathways and signalling cascades. Hence, there is great promise for disease diagnosis, prognosis, and prediction of therapeutic outcome on an individualized basis.

5.1.2 PROTEOMIC HINDRANCES FOR DISCOVERY OF TRUE CANDIDATE BIOMARKERS

Why is there such a disconnection between biomarker discovery using modern proteomic technologies and biomarker qualification requiring much more stringent analytical and clinical criteria? Several major obstacles have been suggested as being responsible for this discrepancy, including:

1. technological variability within/across proteomic platforms;
2. suitable/unsuitable biospecimen collection, handling, storage and processing;
3. capacity/incapacity of credentialing biomarker candidates prior to costly and time-consuming clinical qualification studies using well-established methodologies;
4. necessity for knowledge in the evaluation criteria required for these distinct processes in the pipeline and in regulatory science by the research community;

5. insufficient publicly available high-quality reagents and data sets to the cancer research community;
6. need for improved data analysis tools for the analysis, characterization, and comparison of large datasets and multi-dimensional data;
7. necessity for proper experimental study design when performing studies involving clinical samples in biomarker studies.

If proteomics is to be successfully introduced into clinical diagnostics, universally accepted metrics will be necessary at many steps along the way, to ensure that changes observed are attributable to biological states, not workflow variability. In addition, with the combination of different OMICS- technologies, more reliable data can be achieved. A high number of OMICS-combination-approaches are available for clinical research. It is always necessary to test different tools in order to raise a greater level of efficiency for your clinical study [30]. Figure 1 illustrates the proteomic hindrances for discovery of true (as opposed to surrogate) candidate bio-markers.

With regards to discovery, semi-quantitative proteomic methodologies routinely used for biomarker research between normal and diseased states are differential two-dimensional gel electrophoresis (2DGE), comparative label-free and labelling approaches [e.g., ^{18}O labelling, Isotope Coded Affinity Tags, Isobaric Tag for Relative and Absolute Quantitation (iTRAQ), Stable Isotope Labelling with Amino Acids in Cell Culture (SILAC), Absolute Quantitation (AQUA), Multiple Reaction Monitoring (MRM)] followed by liquid chromatography mass spectrometry (LC-MS). Although such comparative analysis yields important information on possible changes as a result of disease, these current methods in clinical proteomics based, for the most part, on MS and its combination with 2DGE, chromatography or biobead technology, might have limitations related to the sensitivity concentration level.

5.1.2.1 SAMPLE PREPARATION

When using the previously mentioned proteomic tools, sample preparation is one of the most crucial processes in proteomic analysis and biomarker

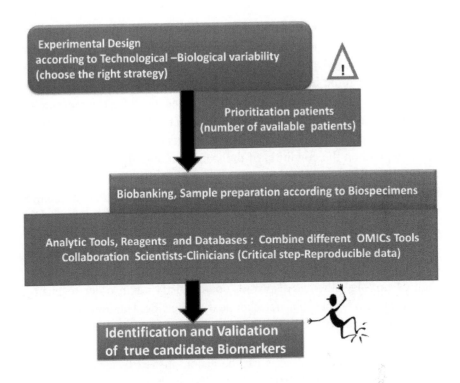

FIGURE 1: Proteomic hindrances for discovery of true candidate biomarkers. This figure illustrates, in a simple manner, relevant discovery aspects of true candidate biomarkers. Points to be considered are: (a) technological and biological variability within/across proteomic platforms; (b) suitable/unsuitable biospecimen collection, handling, storage and processing; (c) capacity/incapacity of credentialing biomarker candidates prior to costly and time-consuming clinical qualification studies using well-established methodologies; (d) the necessity for knowledge in the evaluation criteria required for these distinct processes in the pipeline and in regulatory science by the research community; (e) insufficient publicly available high-quality reagents and data sets to the cancer research community; (f) need for improved data analysis tools for the analysis, characterization, and comparison of large datasets and multi-dimensional data; and (g) necessity for proper experimental study design when performing studies involving clinical samples in biomarker studies. This implies a network-connectivity in relation to: (h) ensuring the choice of the correct strategy, (i) conclusion of the clinical proteomic research study when reaching a reprensative number of patients in order to achieve reliable data, (j) to always carry out inter- and intra-assays of your sample-preparations in order to reproduce your data, (k) to combine different OMIC-Tools to complement and verify the efficiency of your results, (l) Collaboration between clinicians and expert OMIC-scientists is necessary for success.

discovery in solubilized samples. Chromatographic or electrophoretic proteomic technologies are also available for separation of cellular protein components. There are, however, considerable limitations in currently available proteomic technologies as none of these allows for the analysis of the entire proteome in a simple step because of the large number of peptides, and because of the wide concentration dynamic range of the proteome in clinical blood samples. The results of any experiment undertaken depend on the condition of the starting material. Therefore, proper experimental design and pertinent sample preparation are essential for obtaining meaningful results, particularly in comparative clinical proteomics in which one is looking for minor differences between experimental (diseased) and control (non-diseased) samples [31].

Homogenization is one of the preparative steps employed for preparation of biological samples for proteomic analysis, and includes processes such as mixing, stirring, dispersing, or emulsifying in order to change the sample's physical, but not chemical properties. Homogenization for proteomics incorporates five main categories: mechanical, ultrasonic, pressure, freeze-thaw, and osmotic/detergent lyses. Mechanical homogenization for tissues and cells can be accomplished by devices such as rotor–stator, and open blade mills (e.g., Warring blender and Polytron), or pressure cycling technology (PCT) such as French presses. Rotor–stator homogenizers can homogenize samples in volumes from 0.01 mL to 120 L depending on the tip and motor used.

For optimum results, the tissue should be cut into slices, the size of which is slightly smaller than the diameter of the applied stator, as larger samples may clog the generator's inlet, making it impossible to achieve effective homogenization. Depending on the chemical resistance of a cutting tool, it is possible to homogenize samples under acidic or basic conditions in order to prevent degradation by endogenous enzymes. Heat transfer to the processed mixture is low to moderate and the process usually requires external cooling. Sample loss is minimal compared to PCT, where by means of a pressure-generating instrument (Pressure Bioscience, West Bridgewater, MA) alternating cycles of high and low pressure are applied to induce cell lysis [32,33].

In relation to protein solubilisation, proteins in biological samples are generally found in their native state associated with other proteins and often integrated as a part of large complexes, or into membranes. Once isolated, proteins in their native state are often insoluble. Breaking interactions involved in protein aggregation (e.g., disulfide hydrogen bonds, van der Waals forces, ionic and hydrophobic interactions) enables disruption of proteins into a solution of individual polypeptides, thereby promoting their solubilisation. However, because of the great heterogeneity of proteins and sample-source related interfering contaminants in biological extracts, simultaneous solubilisation of all proteins remains a challenge. Integration of proteins into membranes, and their association and complex formation with other proteins and/or nucleic acids hamper the process significantly. No single solubilisation approach is suitable for every purpose, and each sample and condition requires unique treatment. Sample solubilization can be improved by agitation or ultrasonification, but an increase in temperature should be avoided. The selection of the appropriate solubilisation protocol and buffers has specially been facilitated by the availability of commercial kits, although it is somewhat more expensive than routine reagent methods [34,35].

To avoid protein modifications, aggregation, or precipitation resulting in the occurrence of artifacts and subsequent protein loss, sample solubilization process requires the use, in the sample buffer of: (1) chaotropes (urea, thiourea, charged guanidine hydrochloride, for ex.) that disrupt hydrogen bonds and hydrophilic interactions enabling proteins to unfold with ionizable groups exposed to solution; (2) ionic, non-ionic and zwitterionic detergents (SDS, CHAPS, or Triton X-100); (3) reducing agents that disrupt bonds between cysteine residues and thus promote the unfolding of proteins (DTT/dithioerythritol (DTT/DTE) or tributylphosphine (TBP) or tris-carboxy ethyl phosphine (TCEP)) and (4) protease inhibitors [36].

Although there is no general procedure to select an appropriate detergent, nonionic and zwitterionic detergents such as CHAPS and Triton X series are less denaturing than ionic detergents, and have been used to solubilise proteins for functional studies. On the other hand, ionic detergents are strong solubilizing agents that lead to protein denaturation.

However, sodium cholate and deoxycholate are soft detergents compatible with native protein extraction, although variables like buffer composition, pH, salt concentration, temperature, and compatibility of the chosen detergent with the analytical MS procedure, and the way in which to remove it (by dialysis for example) are all crucial factors that need to be considered. Usually, tissue disruption and cell lyses require the combination of detergent and mechanical methodologies [35]. The proper use of the above reagents, together with the optimized cell disruption method, dissolution, and concentration techniques collectively determines the effectiveness of proteome solubilization methodologies.

All the previously detailed information, coupled to the use/study of blood, as a biospecimen in discovery research (a commonly used biospecimen which is highly complex and which has a wide dynamic range of protein concentrations), makes it is very difficult to discover (measure) low abundance proteins (potential biomarkers). One solution to this problem is to develop and apply nanotechnology in clinical proteomics, as well as the throughput of analytical measurement systems while lowering their cost. Not only does nanotechnology have the potential of fulfilling many criteria required for the advancement of clinical proteomics, essential changes in the physicochemical properties of substances on their conversion to the nanostructured state, but it has also made it possible to create efficient systems for drug delivery to targets.

In addition, blood cells offer unique insights into disease processes. Therefore, erythrocytes, granulocytes, monocytes, lymphocytes, and platelets are of special interest for clinical proteomics. Cytometry is currently widely used as an analytical tool for clinical cell analysis directly from anticoagulated whole blood and also for cell sorting to generate pure populations of cells from heterogeneous and highly integrated mixtures as are found in the majority of biological environments. Elispot, slide based cytometry, and tissue arrays together with high-content screening microscopy are further upcoming techniques in cytoproteomics. The major challenge for this type of preanalytical standardization is related to the use of fresh samples, either for direct multiparameter analysis of cellular proteomics in whole blood or body fluids without pre-separation, or for cell sorting and enrichment strategies for subsequent proteomic and functional genomic analysis [37].

5.1.3 NANOTECHNOLOGY TO COMPLEMENT CLINICAL PROTEOMICS

The identification of unique patterns of protein expression, or biomarkers, associated with a specific disease is one of the most promising areas of clinical proteomics. There is an urgent need to discover new biomarkers that are useful for early disease diagnosis. Recently, it has been recognized that the measurement of a panel of multiple biomarkers has the potential to achieve a much higher sensitivity and specificity compared with any single biomarker in the past. Moreover, the highest informative content is thought to reside in the low molecular weight (LMW), low abundance fraction of biological fluids. Nanotechnology offers new approaches to harvest low abundant panels of biomarkers. For cancers, if the disease can be detected prior to the onset of metastases, this can lead to a significant reduction in cancer deaths [38].

The envisioned role of nanotechnology is twofold:

1. to provide access to previously inaccessible data as related to "-omic" technology components with unparalleled efficiency and resolution;
2. to enable innovative therapeutic modalities that leverage the validated system biology outputs for exquisitely specific individualized therapy.

Systems biology has the potential for utilizing subtle biological clues (e.g. "-omic" technology components) for early detection of disease, predicting patient response to therapy, and identifying biomarkers to enable effective targeting of drug- delivery modalities to the disease site. The field of systems biology is still evolving, however there is strong evidence in scientific literature supporting the promise of nanotechnology as an enabling contributor for extracting the elusive "-omic" data for clinical analysis.

For example, investigators have recently shown the ability to reproducibly enhance the presence of the low molecular weight proteome from serum and plasma samples to differentiate the stages of disease as well as predict a patient's response to therapy. As the utility of nanotechnology expands to other "-omic" technologies, the ability to compare and integrate multiple panels of data subsets will tremendously strengthen the validation process for biomarker identification. Furthermore, nanotechnology has already demonstrated a clinical impact upon drug-delivery strategies for a variety of ailments, particularly cancer indications.

The inherent scale of nanotechnology enables a library combining surface modifications (e.g. targeting moieties, charge modifications, stealth) of nanoparticulates, as well as control over size, shape, and other particle characteristics pending on particle material. This variety of options allows the rational design of personalized therapies that are predicated upon established biomarker evidence through system biology discovery, image analysis, mathematical modeling and access to effective chemotherapeutics and other agents [39]. The development of nanotechnology presents an unprecedented opportunity for point-of-care testing devices by enabling both greater analytical sensitivity and the ability to multiplex protein and nucleic acid marker evaluations in the same assay [40]. It is certain that nanotechnology has yet to impart an enabling contribution towards the overall movement to individualized medicine; thus, the potential of nanomedicine coupled to clinical proteomics remains undeniable.

Currently, one of the most promising nanotechnological proteomics under development for medical research is biosensor-based nanodiagnostics. An example of this is the development of a magneto-nano sensor protein chip and a multiplex magnetic sorter based on magnetic nanoparticles that allow rapid conversion of discrete biomolecule binding events into electrical signals, which can detect target molecules down to the single molecule level in less than an hour [41,42]. In consequence, nanotechnology in clinical proteomics today, implies a new medical research direction, dealing with the creation and application of nanodevices for carrying out proteomic analyses in the clinic. Nanotechnological progress in the field of atomic force microscopy facilitates clinical studies on the revelation, visualization and identification of protein disease markers, in particular of those with sensitivity of 10–17 M, much greater than the sensitivity of commonly adopted clinical methods. Also, at the same time, implementation of nanotechnological approaches into diagnostics permits the creation of new diagnostic systems based on the optical, electro-optical, electromechanical and electrochemical nanosensoric elements at high operating speed [42].

In summary, nanobiotechnology is a new focus in technological science. It plays a key role in the creation of nanodevices for the analysis of living systems on a molecular level. Moreover, nanomedicine allows for improved understanding of human life while using the knowledge on human organism at a molecular level. The use of nanotechnological approaches and nanomaterials opens new prospects for the creation of drugs and systems for their directed transport. Implementation of optico-biosensoric, atomic-force, nanowire and nanoporous approaches into genomics and proteomics will significantly enhance the sensitivity and accuracy of diagnostics and will shorten the time for diagnostic procedures, thus undoubtedly improving the efficiency of medical treatment.

5.1.4 BIOINFORMATICS: USEFUL FOR CLINICAL PROTEOMICS

Computational biology covers a wide spectrum of techniques devoted to the generation and use of useful information from structure, sequence or relationships among biological analytes (DNA, RNA, proteins, macromolecular complexes, etc.). Those methods most useful in clinical studies, including biomarkers research, are chiefly the following:

- Next Generation Sequencing (NGS) is recently being used in a detailed study of genes involved in ColoRectal Cancer (CRC).The authors demonstrated that sequencing of whole tumour exomes allowed prediction of the microsatellite status of CGC, and also, facilitating the putative discovery of relevant mutations. Additionally, NGS is applicable to formalin-fixed and paraffin embedded material, allowing the renewed study of relevant clinic material in the pathology departments [43, 44].

- Once modified residues have been found in sequencing or proteomic studies, routine sequence-to-sequence and sequence-to-structure comparisons (MSA: multiple sequence analysis) allow to obtain valuable information about the functional implications related to the mutated residues in the protein context. Multiple alignments of proteins, and chiefly those based on the comparison of experimentally obtained three-dimensional atomic structures (structural alignments), are a very valuable source of information related to the evolutionary strategies. This is then followed by the different members of a family of proteins to conserve or modify their functions and structures [45].

The analysis of structural alignments allows the detection of at least three types of regions or multiple alignment positions according to conservation: (a). Conserved positions, usually the key for function or structure maintenance. (b). Tree-determinant residues, conserved only in protein subfamilies and related to family-specific active sites, substrate binding sites or protein-protein interaction surfaces. These sites contain essential information for the design of family-specific activator or inhibitor drugs [46]. And (c), positions that correspond to compensatory mutations that stabilize the mutations in one protein with changes in the other (Correlated mutations). These sites are very effective for the detection of protein- protein interaction contacts [47]. These last ones allow for the selection of the correct structural arrangement of two proteins based on the accumulation of signals in the proximity of interacting surfaces.

- Because of the sequence-to-structure comparison, and in absence of experimental crystal structures, the homology modelling methods, (also called comparative modelling or knowledge-based modelling), can develop a 3D model from a protein sequence based on the structures of a crystallized homologous protein. The method can only be applied to proteins with a common evolutionary origin: as only for proteins that are hypothesized to be homologous, this assertion implies that their three-dimensional structures are conserved to a greater extent than their primary structures. In the event where good homology hypothesis cannot be seconded, alternative methods can be applied in order to obtain a putative 3D structure. These procedures, known as "far-homology modelling" or "threading" methods, provide structures with lesser confidence compared to those generated using homology modelling methods.

- Data on the 3D structure of the active centre of a protein of interest and/or its natural ligands can be used as a basis for the design of effective drugs. This rational drug design is usually performed via multiple docking experiments in the active centre of the protein of interest. This requires the use of advanced software such as Autodock-4 [48]. Algorithms such as Autodock-4 allow the evaluation of not only the docking to a rigid model of the active centre, and also a certain mobility and adaptation of the side chain of enzyme residues to the ligand shape. Commonly, all the calculated binding conformations to the target protein obtained in every docking run are clustered according to scoring criteria (as "the lowest binding energy model" or "the lowest energy model representative of the most-populated cluster") and sorted according to their estimated free energy of binding. These computer strategies are a useful cost-reducing tool to prospect and model new molecules with potential inhibiting properties or even successful future drugs. Lately, the rational drug design approach has been used for putative cancer therapies, in particular the pharmacological reactivation of mutant p53 [49]. This promising strategy implies the simultaneous use of several ways for the identification of small molecules that target mutant p53, including "de novo" design and screening of chemical libraries.

- To conclude this section, molecular dynamics (MD) techniques are routinely used to obtain refined models for protein structure, protein-protein and protein- ligand interactions. MD is a computational simulation technique in which atoms within molecules are allowed to interact for a period of time according to the principles of physics. In the case of proteins, the relevant forces taken into account are the electrostatic interactions (of attraction or repulsion), Van der Waals interactions, and the properties of the covalent bond (length, angle, and dihedral angle). As a rule, simulation times for macromolecular protein complexes are up to 20 ns and the number of atoms of the simulated systems is in the order of up to 250,000, including solvent molecules. MD tools have been used to simulate the individual behaviour of small protein or peptides [50], protein-protein interfaces and ligand-protein relationship in catalytic macromolecular complexes with GTPase activity [51,52] or kinases involved in cell signalling pathways (e.g. Src tyrosine kinase [53] or the protein kinase B/Akt [54]).

5.1.5 SAMPLE BIOBANKING OMPLEMENTATION NECESSARY FOR CLINICAL PROTEOMICS

A Biobank contains several hundred thousand samples from a broad range of anatomic sites, diseases and with diverse ethnic representation. All biospecimens are obtained using stringent standard operating procedures and ethical protocols to provide assurance to the researcher that the materials will meet their scientific needs.

In order to require tissue samples with accompanying clinical outcome data, it is necessary to maintain a BioReserve repository of frozen and fixed tissues with patient follow-up data.

Through Biobank and BioReserve repositories, it is, thus, possible to provide a rapid delivery of human tissue, biofluids and tissue derivatives that best meet the research requirements. Moreover, during a standard collection protocol for sample biobanking, human tissues or bodyfluids and clinical data can also be custom collected to meet unique requirements. Each donor site uses a standardized clinical data form and pathologic data is classified using codes for anatomic site, morphology and behaviour [55].

As a final check, clinical data management associates review records for each case to ensure complete and consistent data. The stringent evaluation and classification process ensures that scientists receive clinically relevant data to help them in their research. Each biospecimen is assessed using uniform quality assurance tests. The pathologists independently confirm the anatomic site and diagnosis for each tissue procured. In addition, the lab researchers assess the RNA, DNA, proteins etc. integrity of each tissue received. This information is pr ovided to scientists before purchase so they can accurately select the samples which will best meet their needs. This reduces the number of failed OMICS and clinical experiments due to inappropriate or poor quality samples [56].

5.2 CONSIDERATIONS AND FUTURE NEEDS

5.2.1 THE CONVENTIONAL BIOMARKER DEVELOPMENT PIPELINE

It is necessary to integrate genomics, proteomics, nanobiotechnology/ nanomedicine, bioinformatics and biobanking-sample methodologies with clinicians. The mapping of the human genome represents a real milestone in medicine and has led to an explosion in discoveries and translative research in life sciences. Indeed, this important knowledge base has enabled rapid development in the areas of diagnostics, gene therapy, new drug targets discovery, and personalized therapies [57,58]. The expansion of biological knowledge through the Human Genome Project (HGP) has also been accompanied by the de velopment of new high throughput techniques, providing extensive capabilities for the analysis of a large number of genes or the whole genome. The completion of the human genome, however, has presented a new and even more challenging task for scientists: the characterization of the human proteome. Unlike the genome project, there are major challenges in defining a comprehensive Human Proteome Project (HPP) due to (a) a potentially very large number of proteins with PTMs, mutations, splice variants, etc.; (b) the diversity of technology platforms involved; (c) the variety of overlapping biological "units" into which the proteome might be divided for organized conquest; and (d) sensitivity limitations in detecting proteins present in low abundances.

The conventional biomarker development pipeline involves a discovery stage followed by a qualification stage (commonly known as biomarker validation) on large cohorts, prior to clinical implementation and designing complementary OMICs strategies. In common practice, the discovery stage is performed on a MS-based platform for global unbiased sampling of the proteome, while biomarker qualification and clinical implementation generally involve the development of an antibody-based protocol, such as the commonly used enzyme linked ELISA assays. Although this process is potentially capable of delivering clinically important biomarkers, it is not the most efficient process as the latter is low-throughput, very costly and time-consuming. In many cases, affinity reagents for novel protein candidates do not even exist and it is difficult to multiplex targets without creating significant interferences and cross-reactivity. These limitations of immunoassays have called for the development of alternative approaches. The recent surge in the advance of proteomic technologies centering on targeted MS and protein microarrays has provided great opportunities for researchers to use them as "bridging technologies" for clinical proteomic and OMICS investigation of disease-relevant changes in tissues and biofluids.

Some recent studies that combine rigorous study design with a focused mass spectrometry approach, promise to streamline the discovery and validation process [59,60]. These studies deviate from the traditional brute-force discovery efforts, geared to find minute differences between often complex samples, to employ pre-selection and MRM-based quantification strategies. This approach significantly enhances the fidelity of detecting significant differences between even low abundance biomarkers. To put it into the perspective of the proverbial "needle in a haystack" analogy the "haystack" has not become smaller; however, the pre-selection of potential biomarkers of significance has provided the research community with a "magnet" to make the quest for finding the needle more efficient.

On the other hand, apart from restructuring the biomarker development pipeline, it will now become critical to introduce regulatory science to the proteomics together with nanotechnology/nanobiomedicine and bioinformatic research (OMICS technologies in general) with clinical chemistry community so that all these technologies can be translated from the laboratory to the clinic.

5.2.2 THE RELEVANT ROLE OF CLINICAL LABORATORIES

Clinical laboratories have an important role, and clinical scientists undoubtedly play an important part in the analytical validation of diagnostic tests and are thus required to routinely verify (confirm) previously cleared/approved tests by the regulatory agency in their facilities. Post-market analytical validation is routinely performed by clinical researchers via evaluation tests (strategies, instruments, positive and negative controls, reagents, etc.), which complies with regulations, specifications, or conditions. These tests typically involve precision, accuracy, linearity and lower limits of detection and quantification.

On setting up a method for an approved multiplex protein assay using a patient -specific "score", clinical scientists should consider the way in which to perform studies to validate the score. One approach may involve running an adequate number of positive and negative patients to assess the performance of such a "score" in their diagnosis in comparison with their medical charts and final clinical diagnosis. Additionally, international collaboration provides an effective means by which to educate key clinical laboratory audiences about the need for and use of common technologies and standards in proteomic and OMICS workflows and to share knowledge and experience on commonly interesting targets, assays and new technologies.

On the other hand, the reduction universally observed in test development and research activities represents, in part, a shift from laboratories making their own reagents and immunoassays to the purchase of the majority of them from an in vitro diagnostics company.

This is not an entirely negative development. External quality assessment and proficiency testing data clearly demonstrate the benefits of automation, including much improved precision, and there are benefits of scale in centralizing test development processes. Nevertheless, and as we previously mentioned, clinical laboratories should play an active role in the final evaluation of assays and in the study of their clinical utility in relation to their patients.

When considering requirements for the successful introduction of new diagnostic tests, it is helpful to review the general criteria that must be met (see Table 1), focusing on the roles of both research and specialist laboratories and the somewhat different requirements of high- throughput routine laboratories [61-64].

TABLE 1: Tips for the discovery of true candidate biomarkers at clinical laboratories

Necessity	Suggestion
Clinically clearly understood	Direct comparison with the existing best practice in the population for which it is intended
Well-characterized clinical specimens for discovery the relevant clinical population	Several factors have to be taken into account when collecting specimens for the studies of new biomarkers, whether for a specific clinical study or for a biobank in order to enable interpretation of results and ensure appropriate matching of patient and health controls
Well-validated discovery platform which is robust and reliable	The use of internal standards for identifying specific components and quality control via proteomic –mass spectrometry and OMICS strategies is critical.
Clinical evidence for the true candidate biomarker	Take into account: (a) which is the association of our candidate-biomarker with the relevant disease, (b) which is the assessment of clinical utility and impact, (c) which are the circumstances where use of the test would be unjustified and (d) Make a rigorous early investigation of the specific pre-analytical factors which might influence interpretation of the resulting data

This table illustrates the necessities for the successful transition when discovering true biomarkers from the research environment (lab) to the clinical applications and utilities.

5.2.3 INTEGRATION OF OMIC-SCIENTISTS EXPERTS WITH CLINICIANS

The ultimate goal for translational medicine is its capacity to perform assays in various clinical samples at multiple levels: DNA (genome), RNA (transcriptome) and protein (proteome) coupled to bioinformatics and nanotechnology/nanobiomedicine and others, using the knowledge and technologies resulting from large-scale projects. This workflow provides a genetic basis and a good opportunity for the community to characterize and quantify proteins (reflecting genetic alterations if detectable) and their alterations and PTMs in the cell.

It is critical to define the final purpose of a biomarker or biomarker pattern at the onset of the study and to select the case and control samples accordingly. This is followed by the experiment design, starting with the sampling strategy, sample collection, storage and separation protocols, choice and validation of the quantitative profiling platform followed by data processing, statistical analysis and validation workflows. Biomarker candidates arising after statistical validation should be submitted for further validation and, ideally, be connected to the disease mechanism after their identification. Since most discovery studies work with a relatively small number of samples, it is necessary to assess the specificity and sensitivity of a given biomarker-based assay in a larger set of independent samples, preferably analyzed at another clinical centre. Targeted analytical methods of higher throughput than the original discovery method are needed at this point and LC-tandem mass spectrometry is gaining acceptance in this field [65,66].

The resulting proteomic evidence will corroborate or complement the genetic aberrations detected in samples, such as tumours, providing deeper understanding of cancer and other diseases in the context of biological and clinical utility. The integration and interrogation of the proteomic and genomic data (and OMICS data in general) will provide potential biomarker candidates, which will be prioritized for downstream targeted proteomic analysis. These biomarker targets will be used to create multiplex, quantitative assays for verification and pre-screening to test the relevance of the targets in clinically relevant and unbiased samples. The outcomes from this approach will provide the community with verified biomarkers which could be used for clinical qualification studies; high quality and publicly accessible datasets; and analytically validated, multiplex, quantitative protein/peptide assays and their associated high quality reagents for the research and clinical community.

It is also important to state that in order to develop clinical proteomic and OMICS applications using the identified proteins (with and without

PTMs), collaboration between research scientists, clinicians, diagnostic companies and industry, and proteomic experts is essential, particularly in the early phases of the biomarker development projects. Also, complementing the data with other OMICS tools is crucial. The proteomics modalities currently available have the potential to lead to the development of clinical applications, and the channelling of the wealth of the information produced towards concrete and specific clinical purposes is urgent [65-67].

New biomarkers can be taken from research by experts in OMICS and clinicians into routine practice, provided there is sound evidence of clinical utility, funding can be assured, mechanisms are in place to ensure that the test is done only for those likely to benefit, analytical procedures are simple and robust, and quality is verified through internal quality control and efficiency testing procedures. For these requirements to be met in a timely manner for a specific biomarker, it is essential to learn from past mistakes and perhaps to think differently in the future.

For the future, greatly improved involvement and collaboration from all interested parties—including experts in discovery and assay development, in health policy, in clinical trial units, in the diagnostics industry and in laboratories responsible for providing clinical testing—will almost certainly lead to earlier identification and implementation of promising new biomarkers [68-71].

5.3 SUMMARY OF IMPORTANT CLUES WHEN APPLYING CLINICAL OMICS STRATEGIES FOR TRANSLATIONAL MEDICINE RESEARCH

1. Standardizing sample preparation procedures for each sample (e.g. blood, plasma/serum, etc.), is critical for obtaining reliable biomarkers and building a biomarker pattern, since slight changes in a given sample preparation could lead to very different protein profiles.
2. Clinical Proteomics and Bioinformatics for Translational Medicine research studies include steps for improvements that should be made and well-controlled in: (a) analytical tools and biobanking-

samples, (b) discovery, (c) validation, (d) clinical application, and (e) post-clinical application appraisal. It is likely that most, if not all, of the components that are necessary for clinical success are either readily available, or could be allocated with more rigorous research standards and efforts supported by our scientific community, clinicians, health agencies including hospitals, diagnostic companies, and industry. Enthusiasm for the clinical impact of proteomics may need to be tempered, at present, until robust evidence can be obtained, but some clinical successes will eventually be feasible.

3. The rapid proliferation of Nanotechnology/nanobiomedicine and the implementation of sample-Biobanking are revolutionizing science and technology. There is marked interest regarding the use of nanotechnologies in medicine coupled to clinical proteomics, and to complement OMICS tools in general. Therefore clear advances are appearing for the discovery of true candidate biomarkers.

4. However, and as a general rule, it must be taken into account as a very important conclusion, that without: (a) the correct study design, (b) the correct and complementary strategies (c) implementation of robust analytical methodologies and (d) the necessity for collaboration among expert OMICS scientists together with clinicians and the industry, the efforts, efficiency and expectations to make true candidate biomarkers a useful reality in the near future can easily be hindered.

REFERENCES

1. Hassanein M, Rahman JS, Chaurand P, Massion PP: Advances in proteomic strategies toward the early detection of lung cancer. Proc Am Thorac Soc 2011, 8(2):183-188.
2. Anderson L: Candidate-based proteomics in the search for biomarkers of cardiovascular disease. J Physiol 2005, 563:23-60.
3. Rifai N, Gillette MA, Carr SA: Protein biomarker discovery and validation: the long and uncertain path to clinical utility. Nat Biotechnol 2006, 24:971-983.
4. García-Foncillas J, Bandrés E, Zárate R, Remírez N: Proteomic analysis in cancer research: potential application in clinical use. Clin Transl Oncol 2006, 8:250-261.
5. Bouchal P, Roumeliotis T, Hrstka R, Nenutil R, Vojtesek B, Garbis SD: Biomarker discovery in low-grade breast cancer using isobaric stable isotope tags and two- di-

mensional liquid chromatography-tandem mass spectrometry (iTRAQ-2DLC- MS/MS) based quantitative proteomic analysis. J Proteome Res 2009, 8:362-373.

6. Wiener MC, Sachs JR, Deyanova EG, Yates NA: Differential mass spectrometry: a label-free LC-MS method for finding significant differences in complex peptide and protein mixtures. Anal Chem 2004, 76:6085-6096.

7. Geiger T, Cox J, Ostasiewicz P, Wisniewski JR, Mann M: Super-SILAC mix for quantitative proteomics of human tumor tissue. Nat Methods 2010, 7:383-385.

8. Anderson L, Hunter CL: Quantitative mass spectrometric multiple reaction monitoring assays for major plasma proteins. Mol Cell Proteomics 2006, 5:573-588.

9. Wang H, Wong CH, Chin A, Kennedy J, Zhang Q, Hanash S: Quantitative serum proteomics using dual stable isotope coding and nano LC-MS/MSMS. J Proteome Res 2009, 8:5412-5422.

10. Lee J, Soper SA, Murray KK: Microfluidic chips for mass spectrometry-based proteomics. J Mass Spectrom 2009, 44:579-593.

11. Pierobon M, Calvert V, Belluco C, Garaci E, Deng J, Lise M, Nitti D, Mammano E, De Marchi F, Liotta L, Petricoin E: Multiplexed cell signaling analysis of metastatic and nonmetastatic colorectal cancer reveals COX2-EGFR signaling activation as a potential prognostic pathway biomarker. Clin Colorectal Cancer 2009, 8:110-117.

12. Ramachandran N, Raphael JV, Hainsworth E, Demirkan G, Fuentes MG, Rolfs A, Hu Y, LaBaer J: Next-generation high-density self-assembling functional protein arrays. Nat Methods 2008, 5:535-538.

13. Beirne P, Pantelidis P, Charles P, Wells AU, Abraham DJ, Denton CP, Welsh KI, Shah PL, du Bois RM, Kelleher P: Multiplex immune serum biomarker profiling in sarcoidosis and systemic sclerosis. Eur Respir J 2009, 34:1376-1382.

14. Kelleher MT, Fruhwirth G, Patel G, Ofo E, Festy F, Barber PR, Ameer-Beg SM, Vojnovic B, Gillett C, Coolen A, Kéri G, Ellis PA, Ng T: The potential of optical proteomic technologies to individualize prognosis and guide rational treatment for cancer patients. Target Oncol 2009, 4:235-252.

15. Wang P, Whiteaker JR, Paulovich AG: The evolving role of mass spectrometry in cancer biomarker discovery. Cancer Biol Ther 2009, 8:1083-1094.

16. Whiteaker JR, Zhang H, Eng JK, Fang R, Piening BD, Feng LC, Lorentzen TD, Schoenherr RM, Keane JF, Holzman T, Fitzgibbon M, Lin C, Zhang H, Cooke K, Liu T, Camp DG 2nd, Anderson L, Watts J, Smith RD, McIntosh MW, Paulovich AG: Head-to- head comparison of serum fractionation techniques. J Proteome Res 2007, 6:828-836.

17. Ernoult E, Bourreau A, Gamelin E, Guette C: A proteomic approach for plasma biomarker discovery with iTRAQ labeling and OFFGEL fractionation. J Biomed Biotechnol 2010, 2010:927917. Epub 2009 Nov 1.

18. Nirmalan NJ, Hughes C, Peng J, McKenna T, Langridge J, Cairns DA, Harnden P, Selby PJ, Banks RE: Initial development and validation of a novel extraction method for quantitative mining of the formalin-fixed, paraffin-embedded tissue proteome for biomarker investigations. J Proteome Res 2010, 10:896-906.

19. Krishhan VV, Khan IH, Luciw PA: Multiplexed microbead immunoassays by flow cytometry for molecular profiling: basic concepts and proteomics applications. Crit Rev Biotechnol 2009, 29:29-43.

20. Cha S, Imielinski MB, Rejtar T, Richardson EA, Thakur D, Sgroi DC, Karger BL: In situ proteomic analysis of human breast cancer epithelial cells using laser capture microdissection: annotation by protein set enrichment analysis and gene ontology. Mol Cell Proteomics 2010, 9:2529-2544.

21. Anderson KS, Sibani S, Wallstrom G, Qiu J, Mendoza EA, Raphael J, Hainsworth E, Montor WR, Wong J, Park JG, Lokko N, Logvinenko T, Ramachandran N, Godwin AK, Marks J, Engstrom P, Labaer J: Protein microarray signature of autoantibody biomarkers for the early detection of breast cancer. J Proteome Res 2011, 10:85-96.

22. Bateman NW, Sun M, Hood BL, Flint MS, Conrads TP: Defining central themes in breast cancer biology by differential proteomics: conserved regulation of cell spreading and focal adhesion kinase. J Proteome Res 2010, 9:5311-5324.

23. Kristiansen TZ, Harsha HC, Grønborg M, Maitra A, Pandey A: Differential membrane proteomics using 18O-labeling to identify biomarkers for cholangiocarcinoma. J Proteome Res 2008, 7:4670-4677.

24. An HJ, Lebrilla CB: A glycomics approach to the discovery of potential cancer biomarkers. Methods Mol Biol 2010, 600:199-213.

25. Choudhary C, Mann M: Decoding signalling networks by mass spectrometry- based proteomics. Nat Rev Mol Cell Biol 2010, 11:427-439.

26. Madian AG, Regnier FE: Profiling carbonylated proteins in human plasma. J Proteome Res 2010, 9:1330-1343.

27. Iwabata H, Yoshida M, Komatsu Y: Proteomic analysis of organ-specific post- translational lysine-acetylation and -methylation in mice by use of anti-acetyllysine and -methyllysine mouse monoclonal antibodies. Proteomics 2005, 5:4653-4664.

28. Ceroni A, Sibani S, Baiker A, Pothineni VR, Bailer SM, LaBaer J, Haas J, Campbell CJ: Systematic analysis of the IgG antibody immune response against varicella zoster virus (VZV) using a self-assembled protein microarray. Mol Biosyst 2010, 6:1604-1610.

29. Wong J, Sibani S, Lokko NN, LaBaer J, Anderson KS: Rapid detection of antibodies in sera using multiplexed self-assembling bead arrays. J Immunol Methods 2009, 350:171-182.

30. López E, López I, Ferreira A, Sequí J: Clinical and Technical Phosphoproteomic Research. Proteome Sci 2011, 9(1):272.

31. Hernández-Borges J, Borges-Miquel TM, Rodríguez-Delgado MA, Cifuentes A: Sample treatments prior to capillary electrophoresis-mass spectrometry. J Chromatogr A 2007, 1153(1–2):214-226.

32. Guilak F, Alexopoulos LG, Haider MA, Ting-Beall HP, Setton LA: Zonal uniformity in mechanical properties of the chondrocyte pericellular matrix: micropipette aspiration of canine chondrons isolated by cartilage homogenization. Ann Biomed Eng 2005, 33(10):1312-1318.

33. Bodzon-Kulakowska A, Bierczynska-Krzysik A, Dylag T, Drabik A, Suder P, Noga M, Jarzebinska J, Silberring J: Methods for samples preparation in proteomic research. J Chromatogr B Analyt Technol Biomed Life Sci 2007, 849(1–2):1-31.

34. Rabilloud T: Solubilization of proteins for electrophoretic analyses. Electrophoresis 1996, 17(5):813-829.

35. Cañas B, Piñeiro C, Calvo E, López-Ferrer D, Gallardo JM: Trends in sample preparation for classical and second generation proteomics. J Chromatogr A 2007, 1153(1–2):235-258.

36. Görg A, Weiss W, Dunn MJ: Current two-dimensional electrophoresis technology for proteomics. Proteomics 2004, 4(12):3665-3685.

37. Thadikkaran L, Siegenthaler MA, Crettaz D, Queloz PA, Schneider P, Tissot JD: Recent advances in blood-related proteomics. Proteomics 2005, 5:3019-3034.

38. Ray S, Reddy PJ, Choudhary S, Raghu D, Srivastava S: Emerging nanoproteomics approaches for disease biomarker detection: A current perspective. J Proteomics 2011, 74(12):2660-2681. Epub 2011 May 7.

39. Hu Y, Bouamrani A, Tasciotti E, Li L, Liu X, Ferrari M: Tailoring of the Nanotexture of Mesoporous Silica Films and Their Functionalized Derivatives for Selectively Harvesting Low Molecular Weight Protein. ACS Nano 2010, 4(1):439-451.

40. Becker KF, Schott C, Hipp S, Metzger V, Porschewski P, Beck R, Nährig J, Becker I, Höfler H: Quantitative protein analysis from formalin-fixed tissues: implications for translational clinical research and nanoscale molecular diagnosis. J Pathol 2007, 211:370-378.

41. Osterfeld SJ, Yu H, Gaster RS, Caramuta S, Xu L, Han SJ, Hall DA, Wilson RJ, Sun S, White RL, Davis RW, Pourmand N, Wang SX: Multiplex protein assays based on real-time magnetic nanotag sensing. Proc Natl Acad Sci USA 2008, 105:20637-20640.

42. López E, López I, Sequi J, Ferreira A: Discovering and validating unknown phosphosites from p38 and HuR protein kinases in vitro by Phosphoproteomic and Bioinformatic tools. J Clin Bioinforma 2011, 1(1):16.

43. Timmermann B, Kerick M, Roehr C, Fischer A, Isau M, Boerno ST, Wunderlich A, Barmeyer C, Seemann P, Koenig J, et al.: Somatic mutation profiles of MSI and MSS colorectal cancer identified by whole exome next generation sequencing and bioinformatics analysis. PLoS One 2010, 5:e15661.

44. Schweiger MR, Kerick M, Timmermann B, Albrecht MW, Borodina T, Parkhomchuk D, Zatloukal K, Lehrach H: Genome-wide massively parallel sequencing of formaldehyde fixed-paraffin embedded (FFPE) tumor tissues for copy-number- and mutation- analysis. PLoS One 2009, 4:e5548.

45. Zuckerkandl E, Pauling L: Molecules as documents of evolutionary history. J Theor Biol 1965, 8:357-366.

46. López-Romero P, Gómez MJ, Gómez-Puertas P, Valencia A: Prediction of functional sites in proteins by evolutionary methods. In Principles and practice Methods in proteome and protein analysis. Edited by Kamp RM, Calvete J, Choli-Papadopoulou T. Berlin Heidelberg: Springer; 2004::319-340.

47. Carettoni D, Gomez-Puertas P, Yim L, Mingorance J, Massidda O, Vicente M, Valencia A, Domenici E, Anderluzzi D: Phage-display and correlated mutations identify an essential region of subdomain 1C involved in homodimerization of *Escherichia coli* FtsA. Proteins 2003, 50:192-206.

48. Huey R, Morris GM, Olson AJ, Goodsell DS: A semiempirical free energy force field with charge-based desolvation. J Comput Chem 2007, 28:1145-1152.

49. Wiman KG: Pharmacological reactivation of mutant0020p53: from protein structure to the cancer patient. Oncogene 2010, 29:4245-4252.

50. Mendieta J, Fuertes MA, Kunjishapatham R, Santa-Maria I, Moreno FJ, Alonso C, Gago F, Munoz V, Avila J, Hernandez F: Phosphorylation modulates the alpha-helical structure and polymerization of a peptide from the third tau microtubule-binding repeat. Biochim Biophys Acta 2005, 1721:16-26.

51. Mendieta J, Rico AI, Lopez-Vinas E, Vicente M, Mingorance J, Gomez-Puertas P: Structural and functional model for ionic (K(+)/Na(+)) and pH dependence of GT-Pase activity and polymerization of FtsZ, the prokaryotic ortholog of tubulin. J Mol Biol 2009, 390:17-25.

52. Mingorance J, Rivas G, Velez M, Gomez-Puertas P, Vicente M: Strong FtsZ is with the force: mechanisms to constrict bacteria. Trends Microbiol 2010, 18:348-356.

53. Mendieta J, Gago F: In silico activation of Src tyrosine kinase reveals the molecular basis for intramolecular autophosphorylation. J Mol Graph Model 2004, 23:189-198.

54. Calleja V, Laguerre M, Larijani B: 3-D structure and dynamics of protein kinase B-new mechanism for the allosteric regulation of an AGC kinase. J Chem Biol 2009, 2:11-25.

55. Wang W: Role of clinical bioinformatics in the development of network-based Bio-markers. J Clin Bioinforma. 2011, 1(1):28.

56. Baumgartner C, Osl M, Netzer M, Baumgartner D: Bioinformatic-driven search for metabolic biomarkers in disease. J Clin Bioinforma. 2011, 1(1):2.

57. Spitz MR, Bondy ML: The evolving discipline of molecular epidemiology of cancer. Carcinogenesis 2010, 31:127-134.

58. Rosa DD, Ismael G, Lago LD, Awada A: Molecular-targeted therapies: lessons from years of clinical development. Cancer Treat Rev 2008, 34:61-80.

59. Whiteaker JR, Lin C, Kennedy J, Hou L, Trute M, Sokal I, Yan P, Schoenherr RM, Zhao L, Voytovich UJ, Kelly-Spratt KS, Krasnoselsky A, Gafken PR, Hogan JM, Jones LA, Wang P, Amon L, Chodosh LA, Nelson PS, McIntosh MW, Kemp CJ, Paulovich AG: A targeted proteomics-based pipeline for verification of biomarkers in plasma. Nat Biotechnol 2011, 29(7):625-634.

60. Addona TA, Shi X, Keshishian H, Mani DR, Burgess M, Gillette MA, Clauser KR, Shen D, Lewis GD, Farrell LA, Fifer MA, Sabatine MS, Gerszten RE, Carr SA: A pipeline that integrates the discovery and verification of plasma protein biomark-ers reveals candidate markers for cardiovascular disease. Nat Biotechnol 2011, 29(7):635-643.

61. Metcalfe TA: Development of novel IVD assays: a manufacturer's perspective. Scand J Clin Lab Invest Suppl 2010, 242:23-26.

62. Rajappan K, Murphy E, Amber V, Meakin F, et al.: Usage of troponin in the real world: a lesson for the introduction of biochemical assays. Q J Med 2005, 98:337-342.

63. Hlatky MA, Greenland P, Arnett DK, Ballantyne CM, et al.: Criteria for evaluation of novel markers of cardiovascular risk: a scientific statement from the American Heart Association. Circulation 2009, 119:2408-2416.

64. Hortin GL: Can mass spectrometric protein profiling meet desired standards of clini-cal laboratory practice? Clin Chem 2005, 51:3-5.

65. Dowling P, Meleady P, Henry M, Clynes M: Recent advances in clinical proteomics using mass spectrometry. Bioanalysis 2010, 2(9):1609-1615.

66. Latterich M, Abramovitz M, Leyland-Jones B: Proteomics: new technologies and clinical applications. Eur J Cancer 2008, 44(18):2737-2741. Epub 2008 Nov 1. Re-view.

67. Latterich M, Schnitzer JE: Streamlining biomarker discovery. Nat Biotechnol 2011, 29(7):600-602.
68. Anderson NL: The clinical plasma proteome: a survey of clinical assays for proteins in plasma and serum. Clin Chem 2010, 56:177-185.
69. Hortin GL, Jortani SA, Ritchie JC Jr, Valdes R Jr, Chan DW: Proteomics: a new diagnostic frontier. Clin Chem 2006, 52:1218-1222.
70. Beastall GH: The Modernisation of pathology and laboratory medicine in the UK: networking into the future. Clin Biochem Rev 2008, 29:3-10.
71. López E, Muñoz SR, Pascual JL, Madero L: Relevant phosphoproteomic and mass spectrometry: approaches useful in clinical research. Clinical and Transl Med Jour 2012, 1:2-9. (http://www.clintransmed.com/content/pdf/2001-1326-1-2.pdf).

CHAPTER 6

GENOMES2DRUGS: IDENTIFIES TARGET PROTEINS AND LEAD DRUGS FROM PROTEOME DATA

DAVID TOOMEY, HEINRICH C. HOPPE, MARIAN P. BRENNAN, KEVIN B. NOLAN, AND ANTHONY J. CHUBB

6.1 INTRODUCTION

The modern molecular biologist is confronted with increasingly large datasets. Genome sequencing data, proteomics data and microarray data are increasingly accessible, but difficult and laborious to interpret. Considering the investment cost of target validation, one needs to rank genome-sized output data in favour of proteins that can readily be modelled using homology modelling, as these structural models can be used in virtual high throughput screening (vHTS) of large compound libraries [1]–[3]. Microbiologists designing antibiotics need to rank their candidate proteins for lack of similarity with any human protein, to reduce the possibility of potentially toxic off-target side effects due to cross-reactivity between inhibitors and patient host proteins. In addition, it is now possible to screen the proteome for homology to targets of known drugs, using the Drug-Bank dataset [4], and propose FDA-approved drugs for rapid development to Phase IV clinical trials as these compounds are all defined as safe for human consumption. Much of the necessary search functionality is already available online [4]–[7]. However, the assimilation of this data into a cohesive table for analysis is non-trivial for molecular biologists unskilled in programming languages or database management. By providing a convenient online interface and summary table output, we hope to make this analysis open to a wide research audience.

6.2 MATERIALS AND METHODS

Genomes2Drugs was developed using open source Java Enterprise Edition in the NetBeans IDE 6.0 programming environment and deployed on Sun Application Server [8]. The Basic Local Alignment Search Tool (BLAST) program 2.2 was obtained from the USA National Center for Biotechnology Information (NCBI). The human genome protein sequences and PDB protein sequences were also obtained from NCBI. Drug target protein sequences were obtained from the University of Alberta DrugBank website [4]. Output data files are parsed using BioJava 1.6 and the data entered into an open source MySQL 5.1 database. The test genome *Plasmodium falciparum* 3D7 protein sequences were obtained from the European Molecular Biology Laboratory - European Bioinformatics Institute (EMBL-EBI) Integr8 website (493.P_falciparum, [9]).

6.3 RESULTS

Genomes2Drugs is a freely available web-based search engine that simultaneously searches each input protein sequence against the protein sequences of the human genome, the DrugBank dataset drug targets and the PDB protein structure database [http://mmg.rcsi.ie:8080/g2d/]. The schema for information processing is shown in Figure 1. Users can input either a single FASTA formatted protein sequence [10] or multiple sequences, either in an input box or an uploaded text file. For instance, complete proteome sequences can be downloaded from the EMBL-EBI Integr8 website [9], and uploaded into Genomes2Drugs. Screen shots of the input and output screens are shown in supplementary Figure S1 online. Users need

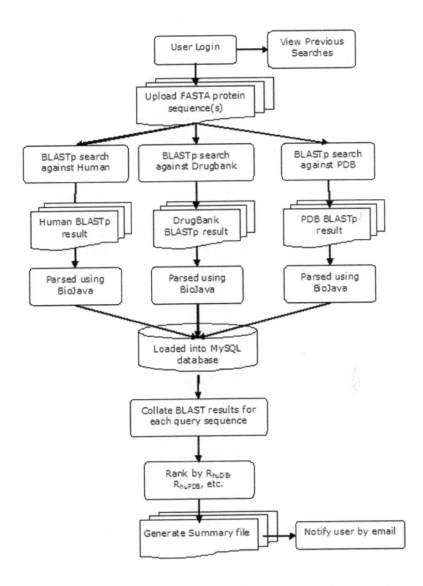

FIGURE 1: Schema of data processing. Genomes2Drugs is a free online resource. The web interface was written using open-source Java Enterprise Edition, BioJava 1.6 and NetBeans IDE 6.0. Input sequences are aligned against the human proteome, the PDB dataset and the DrugBank target proteins dataset. Only the best results are preserved. The resulting output files are parsed using BioJava and entered into a MySQL 5.1 database, where the results are sorted and ranked. Output XML files are generated from this data.

to register and submit an email address, as processing occurs in the background. User information will remain private and will not be given to any third party. The user will be emailed when the job is complete, and can then login to download the result XML file which can be imported into Microsoft Excel as a 'As an XML list', provided the user has downloaded the 'g2d.xsd' file (available online) into the same directory. The results from a few input polypeptides can be opened in Excel, while larger genome wide searches should be opened in a database viewer like Microsoft Access, for which a viewing form is included (see supplementary Figure S1B). For easy of access to the data in Access we have included a template MDB file and XSD schema file which need to be downloaded to the same directory as the XML file. The output terms are described in Table 1. Each E_{BLASTp} value is derived from the optimal alignment across the genome using default settings of NCBI's freely available BLASTp algorithm [5], [6]. As the best alignment score is recorded for each input protein, it follows that a poor score indicates that there is no matching protein in the comparator set. Thus a large E_{BLASTp}[query vs human genome] value indicates that there is likely no match for that query protein in the human genome. Similarly, good sequence identity, with a small E_{BLASTp}[query vs PDB] value indicates that the query sequence has a close homologue in the PDB structural database. No lower limit is set for any E value during the alignment calculation and only the best results are shown.

The <human expect> and <PDB expect> columns can be used individually to rank the whole input genome for proteins showing little homology to the human genome or good homology to a protein for which the crystal structure has been determined, respectively. More conveniently, the ratio of these expect values can be used to rank the output list according to proteins that would be readily structurally modelled, while also showing little identity to any human proteins. This ratio is provided in the logarithmic (base 10) form, in the column R_{huPDB} (2), which has been ranked by descending value.

The ratio values are calculated as follows:

$$R_{huDB} = \log_{10}\left(\frac{E_{BLASTp}[query\ vs.\ human\ genome]}{E_{BLASTp}[query\ vs.\ DrugBank]}\right) \qquad (1)$$

TABLE 1: Key for output file column headings.

Column Title	Explanation
query_id	Unique query entry number
query_accession	First word of input protein title
query_title	Input protein title after '>'
query_length	Number of residues in input sequence
RhuDB	Logarithm (base 10) of the ration of <human expect> and <pdb expect>
RhuDBRank	Entries ranked by descending R_{huPDB}
RhuPDB	Logarithm (base 10) of the ration of <human expect> and <PDB expect>
RhuPDBRank	Entries ranked by descending R_{huPDB}
RDBPDB	Logarithm (base 10) of the ration of <drugbank expect> and <pdb expect>
RDBPDBRank	Entries ranked by descending R_{huPDB}
human_accession	First word of human protein title
human_title	Extracted from target sequence name in BLASTp output
human_expect	Only optimal human/query alignment is returned, i.e. lowest BLASTp E value
human_rank	Query vs. human genome alignments are ranked by descending <human_expect>. I.e. poor/no match to the human genome is scored well and given a low rank number
human_identities	Number of identical residues in query and human sequences
human_percent_identites	(<human identites>/<query length>)*100
human_positives	Number of homologous residues in query and human sequences
human_percent_positives	(<human positives>/<query length>)*100
pdb_accession	Protein Data Bank accession number: pdb|xxxx|x
pdb_title	Name of protein 3-D structure
pdb_expect	Only optimal PDB/query alignment is returned, i.e. lowest BLASTp E value
pdb_rank	Query vs. Protein Data Bank sequence alignments are ranked by ascending <pdb_expect>. I.e. excellent matches with very low E values are scored well and given a low rank number.
pdb_identities	Number of identical residues in query and PDB sequences
pdb_percent_identities	(<pdb_identities>/<query length>)*100
pdb_positives	Number of homologous residues in query and PDB sequences
pdb_percent_positives	(<pdb_positives>/<query length>)*100
drugbank_accession	DrugBank accession number of target protein: nnnn_all_target_protein.fasta

TABLE 1: *Cont.*

Column Title	Explanation
drugbank_title	Name of DrugBank target protein, including target drug accession numbers in parentheses: (DBnnnnn)
drugbank_expect	Only optimal DrugBank/query alignment is returned, i.e. lowest BLASTp E value
drugbank_rang	Query vs. DrugBank sequences alignments are ranked by ascending <pdb_expect>. I.e. excellent matches with very low E values are scored well and given a low rank number
drugbank_identities	Number of identical residues in query and DrugBank sequences
drugbank_percent_identities	(<drugbank_identities>/<query length>)*100
drugbank_positives	Number of homologous residues in query and DrugBank sequences
drugbank_percent_positives	(<drugbank_positives>/<query length>)*100

$$R_{huPDB} = \log_{10}\left(\frac{E_{BLASTp}[query\ vs.\ human\ genome]}{E_{BLASTp}[query\ vs.\ PDB]}\right) \qquad (2)$$

$$R_{huDBPDB} = \log_{10}\left(\frac{E_{BLASTp}[query\ vs.\ DrugBank]}{E_{BLASTp}[query\ vs.\ PDB]}\right) \qquad (3)$$

Where $E_{BLASTp}[]$ is the expect value extracted from the BLASTp alignment output file using open-source BioJava [8]. The BLASTp algorithm approximates the best alignment (E value = 1e-180) to zero. To include these data in the ratios, we set E = 0.0 back to E = 1e-180. To include the important 'NULL' results from the human search in our ratio calculations, we arbitrarily set this to 1000. The full range for the R_{huDB} and R_{huPDB} values is thus −183 to +183. However, a 'NULL' result from the PDB and Drug-Bank database searches needs to be flagged, as these query proteins are likely to be more difficult to homology model, and do not show homology to targets of known drugs. Error messages from these ratios are defined in

Table 2. The negative numbers used will rank these queries to the bottom a descending list.

Query sequences that show good homology to crystal structure template sequences, but poor/no homology to any protein within the human genome, will have high R_{huPDB} values. The researcher may be particularly interested in the "hypothetical" or "unknown" query proteins that are ranked well according to R_{huPDB} (in the top ~100) as these may make excellent targets for novel research into characterisation, validation, crystallography/modelling and virtual high throughput screening.

TABLE 2: Definition of ratio ranges and error codes.

	R_{huDB}	R_{huPDB}	R_{DBPDB}
$E_{BLASTp}[hum]^{\psi}$ vs. $E_{BLASTp}[DB/PDB]^{\xi}$	−183 to 183	−183 to 183	−7000
$E_{BLASTp}[hum]^{\psi}$ vs. 'Null' DB/PDB$^{\varphi}$	−2000	−5000	−8000
'Null' DB/PDB$^{\varphi}$ vs. $E_{BLASTp}[hum]^{\psi}$	−3000	−6000	−9000

$^{\psi}$BLASTp expect value of the best query/human genome alignment (null = 1000)
$^{\xi}$BLASTp expect value of the best query/DrugBank alignment or query/protein data bank alignment (not null)
$^{\varphi}$No alignment found between query and either DrugBank or PDB databases (null)

A sample output from a search using the full proteome of the malaria parasite, *Plasmodium falciparum*, is shown in supplementary Table S1 online. The 5283 FASTA formatted protein sequences in the malarial genome were downloaded from the EMBL-EBI Interg8 website [9] and used as a test set. Of the top 50 entries as ranked by R_{huPDB}, the majority (68%) showed previous investigation and/or homology to crystal structures of *Plasmodium falciparum* proteins, indicating that this simple ranking system highlights good candidate drug targets (see Figure 2). This is further illustrated over the full genome test set in Figure 2. A query entry was defined as a 'hit' if the PDB title contained keywords associated with malaria. After ranking all 5283 test set entries according to R_{huPDB}, the percentage of hits found is plotted as a function of rank number. Thus in the insert in Figure 2 it is clear that ~80% of the hits are recovered within the

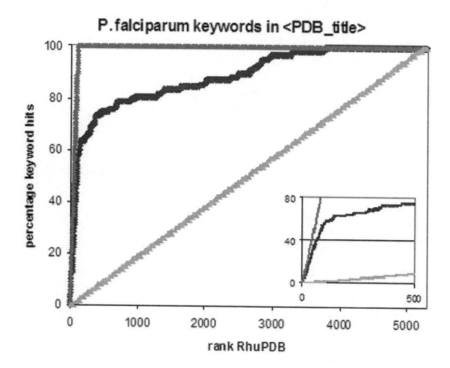

FIGURE 2: Enrichment of *P. falciparum* proteome by R_{huPDB} – PDB targets. Enrichment curves plot the accumulation of user-defined 'hits' as a function of rank number. Thus in an ideal case (medium gray line), each consecutive entry in the ascending ranked list will be a hit. Alternatively, if ranking provides no selection the hits will be distributed randomly across the genome (light gray line). The enrichment percentage as a function of rank are shown in black. The 5283 proteins in the *P. falciparum* 3D7 strain test set were searched using Genomes2Drugs and ranked by R_{huPDB}. *P. falciparum* and malaria related hits from PDB were identified using keyword searching of the <pdb_title> field, and their position in the ranked list identified. The insert, which highlights the first 500 entries, shows that almost 80% of the entries with close homology to known *P. falciparum* crystal structures were identified in the first 10% of the genome.

first 500 entries, or 10% of the genome. The red line in Figure 2 shows an ideal case where each consecutive entry is a hit, while the light blue line shows a random distribution of hits. Interestingly, 25 of the top 50 entries are uncharacterised "hypothetical", "putative" or "unknown" proteins, which warrant further investigation as novel drug targets by virtue of the fact that they are (i) pathogen specific and (ii) similar to a structural template for homology modelling.

Similarly, query sequences homologous to known drug targets, as defined by DrugBank [4], but showing poor/no homology to any human protein, will have high R_{huDB} values. In Figure 3, the full *P. falciparum* proteome test set was ranked according to R_{huDB} and hits identified as having malaria related keywords in the best PDB match title, again indicating that high ranking entries are likely to be well characterised targets for drug discovery and development. Importantly, the same ranking showed good enrichment of known antimalarial drugs, as defined by DrugBank (Figure 4, see listed in supplementary Table S2 online). The DrugBank hits for each query sequence are listed at the bottom of the Microsoft Access form supplied in the output of Genomes2Drugs (see supplementary Figure S1B). These compounds include experimental small molecule drugs as well as FDA (Food and Drug Administration) approved medicinal drugs, which can be purchased and tested for in vitro effectivity [4]. After ranking the *P. falciparum* test set by R_{huPDB}, 8 of the top 50 proteins showed homology to targets of FDA approved drugs. If an FDA approved drug is found to be effective against the pathogen of interest, a 'change-of-application' patent could be sought. As all the necessary toxicology, pharmacology and dosing analysis has already been completed, Phase IV clinical trials to confirm therapeutic use may be more rapidly instigated. This could become an extremely efficient and rapid route for drug development. With a lower financial barrier to entry, this strategy could be especially important in the development of therapeutic drugs against neglected infectious diseases affecting the developing world.

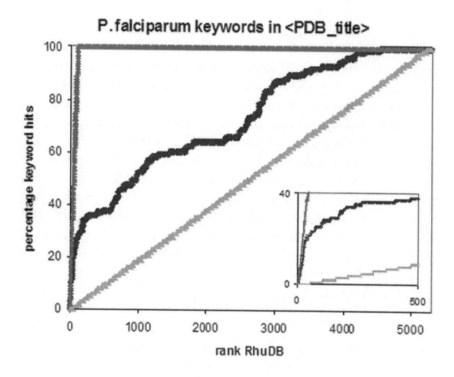

Figure 3. Enrichment of *P. falciparum* proteome by R_{huDB} – PDB targets. Enrichment curves were plotted as described in Figure 2. The 5283 protein malarial proteome was ranked by R_{huDB}. *P. falciparum* and malaria related hits from PDB were identified using keyword searching of the <pdb_title> field. The enrichment percentage as a function of rank are shown in black, while the dark gray line shows an ideal case, and the light gray line indicates a random distribution. The insert highlights the first 500 entries.

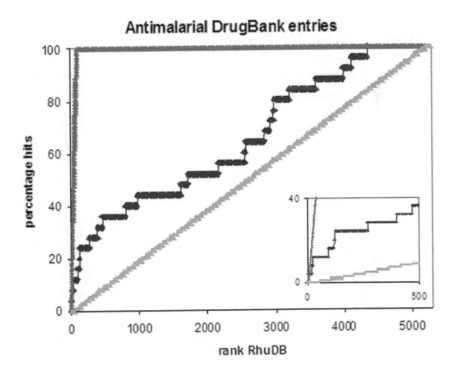

FIGURE 4: Enrichment of *P. falciparum* proteome by R_{huDB} – DrugBank targets. Enrichment curves were plotted as described in Figure 2. The 5283 protein malarial proteome was ranked by R_{huDB}. *P. falciparum* and malaria related hits from DrugBank were identified using keyword searching of DrugBank website [4], as shown in supplementary Table S2 online. The <drugbank_title> field entries were matched to this list of *P. falciparum* or malaria related drug targets. The enrichment percentage as a function of rank are shown in black, while the dark gray line shows an ideal case, and the light gray line indicates a random distribution. The insert highlights the first 500 entries.

6.4 DISCUSSION

We have developed a free online resource that enriches any sized dataset of proteins of interest for those proteins likely to be most usefully in further drug discovery efforts. The program addresses the need to focus drug discovery effort on those protein targets that (i) do not show homology to proteins in the human genomes and (ii) show close homology to proteins for which the 3-dimentional structure is known. As an added feature, each input protein sequence is compared to the DrugBank set of known drug targets, and may identify known drugs that are able to inhibit the protein under investigation.

REFERENCES

1. Kitchen DB, Decornez H, Furr JR, Bajorath J (2004) Docking and scoring in virtual screening for drug discovery: methods and applications. Nat Rev Drug Discov 3: 935–949.
2. Alvarez JC (2004) High-throughput docking as a source of novel drug leads. Curr Opin Chem Biol 8: 365–370.
3. Shoichet BK (2004) Virtual screening of chemical libraries. Nature 432: 862–865.
4. Wishart DS, Knox C, Guo AC, Shrivastava S, Hassanali M, et al. (2006) DrugBank: a comprehensive resource for in silico drug discovery and exploration. Nucleic Acids Res 34: D668–672.
5. Altschul SF, Madden TL, Schaffer AA, Zhang J, Zhang Z, et al. (1997) Gapped BLAST and PSI-BLAST: a new generation of protein database search programs. Nucleic Acids Res 25: 3389–3402.
6. Altschul SF, Wootton JC, Gertz EM, Agarwala R, Morgulis A, et al. (2005) Protein database searches using compositionally adjusted substitution matrices. Febs J 272: 5101–5109.
7. Berman HM, Westbrook J, Feng Z, Gilliland G, Bhat TN, et al. (2000) The Protein Data Bank. Nucleic Acids Res 28: 235–242.
8. Holland RC, Down TA, Pocock M, Prlic A, Huen D, et al. (2008) BioJava: an open-source framework for bioinformatics. Bioinformatics 24: 2096–2097.
9. Kersey P, Bower L, Morris L, Horne A, Petryszak R, et al. (2005) Integr8 and Genome Reviews: integrated views of complete genomes and proteomes. Nucleic Acids Res 33: D297–302.
10. Pearson WR, Lipman DJ (1988) Improved tools for biological sequence comparison. Proc Natl Acad Sci U S A 85: 2444–2448.

There are several supplemental files that are not available in this version of the article. To view this additional information, please use the citation information cited on the first page of this chapter.

PART III

MOLECULAR AND GENETIC MARKERS

CHAPTER 7

PITFALLS AND LIMITATIONS IN TRANSLATION FROM BIOMARKER DISCOVERY TO CLINICAL UTILITY IN PREDICTIVE AND PERSONALISED MEDICINE

ELISABETH DRUCKER AND KURT KRAPFENBAUER

7.1 REVIEW

7.1.1 INTRODUCTION

The strengthening of the robustness of discovery technologies, particularly in genomics, proteomics and metabolomics, has been followed by intense discussions on establishing well-defined evaluation procedures for the identified biomarker to ultimately allow the clinical validation and then the clinical use of some of these biomarkers.

The ability of biomarkers to improve treatment and reduce healthcare costs is potentially greater than in any other area of current medical research. For example, the American Society of Clinical Oncology estimates that routinely testing people with colon cancer for mutations in the K-RAS oncogene would save at least US $600 million a year [1]. On the other side, thousand of papers in the course of biomarker discovery projects have been written, but only few clinically useful biomarkers have been successful validated for routine clinical practice [2]. The following are the major pitfalls in the translation from biomarker discovery to clinical utility:

1. Lack of making different selections before initiating the discovery phase.
2. Lack in biomarker characterisation/validation strategies.
3. Robustness of analysis techniques used in clinical trials.

Each of these details is rarely documented and can dramatically affect the predictive outcome of biomarker results. However, the selection of useful biomarkers must be carefully assessed and depends on different important parameters, such as on sensitivity (it should correctly identify a high proportion of true positive rate), specificity (it should correctly identify a high proportion of true negative rate), predictive value etc. Unfortunately, biomarkers with ideal specificity and sensitivity are difficult to find. One potential solution is to use the combinatorial power of different biomarkers, each of which alone may not offer satisfaction in specificity or sensitivity. Besides traditional immunoassays such as ELISA, recent technological advances in protein chip and multiplex technology offer a great opportunity for the simultaneous analysis of a large number of different biomarkers in a single experiment, which has expanded at a rapid rate in the last decade. However, although many significant results have been derived, one additional limitation has been the lack of characterisation and validation of such technologies. Besides technical characterisation, it also needs quality requirements for correct characterisation of the predictive value of biomarkers. In order to overcome these limitations, some authorities (e.g. Food and Drug Administration (FDA), European Medicines Agency (EMA), European Association for Predictive, Preventive and Personalised Medicine

(EPMA), National Institute of Health (NIH)) already set up recommenda-tions, short proposals and minimum information about a variety of bio-analytical experiments that describe the minimal requirements to ensure that the technical performance as well as the predicted value of biomarkers are correct. For example, EPMA tries to outline a number of key issues in research, development and clinical trial studies, including those associated with biomarker characterisation, experimental design, analytical validation strategies, analytical completeness and data managements [3]. Actual paper follows recommendation presented in the EPMA White Paper [4]. Current recommendations should serve a set of criteria, which will help to carry on to a high-quality data project. Improvements in the quality outcomes are important because without requirements in the improved selection of biomarkers, correct performance of standardisation and validation, the in-terpretation of the results as well as the direct comparisons of the predic-tive value of biomarkers between different research labs or clinical trial studies is not possible. Besides the lack of quality in biomarker selection, a number of other key issues can be identified, which should be addressed in the course of this article. Therefore, the aim of this article is to review and discuss a series of interpretative and practical issues that need to be un-derstood and resolved before potential biomarkers go into the market and become feasible diagnostic tools. The content and structure of the neces-sary information, as well as potential pitfalls and limitations of biomarker research and validation, are discussed briefly in the next subsection.

7.1.2 SHORT OVERVIEW OF DIFFERENT KINDS OF BIOMARKERS

One of the goals of personalising medicine is to use the growing under-standing of biology so that patients receive the right drug for their disease, at the right dose and the right time. Although the definitions of person-alising vary, they all include the use of different biomarkers driven by a decision-making process in which a diagnostic test is pivotal. Biomarkers include gene expression products, metabolites, polysaccharides and other molecules such as circulating nucleic acids in plasma and serum, single-nucleotide polymorphism and gene variants. Ideal biomarkers for use in

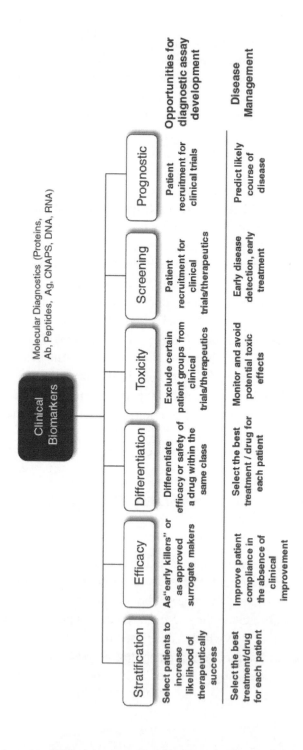

FIGURE 1: Clinical biomarkers: categories/types.

diagnostics and prognostics, and for drug development and targeting, are highly specific and sensitive [5]. Biomarkers can also be categorised as pharmacodynamic, prognostic or predictive [6]:

1. Pharmacodynamic biomarkers indicate the outcome of the interaction between a drug and a target, including both therapeutic and adverse effects [7].
2. Prognostic biomarkers were originally defined as markers that indicate the likely course of a disease in a person who is not treated [8]; they can also be defined as markers that suggest the likely outcome of a disease irrespective of treatment [9,10].
3. Predictive biomarkers suggest the population of patients who are likely to respond to a particular treatment [8,9].

Predictive biomarkers help to assess the most likely response to a particular treatment type, while prognostic markers show the progression of disease with or without treatment. In contrast, drug-related biomarkers indicate whether a drug will be effective in a specific patient and how the patient's body will process it. Figure 1 gives an overview of different biomarker categories and types.

In Figure 1, the clinical biomarkers for diagnostics determine whether a patient is suitable for treatment with a particular drug (by stratification markers), determine the most effective dose for the patient (by efficacy markers), determine the underlying susceptibility of a patient for a particular side effect or group of side effects (by toxicity markers) or evaluate the course and effectiveness end point of a therapy (by surrogate endpoint markers).

Biomarkers can also be used as surrogate end points (end points that substitute for a clinical outcome such as how a patient feels or functions, or how many patients survive) [9,11,12]. Another way of classifying biomarkers is by their role in drug development. Pharmacokinetic or pharmacodynamic biomarkers are involved in early preclinical to phase I studies, and clinical (prognostic, predictive and surrogate) biomarkers play a role in phase II and III trials [10].

The Biomarkers and Surrogate End Point Working Group [13] has defined a classification system that can be used for biomarkers [14]:

1. Type 0 consists of disease natural history biomarkers that correlate with clinical indices;
2. Type I tracks the effects of intervention associated with drug mechanism of action;
3. Type II consists of surrogate end points that predict clinical benefit.

Measurement of different markers (RNA, DNA and/or proteins) needs different diagnostic assays; therefore, different qualification and validation strategies are required.

Pharmaceutical companies are increasingly looking to develop a drug and diagnostic test simultaneously, in a process referred to as drug-diagnostic-co-development so-called companion diagnostic (CDx), to better define the appropriate patient population for treatment. CDx are increasingly important tools in drug development because they lead to the following:

1. Reduced costs through pre-selected (smaller) patient population;
2. Improved chances of approval;
3. Significantly increased market uptake;
4. Added value for core business (late phase);
5. Regulatory trend to have CDx mandatory.

The first drug introduced using the personalised medicine paradigm—Herceptin (Trastuzumab; Roche/Genentech, South San Francisco, CA, USA)—has now been on the market for more than a decade. However, the number of drugs marketed alongside CDx remains small (see Table 1).

Regulatory hurdles have been cited as other main reasons for the slow growth in this area. The differences between the regulatory process in the European Union (EU) and USA and the complexities of the regulatory processes in both regions cause other huge problems for companies. These difficulties affect the preparation of dossiers and their timing and are amplified when considering a CDx project, particularly where more than one company (e.g. pharmaceutical and diagnostic companies) is involved.

Advances in the science underlying drug development have made the discovery of novel biomarkers a real possibility, whilst still challenging, and the use of biomarkers to drive drug development programmes has been increasing steadily over the past decade. Whilst the majority of these biomarkers will not be translated into CDx tests, the growth of biomarker use indicates that the future of the industry will lie in personalised medicine.

TABLE 1: Overview of already approved CDx on the markets

Biomarker	Related drug	Company	Indication	Test
Her-2/neu	Herceptin	Genentech/Roche	Breast cancer	PathVysion®FISH
Kit (CD117)	Gleevec/Glivec	Novartis	Gastrointestinal	c-Kit pharmDx
EGFR	Erbitux/ Tarceva	Bristols-Myers/ Genentech	Colorectal/ NSCLC	EGFR pharmDx kit
CD20	Rituxan/ Bexxar	Genentech/Glaxo	NHL	Flow cytometry
CD25	Ontak/Onzar	Eli Lilly	Lymphoma	Flow cytometry
CD33	Mylotarg	Wyeth	Leukaemia, CML	Flow cytometry
Estrogen receptor	Nolvadex	AstraZeneca	Breast cancer	Hormone receptor assay
HLA A2/HLA C3	Melacine	GlaxoSmithKline	Melanoma	Serology, DNA-based
Philadelphia chromosome	Roferon-A/ Gleevec/Glivec	Roche/Novartis	Leukaemia, CML	BCR-ABL chromosome translocation test
T(15;17) translocation	Trisenox	Cephalon	Leukaemia, CML	Fluorescence in situ hybridisation (FISH)
PML/RAR-α gene expression	Vesanoid	Roche	Leukaemia, CML	

EGFR, epidermal growth factor receptor; CML, chronic myelogenous leukaemia; NSCLC non-small-cell lung carcinoma; NHL, non-Hodgkin lymphoma.

As reflected in Figure 2, the search of the scientific literature indicates that many studies report the discovery of different potential biomarkers, but most of them do not meet the criteria of high sensitivity and specificity. The lack of sensitivity and/or specificity leads to a low number of patent application and, in addition to this, to a low number of successful market applications.

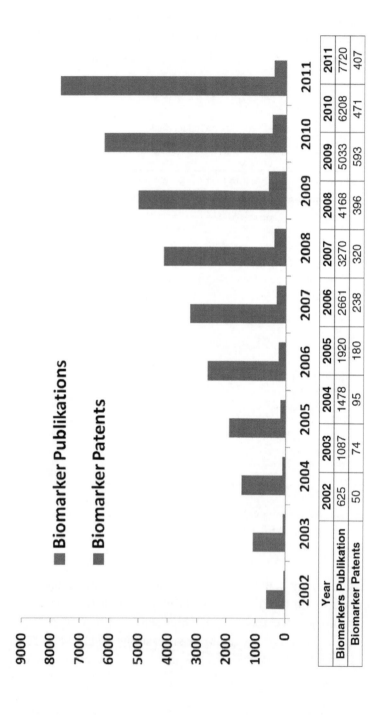

Year	2002	2003	2004	2005	2006	2007	2008	2009	2010	2011
Biomarkers Publikation	625	1087	1478	1920	2661	3270	4168	5033	6208	7720
Biomarker Patents	50	74	95	180	238	320	396	593	471	407

FIGURE 2: Overview of the relationship between publications and patenting of biomarkers.

If the biomarker used for patient selection is known from the earliest stages of the development process, the process of assay development can begin early, and there will be a selection of diagnostic assay used in clinical trials from an early stage. Biomarkers related to response to therapy are often the result of clinical investigations in patients and may not be available until later in the development programme.

Diagnostic development is undertaken in three stages once a biomarker has been identified. Analytical validation ensures the consistency of the test in being able to measure the specific biomarker. Clinical validity relates to the consistency and accuracy of the test in predicting the clinical target or outcome claimed, and clinical utility relates to the fact that the test should improve the benefit/risk of an associated drug in the selected and non-selected groups. Table 2 describes strategic consideration and implication positions of key stakeholders—regulators, pharma and diagnostics companies, patients, physicians and healthcare providers.

TABLE 2: Strategic considerations and implications of personalised medicine

Pharmaceutical companies	• Generate new revenue stream
	• Increased targeted therapies
	• Improve current clinical trials design (kick-off candidates at phase III)
	• Differentiate CM product offerings
	• Shorten clinical trials
	• Improve go-no-go decisions in clinical trials and make it earlier
Diagnostic companies	• New pivotal in the personalised medicine
	• Need to establish relationships with pharmaceutical companies
Payers/health ensurers	• Payers ensure payment of personalised medicine
	• Agree on reimbursement
	• Improve the availability of personalised medicine and their respective diagnostic
	• Have control over escalating healthcare costs
Regulatory authorities	• Clinical trials with improved statistical relevance
	• Will aid co-development programmes
	• Enhance the utility of test information on product labelling

Case studies of drugs and their companion diagnostics that have been approved over the last 10 years indicate that the number of co-developed products is small. The majority of diagnostic tests available to drive patient selection for particular drugs have been added years after the drug's approval. However, experience from the EU and USA also indicates that regulators will not approve targeted drugs in the absence of available, relevant diagnostic tests.

7.1.3 KEY POINTS TO BE ADDRESSED

According to Issaq et al. [5], the failure in finding high-sensitive and high-specific biomarkers may be attributed to the following factors:

1. Small number of samples that are analysed;
2. Lack of information on the history of the samples;
3. Case and control specimens which are not matched with age and sex;
4. Limited metabolomic and proteomic coverage; and
5. The need to follow clear standard operating procedures for sample selection, collection, storage, handling, analysis and data interpretation.

Furthermore, most studies to date used samples with a complex matrix such as serum, plasma, urine or tissue from patients and controls. Another reason for pitfalls in biomarker validation is the usually slow progression of some diseases, requiring high numbers of well-stratified patients who are undergoing long-term treatment when conventional diagnosis and imaging techniques are used. Importantly, there is a lack of sensitive and specific prognostic biomarkers for disease progression or regression that would permit a rapid clinical screening for potential responders and non-responders. Nonetheless, in view of an urgent need for novel therapeutics that have a positive impact on morbidity and mortality of chronic diseases, the field is now moving more quickly towards clinical translation. This development is driven by smart preclinical validation, a better study design and improved surrogate readouts using currently available methodologies and diagnostic techniques. Moreover, upcoming novel biomarkers and diagnostic technologies will soon permit a more accurate and efficient assessment of disease progression and regression.

7.1.4 CONSIDERATIONS BEFORE INITIATING THE BIOMARKER DISCOVERY PHASE

Although some biomarkers have been approved by the FDA as qualitative tests for monitoring specific diseases (e.g. nuclear matrix protein-22 for bladder cancer), unfortunately, the majority of found biomarkers (proteins or metabolites) are not sensitive and/or specific enough to be used for population screening. One of the major reasons that proteomic and metabolomic studies over the past decade have failed to discover molecules to replace existing clinical tests is due to errors in either study design and/ or experimental execution. Werner Zolg wrote in a review [15] that, before initiating the discovery phase, the first step in the process chain of creating new diagnostic content is to make critical decisions on the sample selection that will directly impact the outcome of the identification process. The very selection of the discovery samples and their degree of characterisation of the material, down to the standard operation procedures on how the samples were acquired and stored, can be decisive for success or failure. By selecting tissue as the discovery material for biomarker identification, one must inevitably choose between cultured cells or specimen directly obtained from patients. There are advantages/limitations to either option.

7.1.5 CONSIDERATION ON THE SELECTION AND RANDOMISATION OF PATIENTS FOR BIOMARKER STUDIES: LOOKING FOR THE 'IDEAL' PATIENTS

The optimal selection and randomisation of patients is essential and has to be included in each clinical trial, testing the efficacy of drugs and biomarkers. In particular, given the variant course of disease progression even in well-selected patients with a dominant single aetiology, subjects should be well matched according to factors such as the following lifestyle risk factors:

(1) alcohol and tobacco consumption, (2) body mass index, (3) physical activity, (4) signs of the metabolic syndrome or (5) use of (over-the-counter) medications. As in other studies, age and sex should be balanced. In addition, stratification of patients as to their genetic risk of developing a specific disease, (e.g. using a score) will be central to obtaining a balanced randomisation of the placebo vs. the treatment group. These facts alone should significantly reduce the number of patients and the duration of the trial needed to demonstrate a significant reduction of disease progression or induction of regression. Histological end points in proof-of-concept trials will still be required by regulatory authorities, apart from long-term hard end points, such as morbidity and mortality in phase III trials. At present, it is not possible to exactly predict the number of patients and the time on treatment that are needed to demonstrate the clinical benefit of a drug agent or biomarker. This is one major reason that companies have been reluctant to enter this difficult field.

7.1.6 THE CURRENT STATE OF BIOMARKER DISCOVERY

The search of the scientific literature clearly indicates that most published biomarkers are inadequate to replace an existing clinical test or that they are only useful for detecting advanced disease stage, where the survival rate is low. Many molecular or genetic biomarkers have been suggested for the detection of different diseases; however, most of them do not possess the required sensitivity and specificity. Another reason why most proposed metabolomic and proteomic biomarker results that have not progressed from the laboratory to the clinic study is that the majority stopped at the first phase of biomarker discovery. According to other studies [5,16,17], there are five phases that a protein or a metabolite has to go through to become a biomarker. Phase I is preclinical exploratory studies to identify potentially useful markers, phase II is clinical assay development for clinical disease, phase III is retrospective longitudinal repository studies, phase IV is prospective screening studies and phase V is control studies [5].

Listed examples of already approved biomarkers in Table 1 show that there are no 100% sensitive and specific biomarkers for different types of diseases to date. A biomarker with a high sensitivity has a low specificity and vice versa. Unfortunately, biomarkers with ideal specificity and sensitivity are difficult to find. One potential solution is to use the combinatorial power of a number of different biomarkers, each of which alone may not offer satisfactory in specificity. For example, Horstmann et al. [18] studied the effect of using a combination of bladder cancer biomarkers on sensitivity and specificity. Although none of the combinations resulted in 100% sensitivity and specificity, the sensitivity improved from 91% (using two biomarkers) to 98% using a combination of four different biomarkers.

7.1.7 PITFALLS AND LIMITATIONS

However there exist different reasons why most potential biomarkers failed in achieving adequate sensitivity and specificity and are not accepted as clinical tests. One main reason is that most biomarkers are dealing with detecting diseases at an early stage in humans that have different age, sex and ethnicity. Other important fact is to find a protein or a metabolite at an extremely low concentration level among thousands of other proteins and metabolites. To improve sensitivity and specificity, there are different strategies: potential solutions are listed as follows:

1. Improve the assay (e.g. antibody with a higher specificity and/or in combination with a detection conjugate with a higher sensitivity),
2. Combine several markers,
3. Check for subpopulations and stratify population (e.g. matched by gender, age, pathology).

The current procedure for the search of biomarkers is dealing with potential errors in the study design that can be avoided in future studies.

Figure 3 gives an overview about two main reasons why most potential biomarkers failed in achieving adequate sensitivity and specificity and are not accepted as clinical tests. One main reason is pitfalls and limitations in biomarker discovery and second main reason is pitfalls in biomarker validation.

7.1.8 AGE, SEX AND RACE

Biomarker studies are normally carried out using body fluids or tissues collected from patients and healthy subjects of different ages, sex and race. Using samples from patients and controls that are of different ages and sexes can dramatically influence the results. In a recent study, Lawton et al. used 269 subjects, 131 males and 138 females, to study the effects of age, sex and race on plasma metabolites. The patients were of Caucasian, African-American and Hispanic descent and ranged in age from 20 to 65 years. The subjects were divided into three different age groups; 20–35, 36–50 and 51–65. Using GC/mass spectrometry (MS) and high-performance liquid chromatography (HPLC)/MS, they reported that 'more than 300 metabolites were detected of which more than 100 metabolites were associated, with age, many fewer with sex and fewer still with race' [19].

7.1.9 SELECTION OF PATIENTS AND CONTROLS

Patients for biomarker studies should be carefully selected by a specialist (e.g. oncologist for cancer studies or a pathologist for tissue samples) to insure the presence or absence of diseases. Unfortunately, predictive curve values of biomarkers with no or less overlapping of diseased vs. non-diseased cohorts are difficult to find. There exist always more or less overlapping areas between healthy and diseased cohort. The overlapping area allows the analyst to calculate the proportion of patients whose diagnosis was correctly predicted by the model (true positives for sick patients and true negatives for healthy patients) or false negative or false positive values [3].

Generally, the number of patients and control subjects in published studies is very small to give an acceptable statistical value. Also, many of the potential proposed markers have not been confirmed or validated in a high-quality manner. Body fluids and tissues are collected from a group of patients of different disease stages, and results are compared with a group of healthy persons. The effect of a disease stage on sensitivity of a single biomarker should be taken into consideration as mentioned previously because sensitivity improves with increase in disease stage. Grossman

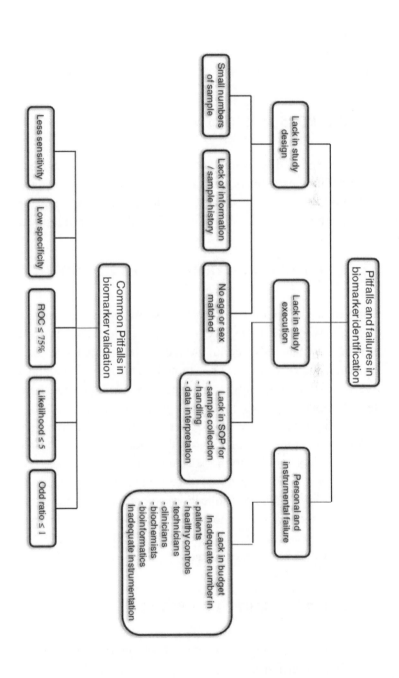

FIGURE 3: Pitfalls and failures in biomarker identification.

[20] adequately summarises the importance of consistency through his observation that 'the contradicting published reports likely [resulted] from studies testing different patient populations, using different methodologies, and applying different [cut-offs] for a positive test'.

7.1.10 ERRORS IN STUDY EXECUTION

Study execution deals with experimental parameters that need to be considered. These parameters include many different variables, such as sample collection, handling and storage, sample comparison, number of samples, sample preparation, methods of analysis and number of replicates.

7.1.11 SAMPLE COLLECTION, HANDLING AND STORAGE

Samples are collected from a person who passed a physical exam by a physician who determines that the person of interest has a concrete disease or is healthy. Samples (serum, plasma, urine, saliva, tissue etc.) should be collected in freezer-type tubes, immediately snap frozen and stored in a freezer until time of analysis. It is recommended that, for short-time storage (less than 1–2 weeks), storage condition should be at −20°C, and for long-term storage (more than 2 weeks), storage condition should be at −70°C. At the time of analysis, samples should be thawed at 4°C or on ice and prepared according to the selected method of analysis. The history of the sample is very important and may have been obtained from sample storage banks with proper collection, storage information about the stage of disease, medication, pathology, age, gender and condition of patients. A lack of consistency in sample collection and storage can doom a study before any data are even collected.

7.1.12 DIRECT SAMPLE COMPARISON

If this option is chosen, the degree of sample characterisation is critical. It is of importance that the specimens used in the diseased cohort are not simply classified as 'diseased' (if possible, together with the stage of the disease) but that a detailed histopathological assessment of the distribution of cell types (e.g. tumour cells, necrotic cells, stroma) in the diseased specimens is carried out [5]. This distribution should be as uniform as possible in all samples, and it should represent the correct disease/healthy state. Otherwise, normalising the analytical outcome becomes very difficult.

7.1.13 NUMBER OF SAMPLES

The number of samples that have to be placed in the diseased and healthy control groups in order to be compared with a variety of analytical approaches remains a matter of discussion. A minimum of 15 samples in the discovery phase is necessary to get a reasonable representative selection basis for marker candidates. If the number is for practical reasons (resources, cohort and time lines), which is very small (e.g. less than 10 per group), then the observed differences between the two sets of specimens are in danger of being over-interpreted when extrapolated to generalised cohorts. Low sample sizes make the correct identification of those differences increasingly difficult. To overcome these limitations, Zolg [15] recommended running second and third discovery rounds to complement the results of the first round. Ideally, the sample number analysed should not only allow stating the presence or absence of a given protein, but should also give the opportunity to identify trends. Another opportunity is to pool the samples, i.e. to physically combine several of the extracts to create fewer samples, to be put through the entire analytical process. Pooling of samples inevitably leads to a loss of information. The distribution of proteins is averaged by the very pooling process with the prospect that individual proteins are pushed below the detection limit by one member of the pooling cohort not expressing the protein in question. At any rate, somewhere in the selection process, the individual spectrum of proteins has to be established. Therefore, the pooling process just shifts the workload to a later point in the process chain, and very good arguments have to be found to deliberately increase the complexity of the data sets by pooling.

7.1.14 SAMPLE PREPARATION

Preparations of the sample for proteomic and metabolomic analyses prior to analysis are very important and can introduce errors that always will affect the final results [3,5]. The search for biomarkers in biological samples involves different steps depending on the sample type and if it is analysed for metabolites or proteins. Extraction of metabolites from the blood, urine or tissue required multiple purification and extraction procedures using different solvent systems as discussed by Want et al. [21] and Issaq et al. [5]. It is not always possible to extract or to isolate all the metabolites from a sample with a single solvent since metabolites have different chemical and physical properties and are present in a wide dynamic range of concentrations. The search for a protein biomarker involves extraction of the proteins followed by fractionation, purification, specific enrichment and then analysis by different analytical methods (e.g. 2DE-PAGE, immunoassays, Western blot, HPLC/MS/MS). Analysis of the blood as well as the serum is more complicated than that of urine or saliva as it contains fewer proteins, and high-abundant protein must be depleted prior to analysis. Approximately 99% of the protein content of the blood (both serum and plasma) is made up of only about 20 proteins (http://www.plasmaproteome.org) [22]. While depletion of these proteins will allow the detection of low-abundant proteins, it may remove proteins that are bound to these 20 proteins, resulting in a loss of potentially important information [23]. Tissues are homogenised first followed by metabolites, and proteins are extracted and analysed. Incomplete homogenisation can lead to losses that can affect the accuracy of the results. In addition, one cannot ignore human errors in sample collection, storage, weighing, extraction etc.

7.1.15 METHODS OF ANALYSIS

Choosing the optimal analysis method is critical in biomarker search by proteomics and metabolomics. For example, analysing the plasma proteome involved protein precipitation and solubilisation; therefore, the downstream fractionation method must be either electrophoresis or a liquid-phase method. Unfortunately, studies have shown that the proteome analysis by groups using different methods resulted not only in different numbers of protein identifications, but also in poor overlap between the results [5,23,24]. These results prove that the selected method of analysis is an important parameter.

7.1.16 NUMBER OF REPLICATES

Sample should be analysed in triplicate and report the mean and standard deviation. Unfortunately, most published proteomic and metabolomic studies only analyse each sample once, which does not permit the deviation from the mean (i.e. the error in the measurement) to be calculated. Proteomic analysis of a biological sample involves different analytical steps in the course of sample preparation. Each one of these steps can introduce an error. Due to difficulties either in sample preparation, in protein preparation or in assay or protein chip hybridisation, the amount of replicas varied from zero to six. Thus, implicating different optimal statistical tests were necessary for the various settings.

7.1.17 CONSIDERATION ON THE IMPROVEMENT OF CURRENT EFFICACY READOUTS BY DEVELOPMENT OF NON-INVASIVE DIAGNOSTIC TOOLS

Further improvement is desirable to reduce the number of study patients, trial duration, costs and, most importantly, possible risks for individuals. Thus, new innovative diagnostic techniques are needed that allow an exact assessment of the degree of disease and, more importantly, of the dynamic processes underlying the diseases. Such biomarkers and technologies will have to be specific for the targeted structure, i.e. the cells or key molecules involved in the development of the disease. Ideally, sensitive and specific markers/imaging methodologies will allow a rapid and mechanism-based screening for and efficacy monitoring of treatments. Additionally, there is a need for universal-standardised reporting methods to aid interpretation and comparison of potential clinical biomarker trails. All current non-invasive methodologies (serum markers, serum marker algorithms, contrast imaging etc.) yield a sufficient to excellent diagnostic accuracy for the detection (or exclusion) of an upcoming or current disease.

7.1.18 REGULATORY OUTLOOK AND FUTURE ASPECTS

The regulatory landscape for biomarker discovery and validation projects (especially for drug-diagnostic co-development = companion diagnostic) is evolving and getting more important to the upcoming clinical trial studies. In the past few years, available data have been reviewed by FDA and EMA, and experience from some exploratory data submission process was used to create a formal biomarker qualification purpose [25].

Both the FDA and EMA have similar biomarker qualification processes in place that enable research institutes and pharmaceutical companies to obtain advice or qualification of the biomarker in question. In both cases, similar guidance concepts were developed that are very clear on the fact that biomarker qualification does not constitute a review of a diagnostic for commercialisation. Nevertheless, for the future, biomarker qualification submissions are strongly recommended by US and EU authorities and will be more and more required for drug/diagnostic co-development projects in both regions [25]. Further guidance on clinical trial enrichment and internal standard operating procedures for cross-labelling efforts are also expected and will improve the penetration of personalised medicine in clinical practice.

The FDA's first guideline was finalised in 2005, and it is based on the fact that many clinical trial studies were utilising biomarkers but that these data were often exploratory and that their regulatory submission was not required [25]. However, the US regulatory agency regarded the submission of these data as beneficial for both the industry and the FDA to ensure that regulatory scientists are familiar with and are prepared to evaluate future submissions. This data mainly includes pharmacogenomic information, and the programme is referred to as a voluntary exploratory data submission (VXDS). The success of this VXDS programme has led to the development of a number of new (draft) guidance documents including those related to the biomarker qualification process and to clinical pharmacogenomics in the early phases of drug development. Further guidance on clinical trial enrichment and internal standard operating procedures for cross-labelling efforts within FDA offices is also expected and is continuously under discussion.

Since the FDA's initial publication, the International Committee on Harmonisation (ICH) has published a guidance relating to pharmacogenomic data (ICH E15) that defines pharmacogenomics, pharmacogenetics, genomic biomarkers, and relevant sample and data coding. Standardised terminology is presented for incorporation in future regulatory documents related to pharmacogenetics and pharmacogenomics. Further ICH guidance, topic E16, on the information required for biomarker qualification was published in 2010. In addition, the FDA has established processes for working jointly with EMA on the review of exploratory information. A review of their experience and the impact of the guidance were published in 2010 [16].

7.2 CONCLUSIONS

While application of potential biomarkers in preclinical development is far advanced, only a handful have passed clinical trials (see Table 1) and are already commercially successful on the market (see Table 3). Reasons for the pitfalls are manifold, including difficult validation strategies and the usually slow disease progression, requiring high numbers of well-stratified patients undergoing long-term treatment when conventional diagnostic parameters or related end points are used. Importantly, there is a notorious lack of sensitive and specific surrogate biomarkers for disease progression or regression that would permit a rapid clinical screening for potential drug candidates. Nonetheless, in view of an urgent need for new drugs that positively impact morbidity and mortality of different diseases, the biomarker field is now moving more quickly towards clinical translation. This development is driven by thoughtful preclinical validation, a better study design and improved surrogate readouts using currently available methodologies. Moreover, upcoming novel biomarkers and imaging technologies will soon permit a more exact and efficient assessment of disease diagnosis, disease progression and disease regression as already published in other works [26,27].

TABLE 3: Molecular diagnostic players with approved tests

Manufacturer	Headquarter	Number of tests	Global sales IVD
Roche Molecular Diagnostics	Switzerland	24	20%
Gen-Probe	CA, USA	18	
Cepheid	CA, USA	13	
Becton, Dickinson and Company	NJ, USA	11	
AdvanDx	MA, USA	10	
Abbott Molecular	IL, USA	8	15%
Hologic	MA, US	7	
Nanosphere	IL, USA	7	
Qiagen	Germany	7	
Idaho Technology	UT, USA	5	
AutoGenomics	CA, USA	4	
bioMerieux	France	4	
Luminex Molecular Diagnostics	TX, USA	4	
Siemens Healthcare Diagnostics	IL, USA	3	15%[a]
Others		28	
Total		153	

[a]*Mainly imaging technology. After Datamonitor; adapted from the Association of Molecular Pathology.*

REFERENCES

1. Javle M, Hsueh CT: Recent advances in gastrointestinal oncology—updates and insights from the 2009 annual meeting of the American society of clinical oncology. J Hematol Oncol 2010 2009, 23(3):11-23.
2. Poste G: Bring on the biomarkers. Nature 2011, 469:156-157.
3. Waerner T, Urthaler J, Krapfenbauer K: The role of laboratory medicine in healthcare: quality requirements of immunoassays, standardisation and data management in prospective medicine. EPMA J 2010, 1:619-626.
4. Golubnitschaja O, Costigliola V, EPMA: General reports & recommendations in predictive, preventive and personalised medicine 2012: white paper of the European Association for Predictive, Preventive and Personalised Medicine. EPMA J 2012, 3:14.

5.   Issaq HJ, Waybright TJ, Veenstra TD: Cancer biomarker discovery: opportunities and pitfalls in analytical methods. Electrophoresis 2011, 32(9):967-975.

6.   Amur S, Frueh FW, Lesko LJ, Huang SM: Integration and use of biomarkers in drug development, regulation and clinical practice: a US regulatory perspective. Biomark Med 2008, 2(3):305-311.

7.   Feuerstein GZ, Dormer C Jr, Rufollo RR, Stiles G, Walsh FS, Rutkowski JL: Translational medicine perspectives of biomarkers in drug discovery and development. Part I. Target selection and validation - biomarkers take center stage. Int Drug Discovery 2007, 2(5):36-43.

8.   Bruenner N: What is the difference between "predictive and prognostic biomarkers"? Can you give some examples? Connection 2009, 13:18.

9.   Buyse M, Sargent DJ, Grothey A, Matheson A, de Gramont A: Biomarkers and surrogate end points - the challenge of statistical validation. Nat Rev Clin Oncol 2007, 7(6):309-317.

10. Buyse M, Michiels S, Sargent DJ, Grothey A, Matheson A, de Gramont A: Integrating biomarkers in clinical trials. Expert Rev Mol Diagn 2011, 11(2):171-182.

11. Rosenkranz B: Biomarkers and surrogate end points in clinical drug development. Appl Clin Trials 2003, 5(2):30-40.

12. Lassere MN, Johnson KR, Boers M: Definitions and validation criteria for biomarkers and surrogate end points: development and testing of a quantitative hierarchical level of evidence schema. J Rheumatol 2007, 34(3):607-615.

13. Atkinson AJ, Colburn WA, DeGruttola VG: Biomarkers and surrogate end points: preferred definitions and conceptual framework. Clin Pharmacol Ther 2001, 69(3):89-95.

14. Frank R, Hargreaves R: Clinical biomarkers in drug discovery and development. Nat Rev Drug Discov 2003, 2(7):566-580.

15. Zolg W: The proteomic search for diagnostic biomarkers: molecular & cellular. Proteomics 2006, 5:10. 1720–1726

16. Goodsaid FM, Mendrick DL: Translational medicine and the value of biomarker qualification. Sci Transl Med 2010, 2(47):47.

17. Pepe MS, Etzioni R, Feng Z, Potter JD, Thompson ML, Thornquist M, Winget M, Yasui Y: Phases of biomarker development for early detection of cancer. J Natl Cancer Inst 2001, 93:1054-1061.

18. Horstmann M, Patschan O, Hennenlotter J, Senger E, Feil G, Stenzl A: Combinations of urine-based tumour markers in bladder cancer surveillance. Scand J Urol Nephrol 2009, 43:461-466.

19. Lawton KA, Berger A, Mitchell M, Milgram KE, Evans AM, Guo L, Hanson RW, Kalhan SC, Ryals JA, Milburn MV: Analysis of the adult human plasma metabolome. Pharmacogenomics 2008, 9:383-397.

20. Grossman HB: Are biomarkers for bladder cancer beneficial? J Urol 2010, 183:11-12.

21. Want EJ, O'Maille G, Smith CA, Brandon TR, Uritboonthai W, Qin C, Trauger SA, Siuzdak G: Solvent-dependent metabolite distribution, clustering, and protein extraction for serum profiling with mass spectrometry. Anal Chem 2006, 78:743-752.

22. Xiao Z, Conrads TP, Lucas DA, Janini GM, Schaefer CF, Buetow KH, Issaq HJ, Veenstra TD: Direct ampholyte-free liquid-phase isoelectric peptide focusing: application to the human serum proteome. Electrophoresis 2004, 25:128-133.
23. Anderson NL, Polanski M, Pieper R, Gatlin T, Tirumalai RS, Conrads TP, Veenstra TD, Adkins JN, Pounds JG, Fagan R, Lobley A: The human plasma proteome: a nonredundant list developed by combination of four separate sources. Mol Cell Proteomics 2004, 3:311-326.
24. Buscher JM, Czernik D, Ewald JC, Sauer U, Zamboni N: Cross-platform comparison of methods for quantitative metabolomics of primary metabolism. Anal Chem 2009, 81:2135-2143.
25. Sleigh S, Barton C: Advances in Drug-Diagnostic Co-Development. London: Scrip Business Insights; 2011.
26. Ausweger C, Burgschwaiger E, Kugler A, Schmidbauer R, Steinek I, Todorov Y, Thurnher D, Krapfenbauer K: Economic concerns about global healthcare in lung, head and neck cancer: meeting the economic challenge of predictive, preventive and personalized medicine. EPMA J 2010, 1(4):627-631.
27. Koehn J, Krapfenbauer K: Advanced proteomics procedure as a detection tool for predictive screening in type 2 pre-diabetes. EPMA J 2010, 1(1):19-31.

CHAPTER 8

HOW BIOINFORMATICS INFLUENCES HEALTH INFORMATICS: USAGE OF BIOMOLECULAR SEQUENCES, EXPRESSION PROFILES, AND AUTOMATED MICROSCOPIC IMAGE ANALYSES FOR CLINICAL NEEDS AND PUBLIC HEALTH

VLADIMIR KUZNETSOV, HWEE KUAN LEE,
SEBASTIAN MAURER-STROH, MARIA JUDIT MOLNБR,
SANDOR PONGOR, BIRGIT EISENHABER,
AND FRANK EISENHABER

8.1 WHEN WILL GENOME SEQUENCES, EXPRESSION PROFILES, AND COMPUTER VISION FOR BIOIMAGE INTERPRETATION BE ROUTINELY USED IN CLINICAL MEDICINE?

There is apparently no doubt for anyone that modern life science research based on the new high-throughput technologies most prominently represented by genomic sequencing together with the increasingly powerful and, at the same time, affordable information technology products will

This chapter was originally published under the Creative Commons Attribution License. Kuznetsov V, Lee HK, Stroh SM, Molnár MJ, Pongor S, Eisenhaber B, and Eisenhaber F. How Bioinformatics influences Health Informatics: Usage of Biomolecular Sequences, Expression Profiles and Automated Microscopic Image Analyses for Clinical Needs and Public Health. Health Information Science and Systems *1,2 (2013). doi:10.1186/2047-2501-1-2.*

dramatically change healthcare. The main idea behind these expectations is that the new availability of data characterizing the patients' individuality at the level of genome, biomolecules and gene/protein networks together with evermore powerful diagnostic, mainly imaging tools at the histological, anatomical and physiological levels allow ever finer stratification of the patients' conditions once the molecular data is integrated with clinical data and, finally, it will lead to the design of personalized treatment regimes.

Unfortunately, the discussion in the media has become hyped with expectations increasingly getting out of touch with the progress that both biomedical science [1] and healthcare at the ground can deliver in the short and medium term. In this discussion and, to some extent, review article, we try to analyze what are major trends in computational biology and bioinformatics that support the advance towards stratified and personalized medicine and what are the fundamental and some of the procedural barriers on the path towards the solution of major healthcare problems such as infections, cancer, metabolic and neurodegenerative diseases, familial disorders, etc.

The article is structured as follows: In the section The hype around genomics and proteomics technologies in the healthcare context and fundamental reasons calling for a temperate view, we look into the general developments that fuel the expectations of revolutionary change in health care and public health; we talk about several roadblocks that have been removed on the path towards personalized/stratified medicine and the possible role of bioinformatics and computational biology in this process. We also emphasize what are the reasons why many of the expectations will not materialize in the short- to medium-term time frame. Section Management of innovation cycles of high-throughput technologies and the role of bioinformatics in this process is dedicated to issues that arise when bioinformaticians/computational biologist actually penetrate into the actual health care provision system under the condition when the application of new computational analysis methods and evaluation protocols is not really routine.

In sections Bioinformatics moving towards clinical oncology: biomarkers for cancer classification, early diagnostics, prognosis and personalized therapy (cancer biomarkers), Sequence-structure-function relationships for pathogenic viruses and bacteria and their role in combating infections (infectious diseases) and Impact of Bioimage Informatics on Healthcare (computerized histopathology), we exemplarily discuss and partially review the progress in application areas that have already or will likely benefit in the near future from interaction with bioinformatics/computational biology approaches. Although often histologically similar, increasingly more cancer subtypes are getting characterized at the level of the specific, individual biomolecular mechanisms that drive the growth of the tumor cell population and, thus, are essentially understood as different diseases. Cancer biomarkers are critical for diagnosis, classification, prognosis and therapy progress evaluation in this concept (section Bioinformatics moving towards clinical oncology: biomarkers for cancer classification, early diagnostics, prognosis and personalized therapy).

Due to their small genome and the possibility to successfully deduce phenotype properties from mutations, viral and bacterial pathogens are thankful objects for computational biology analysis in the clinical context (in contrast to the situation with higher eukaryotes such as human; section Sequence-structure-function relationships for pathogenic viruses and bacteria and their role in combating infections). As example, we review in depth the clinically relevant characterization of patient-specific influenza viral infections. We also show that genome analysis of enterohemorrhagic *E.coli* allows selecting existing FDA approved drugs for treatment.

In section Impact of Bioimage Informatics on Healthcare, we review advances in the automated assessment of histopathological and, to a minor extent, other medical images. Possibly, these developments in this area might have a non-spectacular but a very profound impact on health care delivery very soon since the problems to overcome are more of the engineering type and not of fundamentally scientific origin.

8.1.1 THE HYPE AROUND GENOMICS AND PROTEOMICS TECHNOLOGIES IN THE HEALTHCARE CONTEXT AND FUNDAMENTAL REASONS CALLING FOR A TEMPERATE VIEW

Several roadblocks towards the goal of stratified/personalized medicine have disappeared very recently. The spectacular improvement of nucleic acid sequencing technologies lead to a reduction in costs, both in time and money, at a scale that can only be described as jaw-dropping for the observer. Whereas the first full human genome sequencing absorbed about 3 billion USD in the USA alone and it took about a decade to be accomplished [2], recently offered machines such as Ion Proton™ Sequencer (Life Technology) or HiSeq™ 2500 (Illumina) [3] move these numbers rather close towards 1000 USD and a single day. And this appears not to be the endpoint of the technology development with more progress to be expected in the medium-term future. Naturally, dreams about all kinds of sequencing applications, especially, in clinical contexts and with affluent patients start sprouting. To note, the progress of nucleic acid sequencing is just the most eye-catching; essentially, it hides dramatic progress also in many other areas and high-throughput technologies such as expression profiling, histopathological image processing, etc. We need to acknowledge, that for life sciences, where, historically, getting at least some verifiable, quantified data for their biological system of study was a major difficulty and the setup of experiments and not the analysis of the measurement absorbed most of the intellectual capacity [4], the current deluge of quantified data is really a game changer and puts theoretical analysis detached from experimentation into general importance for the field for the first time.

The second major change is in IT itself. The older among the list of authors still remember their times as PhD students when the access to mainframe machines was cumbersome and heavily restricted and a good desktop computer with graphical interface in the late eighties/early nineties had the price of a luxury sports car. Today, for nominally the same money, one can equip several research teams if not a small institute with computer clusters (e.g., a 64 core computer trades for just about 10000 USD), storage systems and network tools that are more powerful than necessary for about 90% of the tasks in computational biology. Thus, computing and storage opportunities are essentially no longer the limiting factor for life science research compared with just a decade or even a few years ago.

The hype currently accumulating around the new opportunities with sequencing and other high-throughput technologies, maybe, is sensed most directly in the entrepreneurs' and scientists' comments compiled by Bio-IT World at its WWW page dedicated to the 10th anniversary of its own launch [5]. Although there are some minority cautionary notes, one cannot get away with the general impression that concluding from molecular data to clinically important statements is mainly seen as a problem of the scale of data generation. It is expected that the IT-centric efforts of integrating patient-specific sequencing, expression, tissue imaging data with clinical information (whatever might be the exact meaning of this "data integration"; just putting everything into one electronic database) will inevitably lead to significant healthcare outcomes in terms of personalized medicine.

This surprisingly optimistic view remembers the euphoria that, ten years ago, accompanied the presentation of the first draft of the human genome caused by the anticipation that "Genetic prediction of individual risks of disease and responsiveness to drugs will reach the medical mainstream in the next decade or so. The development of designer drugs, based on a genomic approach to targeting molecular pathways that are disrupted in disease, will follow soon after" [6]. With hindsight, we know that the progress in the last decade has not reached the promises, not even nearly [1,7]. The hype in the media is also in suspicious contrast to the recent attempt of certain pharmaceutical companies to slash down their own research force and to promote the idea of open innovation, i.e., essentially unloading research efforts, costs and research risks into the public sphere.

Whereas the general developmental trend appears correctly predicted, the devil is in the detail and the serious disagreement is about timescales and in which areas/applications the healthcare breakthroughs from genomics and other technologies are more likely in the time closer to us. Moving from the scientific laboratory to actual healthcare is also associated with a myriad of additional issues besides the scientific task itself. Apparently boring questions such as predictive power, robustness, standardization, availability and reliability of the new methods in conditions of routine application in regular hospitals, clinics and in the out-patient context by possibly scientifically insufficiently trained personnel become urgent. This includes the comparison of the new methods with more traditional, tested approaches not only from the viewpoint of medical science but also cost-wise (in terms of money and working time for tests and data analyses). Since considerable economic interest is associated with the upcoming healthcare revolution not only from IT equipment and healthcare solution providers but also from charlatans who, for example, try to sell life style advice derived from the customers' own genome sequence already today, it is important to get the discussion away from the level of fairy tale and hyped promises and to assess the current state of the art realistically.

Besides the costs, the most important argument against having genome sequencing and expression profiling from every patient at present is the fact that the overwhelming part of this data cannot be interpreted into biologically and/or medically significant conclusions. Today, ever faster sequencing leads foremost to ever faster growing amounts of non-understood sequence data. To note, we need to know about the biomolecular mechanisms that translate the genome sequence into phenotypes when we wish to interfere rationally at the molecular level. As elaborated elsewhere, the biological functions of about every second human gene are not well or even completely not known [1]. The whole mystery of non-coding RNA function is hardly scratched upon; yet, we know that many, also non-protein-coding regions of the genome are actively transcribed and this expression influences important biological processes [8,9]. Maybe, it was one of the most important insights from the whole human genome sequencing project that we can estimate now how much human biology at the molecular level we do not know, namely most likely (much) more than 50% [1]. To just search for correlations between phenotypic, including clinical conditions and genomic changes will appear insufficient because of several reasons: 1) the path relating genome features and phenotype is extremely complex in many cases. 2) The statistical significance criteria will require impossibly large cohorts. 3) Rationally designed therapy without mechanistic insight is problematic. Given the pace of progress in the area of biomolecular mechanism discovery during the last decade, it is expected that it will take another century until we will understand our own genome. Presumably, scientific, technological and social factors will kick in that will accelerate the advance [1]; yet, it is clear that this is not a short term issue.

Most likely, biomedical applications that rely either on the comparison of DNA or, generally, nucleic acid sequences, without necessarily understanding their biological meaning or on the biomolecular mechanisms that are already more or less known have the greatest likelihood to achieve importance for healthcare, public health and biotechnology. To the first class of applications belong methods for the identification of the human individual's origin and identity, be it in the forensic, genealogy or legal context, but also the diagnostics of hereditary diseases and the characterization of food items in terms of quality and origin. With regard to the latter class of applications, those diseases that require the investigation of less complex gene networks and biomolecular mechanisms will have better chances to benefit from sequencing, expression profiling and histopathological imaging informatics than those with more complex mechanisms. In this light, the perspectives of fighting infections or cancer are more promising than, for example, those of battling obesity since energy metabolism appears to be one of the most complexly regulated systems in humans.

In this context, does the sequencing of patients' DNA in a large scale make sense? In several countries, for example in Norway [10], programs are being implemented that aim exactly at realizing this vision, the sequencing of the patients' genomes and of their cancers. It appears to us that, at this stage, the move may be justified for small, rich countries that have the necessary capacity to finance an extensive follow-up fundamental research effort to study the newly collected data since, in many cases, the clinical outcome for the specific patient might be negligible at present. Thus, sequencing, expression profiling, etc. make sense in a clinical setup where the data can enter into a research environment for proper, non-standard data analysis and where, beyond potential benefit for the specific patient, these expensive laboratory investigations can have serendipitous consequences for the scientific knowledge gain that might benefit many other future patients.

8.1.2 MANAGEMENT OF INNOVATION CYCLES OF HIGH-THROUGHPUT TECHNOLOGIES AND THE ROLE OF BIOINFORMATICS IN THIS PROCESS

In addition to fundamental scientific problems with biomolecular mechanisms discovery, we need to emphasize that high-throughput technologies such as nucleic acid sequencing are far from mature. The renewal cycle involves maximally a couple of years and it might be already tomorrow that, due to some unexpected innovation, the equipment purchased yesterday is hopelessly out of date even if the machines continue to look shiny.

Since the new generation of sequencing, expression profiling and other high-throughput technologies tend to generate the biological data at much lower costs and with higher accuracy than their predecessors, it does not make sense to produce more data than can be properly analyzed within a reasonably short time frame; future researcher will rather look at regenerated data produced with newer technologies available then instead of reviving old data files.

Even for dedicated research institutions with rich budgets, it remains a financial problem to participate in every step of technology development. It is not just the purchase of new pieces of equipment, but also the establishment of subsequent data analysis pipelines, software replacements and the training of the respective staff or even the hiring of new types of professionals. The latter issues might create more headache than the sequencer purchase itself.

Many clinical labs attached to research and other top-end hospitals around the world are thinking about how to prepare for a swift increase in genomics and proteomics analysis needs. Ever since their emergence in 2005, next-generation sequencing (NGS) technologies have proven revolutionary research tools in a variety of scientific disciplines of the life sciences. NGS technologies are now increasingly being applied in clinical environment, which is partly due to the emergence of novel and efficient sequencing protocols and partly to the appearance of smaller, less expensive sequencing platforms. The possibilities of applying NGS in clinical research ranges from full human genome profiling [11], microbiome profiling [12] to biomarker discovery, stratification of patients for clinical trials, prediction of drug response and patient diagnosis. Such applications often involve targeted re-sequencing of genes of clinical relevance whereby not the entire genome is sequenced, only a few dozen PCR-amplified regions or known disease-related genes. These genes harbor diagnostic or causative mutations of diseases including indels and single nucleotide polymorphisms. Individual genes have previously been interrogated in clinical testing using traditional techniques such as Sanger sequencing however NGS technologies have already begun to supplant the previous tools of choice in these areas, offering increased speed and throughput with reduced running costs.

Targeted re-sequencing in the clinical context presents specific requirements and new challenges also for bioinformatics which is aggravated by new computational needs of fast changing sequencing platforms. Just to mention one problem, that of multiplexing: simultaneous analyses of many patients for many diseases require accurate and unequivocal identification of many persons and many genes within an ensemble of many hundred thousand reads. Molecular bar-coding makes this possible, but standard bioinformatics tools are not ready to handle bar-coding information [13,14].

Clinical labs seek the advice of bioinformaticians regarding what kind of software to use. The usual standard answer is to use the current best of genomics software. Unfortunately, it is often found that these tools are not even always capable of doing the clinical application job, for example detecting specific mutation types. The reason is simple: Genome aligners were designed to map short reads to a whole genome, i.e., finding relatively strong similarities in a background of weak or minimal similarities. This scenario has called for specific speed-up solutions and approximations, many of which may not necessarily be true for amplicon sequencing protocols. So, clinicians usually face two problems: i) Buy an expensive hardware and non-transparent, and more often than not, very computer time-consuming commercial software from the platform vendor, or ii) seek advice from trained bioinformaticians who may point them to academic tools developed for genome analysis, but not necessarily suitable for amplicon sequencing. The solution is not easy. Platform vendors cannot be blamed for proposing a technically sound solution which, for the moment, has no chances to follow the exponential growth of clinical analysis needs. So, it is the task of future bioinformatics projects to develop accurate and flexible solutions for clinical applications.

8.2 BIOINFORMATICS MOVING TOWARDS CLINICAL ONCOLOGY: BIOMARKERS FOR CANCER CLASSIFICATION, EARLY DIAGNOSTICS, PROGNOSIS AND PERSONALIZED THERAPY

Losses of human lives and sufferings as a result of cancer remain one of the critical obstacles in prolonging active human life span. Worldwide, cancers are responsible for one in eight deaths [15]. In Singapore, cancers are the major causes of mortality and accounts for about 28.5% of all deaths [16]. In our present understanding, cancer is a disease involving genetic changes in certain cell populations that lead to cellular reprogramming and uncontrolled cell division; in turn, the formation of a malignant mass can create a variety of clinical symptoms. The huge individual genome variation and diversity of cellular phenotypes in cancers often complicates clinical detection, classification, prognosis and treatment of patients. In fact, histologically similar cancers do not necessarily represent the same disease due to differences in the biomolecular mechanisms leading finally to similar clinical outcomes. Consequently, among the list of 10 most important human diseases, the pharmacotherapy efficacy of cancer is very low except for a few rare subtypes [17]. The progress in the early diagnostics/detection and therapy of many cancers is very slow. For instance, for the past 30 years, ovarian cancers (OC) mortality rate has remained very high and unchanged, despite considerable efforts directed toward this disease.

Current clinical oncology needs (i) improvement of disease classification, (ii) increased specificity and sensitivity of early detection instruments/molecular diagnostics systems, (iii) improved disease risk profiling/prediction, (iv) improvement of cancer therapeutic methods including next generation drugs with higher specificity and lowered toxicity (ideally, inhibitors of the exact biomolecular mechanisms that drive individual cancer growth) and generally more stratified or even personalized therapies, (v) understanding of the anti-cancer immune response, (vi) adequate monitoring and rehabilitation during post-treatment recovery period and (viii) patients' social adaptation.

At present, there are two main lines of support for clinical oncology from the side of computational biology fuelled by data generated by genomics and proteomics high-throughput technologies. On the one hand, genome and RNA sequencing as well as expression profiling of cancer biopsy samples opens the possibility to understand the biomolecular mechanisms that are behind the malignant transformation in the individual patient's tumor case. On the other hand, the status of biomarkers can be measured and used to provide more accurate diagnostics of a specific cancer type, prognosis and selection of personalized therapy.

8.2.1 HUNTING AFTER CANCER MUTATIONS IN A CLINICAL SETUP

The problems associated with large-scale sequencing and expression profiling of cancers need to be seen from two sides. Whereas the technical aspects of correct sequence and expression profile determination from generally miniscule biopsy amounts are considerable but manageable (see a recent review of some of the IT and bioinformatics aspects [18]), the evaluation of the data in terms of clinically relevant conclusions for the specific patient is presently impossible in most cases and the clinically relevant effort is centered more around the question whether the actual patient happens to carry a cancer that belongs to one of the better understood subtypes. At the same time, sequencing and expression profiling of carefully selected cohorts of cancer patients are of immeasurable value for biomedical research aimed studying yet unknown biomolecular mechanisms.

Technically, analyzing somatic mutations in complex diseases such as cancer is particularly challenging since the mutant alleles can be easily diluted below detection thresholds due to the presence of wild type non-tumor DNA and the inherent genetic heterogeneity of the tumor itself. The problem is further aggravated by the limited amount of DNA (1-100 ng) available from biopsies on the one hand, and the clinical sample preparation, on the other: For example, clinical samples fixation in formalin randomly breaks DNA into 200-400 bp long fragments.

The current gold standard method tries to circumvent these problems by applying targeted PCR amplification to 100-200 bp long target sequences which is followed by Sanger sequencing of the PCR amplicons. Next generation sequencing (NGS) platforms such as the 454 FLX Genome analyzer (Roche) or Ion Torrent Personal Genome Machine (Life Technology), offer important advantages due to their extremely high (1000-10000 fold) sequence coverage. Thus, sensitivity as compared to Sanger sequencing is increased. This is very important for detecting low frequency mutations, which makes NGS an attractive option for diagnostic sequencing.

For clinical analysis of the transcriptome, deep sequencing technologies (e.g. RNA-seq, etc.) allow detecting low abundant RNA transcripts. Many classes of these transcripts (e.g., long non-coding RNAs) play essential regulatory roles in cancer development and can potentially be used for clinical sub-typing, detection, prognosis and therapy design of cancers. Detection of the rare genome aberrations and low-abundant transcripts in cancers and in human body fluids might be important. However, clinical studies of such data require development of appropriated biomedical research infrastructure, collection of large patients' cohorts, management of well-coordinated interdisciplinary research projects, dynamical and integrative databases, novel IT solutions and massive data analyses within a computational biology research effort.

Another advantage of NGS technology is its ability to deal with parallel sequencing of multiple genes. The widely respected white paper of the American Society of Clinical Oncology [19] suggested that all targeted drugs should be registered based on the molecular profile independently from the tumor type. Recently, researchers of the Massachusetts General Hospital argued that simultaneous analysis of 12 genes is useful for the diagnosis of lung cancer [20]. Therefore, there is a clinical need for targeted re-sequencing of dozens of genes in each cancer patient. There are several, commercially available multiplex re-sequencing assays in clinical use today. A typical analysis for cancer targets may require PCR-based re-sequencing of 10 to 1500, mainly exon-derived amplicons selected from 10 to 400 genes, and a minimum amount of 10 ng DNA [21].

8.2.2 BIOMARKERS FOR CANCER CLASSIFICATION: MUTATIONS IN SIGNALING PROTEINS

A biomarker is a traceable biochemical substance that is informative about the status of a disease or medical condition. For practical purposes, it is sufficient to show a close correlation between the occurrence of the bio-marker and the cancer type and development in model systems and in clinical trials. Yet, the likelihood of the biomarker actually being associated with the cancer subtype considered is dramatically increased if the biomarker plays a role in the biomolecular mechanisms driving the cancer and not just in some secondary or tertiary effects of cancer growth. However, discovery of reliable diagnostic, prognostic and drug response cancer biomarkers faces big challenges due to patient heterogeneity, small sample sizes, and high data noises.

A couple of cancer subtypes well-characterized mechanistically have recently seen spectacularly successful treatment. Mutations in signaling proteins have been found to drive cells into the cancer state and the design of drugs that specifically bind to these mutated forms have been shown to suppress cancer development. For the drugs to be applied, a companion diagnostic test is necessary to verify whether the potential patient has indeed a cancer driven by the target supposed. As a rule, this will dramatically shrink the number of patients but the selected ones have a high chance to receive benefits from the treatment. Three cases illuminating the trend towards mutation-specific targeting drugs are reviewed in some detail below.

Several forms of chronic myelogenous leukemia (CML) and gastrointestinal stromal tumors (GISTs) are characterized by the Philadelphia chromosome, a chromosomal translocation, and the subsequent fusion of genes bcr and abl. As a result, the tyrosine kinase abl is locked in its active signaling state and affecting the downstream pathways Ras/MapK (increased proliferation due to increased growth factor-independent cell growth), Src/Pax/Fak/Rac (increased cell motility and decreased adhesion), PI/PI3K/AKT/BCL-2 (suppression of apoptosis) and JAK/STAT (driving proliferation). The inhibitor Imatinib (STI571, Gleevec) inhibits bcr-abl and, as a result, an originally fatal disease is transformed into a chronically manageable one [22]. The same inhibitor is also active for some sequence variants of c-kit and PDGF-R (platelet-derived growth factor receptor) and, thus, can be applied in a handful of other cancers. Since application of the drug is essentially selectively killing sensitive cells, strains with resistant mutations survive and it might require the application of other batteries of drugs to bring these strains down, too [23].

Another case with some success are melanoma subtypes with the B-RAF mutation V600E that can be treated with vemurafenib (PLX4032, RG7204) [24,25]. In melanomas with mutant B-RAF (V600E), the drug inhibits specifically B-RAF (V600E) monomers. Since the ERK signaling inhibition is tumor-specific, these RAF inhibitors have a broad therapeutic index and a remarkable clinical activity in patients with melanomas that harbor the respective B-RAF mutant (V600E). However, resistance invariably emerges, for example via alternative splicing. The version p61 B-RAF (V600E) shortened by exons 4-8 shows enhanced dimerization in cells with low levels of RAS activation and ERK signalling is resistant to the RAF inhibitor [25].

Certain EGFR (epidermal growth factor receptor, another tyrosine kinase) driven cancers of breast, lung, pancreas, etc. are sensitive to gefitinib (Iressa) or erlotinib (Tarceva). The EGFR class includes Her1 (erb-B1), Her2 (erb-B2), and Her 3 (erb-B3). The EGFRs are hyper-activated due to a mutation in the tyrosine kinase domain and this leads to inappropriate activation of the anti-apoptotic Ras signalling cascade, eventually resulting in uncontrolled cell proliferation [26].

8.2.3 BIOMARKERS FOR CANCER CLASSIFICATION: UP-REGULATED GENES

The literature on cancer biomarkers is enormous and it is beyond this review to be comprehensive. Here, we focus on developments with our authors' involvement.

Lung adenocarcinoma (AC) is the most common type of lung cancer which is the leading cause of cancer deaths in the world. The genetic mechanisms of the early stages and lung AC progression steps are poorly understood. Currently, there are no clinically applicable gene tests for early diagnosis and lung AC aggressiveness assessment. Recently, authors of this review (VK et al.) suggested a method for gene expression profiling of primary tumours and adjacent tissues (PT-AT) based on a new rational statistical and bioinformatics strategy of biomarker prediction and validation, which could provide significant progress in the identification of clinical biomarkers of lung AC. This approach is based on the extreme class discrimination (ECD) feature selection method that identifies a combination/subset of the most discriminative variables (e.g. expressed genes) [27]. This method includes a paired cross-normalization (CN) step followed by a modified sign Wilcoxon test with multivariate adjustment carried out for each variable. Analysis of paired Affymetrix U133A microarray data from 27 AC patients revealed that 2,300 genes can discriminate AC from normal lung tissue with 100% accuracy. Our finding reveals a global reprogramming of the transcriptome in human lung AC tissue versus normal lung tissue and for the first time estimates a dimensionality of space of potential lung AC biomarkers. Cluster analysis applied to these genes identified four distinct gene groups. The genes related to mutagenesis, specific lung cancers, early stage of AC development, tumour aggressiveness and metabolic pathway alterations and adaptations of cancer cells are strongly enriched in the discriminative gene set. 26 predicted AC diagnostic biomarkers (including SPP1 and CENPA genes) were successfully validated on qRT-PCR tissue array. The ECD method was systematically compared to several alternative methods and proved to be of better performance [27]. Our findings demonstrate that the space of potential clinical biomarker of lung cancers is large; many dozens of combined biomarkers/molecular signatures are possible. This finding suggests that further improvement of computational prediction and feature selection methods is necessary in conjunction with systematic integration of massive and complex data analysis.

Similar computational approaches applied on breast cancer patients' expression data allowed important new insights into molecular and clinical classification, tumor aggressiveness grading and identification of novel tumor sub-types. Current statistical approaches for biomarker selection and signature extraction were extended by developing a hybrid univariate/multivariate approach, combining rigorous statistical modeling and network analysis [28]. In this approach, single survival-significant genes can be identified and used to generate important cancer related gene networks. The method also allows estimating the synergistic effect of two or several genes belonging to the same or different networks on the patients' survival. With this analysis, we generated and evaluated several related signature sets which are superior to traditional clinical prognostic markers and existing breast cancer classifications [28-30]. The final groupings have significantly different p53 mutation status, tumor aggressiveness grading and metastasis events. Most importantly, it could be shown that the intermediate class of G2 breast cancers does not have a justification at the level of gene expression. The G2 cases are shown to be either G1-like or G3-like. This implies that G2 patients with a G3-like expression profile are recommended to receive the more aggressive treatment reserved for G3 patients.

Currently, using clinical and molecular markers does not provide specific and reliable ovarian cancer (OC) patients' stratification, prognosis and treatment response prediction. High-grade epithelial ovarian serous carcinoma (HG-EOC), a major type of OC, is poorly detected. At the molecular level, the tumors frequently exhibit altered expressions of many hundreds and thousands features at genome, transcriptome and proteome levels. The specific and reliable biomarkers of this complex disease and appropriate therapeutic targets have not been defined yet. Similar computational approaches as described above in the cases of lung and breast cancers have been used to derive expression signatures for OC and they were found to include the EVI1 gene [31].

It is also notable that non-coding RNAs can also be used as biomarkers [32]. To conclude, the identification of reliable diagnostic, prognostic and drug response-related biomarkers for cancer requires integrative data analysis and understanding of the molecular and cellular basis of genome loci and gene expression and pathways.

8.3 SEQUENCE-STRUCTURE-FUNCTION RELATIONSHIPS FOR PATHOGENIC VIRUSES AND BACTERIA AND THEIR ROLE IN COMBATING INFECTIONS

Whereas the discussion above has highlighted that sequence-function relationships are not well understood and this status will continue for a while, the situation for the small genomes of pathogenic viruses and bacteria is considerably more promising. Their genome size is much smaller (from a handful of genes in the case of viruses to maximum a few thousand genes for bacteria) and their physiology is much more completely understood at the level of biomolecular mechanisms. For example, there is no gene in the influenza virus where at least some mechanistic aspect of its molecular and cellular function is known; a stark contrast to the situation for the human genome where about half of the genes still await their at least initial characterization [1] and even the compilation of the complete proteome is not in sight [33].

With sequencing getting increasingly cheaper and efficient, it became possible to explore the full genome of the set of strains that is actually invading the patient's body. This is important since, to evade the patient's immune system, the pathogen mutates and one or several of the mutants might find the weak spots of the patient and propagate. This allows not only designing efficient patient-specific treatment strategies, for example by deducing certain drug resistances theoretically from the pathogen's genomic sequence before even trying actually the respective drug in the treatment. It provides also much better options for epidemiology and public health since each strain can be individually determined and, thus, the actual spread of the pathogen can be traced geographically and in real time. Measures for preventing and combating epidemics can be designed more rationally and with lower costs for social and economic life.

Most attention with regard to rationally designed strategies for fighting infection so far has been directed towards the acquired immunodeficiency syndrome (AIDS) caused by the human immunodeficiency virus

(HIV) and this can rightly be considered a success story for computational biology. A previously absolutely fatal disease has been transformed into a chronic illness with high quality of life and, for many patients, with apparently zero viral blood counts. Not only have all the drugs against AIDS used in the multi-drug cocktail for high active antiretroviral therapy (HAART) been rationally designed against structures of HIV proteins to interfere into the well-studied life cycle of the virus [34]. New drugs appear all the time and provide new treatment opportunities for patients harboring strains resistant against the standard cocktails [35]. Sophisticated knowledge-based therapeutic algorithms [36] are available to treat AIDS patients optimally depending on the mutation spectrum within the patient's viral load [37,38].

Similar strategies are useful for other pathogens that try to evolve away from the attack of antibiotics/antiviral therapy or the immune system's efforts. *Staphylococcus aureus* causing a wide range of infection from skin to post-operative wound infections has great adaptive potential and can generate forms (best known as methicillin-resistant *Staphylococcus aureus* - MRSA) widely resistant against many available antibiotics. Exact determination of the molecular epidemiology with multi-locus sequence typing and other methods can be the basis for an optimized antibiotics selection for more efficient therapy [39].

In the following, we explore how classical bioinformatics aimed at studying biomolecular sequences and structures can impact infection medicine in context with the influenza virus and the enterohemorrhagic *E. coli* pathogens.

8.3.1 GENOME SEQUENCE STUDIES OF THE INFLUENZA VIRUS AND PUBLIC HEALTH

Besides the occasional pandemics, recurrent seasonal influenza and its ongoing evolution has always been an important topic concerning public health. Whenever a new flu strain emerges and threatens to circle the globe, health authorities and clinicians need to know the characteristics of the new virus including virulence, drug susceptibility and vaccine efficacy. The recent swine flu pandemic from 2009 is an excellent example how computational methods can provide crucial support not only in the early molecular characterization [40-42] but also to follow the still ongoing evolution of the virus. Modern sequencing technology and increased preparedness resulted in a significant worldwide increase of institutions and hospitals that can generate molecular sequence data from patient samples. But when the patient-specific strain sequences are available after sequencing ordered by hospitals or ministries, it appears that the institution cannot properly handle them. The expertise for the subsequent steps of computational analysis to connect the genotype to possible phenotypes is often sparse. Bioinformatics can be used to rapidly screen influenza sequences for potentially interesting mutations, for example, through comparative genomics, 3D structural modeling, literature text mining and plotting geotemporal occurrence patterns for epidemiological significance.

While this sounds exciting, are we really in a state that we can reliably predict relevant phenotypic changes from sequence mutations? First, the influenza genome is small and codes for only 10-13 proteins all of which are well characterized in their functions and there exists a mechanistic understanding how they work together as well as how they interact with the infected host. Second, there is wide interest in influenza research and the amount of available sequences, crystal structures, experimental data and associated literature is enormous which allows transferring information and annotations if very closely related strains are compared. For example, the typical Tamiflu resistance mutation H274Y in the neuraminidase protein has the same effect on equivalent positions in seasonal H3N2, old seasonal H1N1, pandemic H1N1, avian H5N1, etc.

But what can be said about "new" mutations? In the second wave of the 2009 H1N1 pandemic, a Norwegian team reported a high frequency of a new hemagglutinin mutation D222G in severe cases [43]. The power of bioinformatics for linking genotype to phenotype for influenza mutations can be shown for this example, as within a few hours from first reports of the mutation one could find a possible mechanistic explanation on how this mutation could possibly exert its severity using computational tools and databases alone. The first obstacle is the numbering, different groups prefer to use old seasonal H3N2 based numberings also for H1N1 pandemic strains but it is important to know that D222G is actually corresponding to the mutation D239G in the literal sequence numbering of circulating pandemic strains which is necessary to find and count appearances of this mutation in available influenza surveillance sequences. This can easily be resolved computationally by aligning with respective reference strains with defined numbering. Sequence alignments to strains with known structure can also be used to build homology models and find the corresponding position of the mutation in the 3D structure. It turns out that D222/239G was located within the receptor binding pocket which determines the type of sugar-linked sialic acids recognized on human host cells but the precise effects on substrate specificity is still challenging to predict in detail by docking and modeling alone. Being able to switch between numbering schemes is also important to find prior work on related mutations in the literature. Indeed, a corresponding position in avian H1N1 has previously been investigated [44] as mutation G225D which is exactly equivalent to the new D222/225/239G but with inverted direction. The paper had found that G at this position is associated with preference for α2-3 avian-like receptor specificity while D would bind better to α2-6 human-like receptors. By analogy, it was possible to deduce that the new D222/225/239G mutation in the pandemic H1N1 could possibly shift the receptor preference to avian-like α2-3 receptors. The next important additional hint from the literature was that also humans have some α2-3 receptors but they are found deeper in the lungs, notably in the bronchiolae [45]. Finally, everything comes together and a hypothetical mechanism on how the new mutation could be related to severity is apparent where the D239G would change the receptor specificity to allow infections deeper in the lungs (Figure 1). More than a year later, this exact mechanism of the D222/225/239G mutation was studied in detail [46] and the experiments verified what could be suggested already much earlier by computational and literature analysis by a bioinformatics expert within a few hours. Many of the functions described here, have now been implemented in the WWW-based FluSurver that can accept patient-specific virus genome information and generate a clinical relevance report automatically (SMS et al., to be published).

There are many more examples where Bioinformatics analysis helped to elucidate phenotypic roles of new influenza mutations such as marker mutations of new variants rising in occurrence [47], changes in hemagglutinin surface epitopes [48] and glycosylation sites as well as detect known

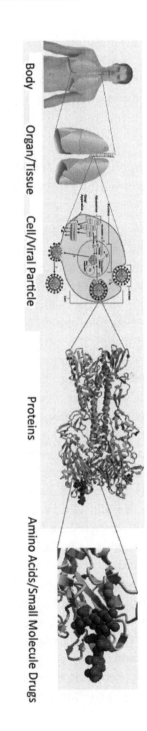

FIGURE 1: The link between an influenza virus mutation and the altered course of infection. Schematic representation showing how a single viral amino acid mutation (right, red balls) can affect host cell receptor (blue balls) interaction, which can alter viral localization and where the infection takes place, which in turn can affect severity and symptoms for the patient (left). A thorough understanding of the effects of mutations on biological mechanisms is also important for other human diseases such as cancer as well as patient-specific response to different treatments. Attribution of images: The 3 left-most images of the composed figure are public domain or under free-to-use licenses at Wikimedia commons from the following sources: patient body and organ [118] and infected cell [119].

[49] and novel [50-52] mutations in the neuraminidase drug binding pocket that alter antiviral drug efficacy. While the wealth of prior work on influenza is crucial for the ability to make relevant computational predictions, it shows that, with a concerted effort, similar successes may be achieved in other areas of high interest.

8.3.2 CONCLUSIONS FROM THE SEQUENCE OF THE ENTEROHEMORRHAGIC O104:H4 E. COLI STRAIN

Next generation sequencing has dramatically brought down the cost of genome sequencing but the current reality is that there usually is a long way from the initial genomic data to information relevant for clinicians. However, there are exceptions. When an enterohemorrhagic O104:H4 *E. coli* strain caused a major outbreak in Germany [53] in 2011, the genome sequence was rapidly available through next generation sequencing [54]. At the same time, the Robert Koch Institute provided the microbial characterization including the clinically important antibiotic susceptibility profile [55]. In principle, the information if a specific antibiotic drug is effective against an organism should be encoded in its genome by the presence of the known target gene of the respective drug as well as the absence of associated drug resistance factors. Clearly, the prerequisite for computationally deriving an antibiotic susceptibility profile depends not only on the availability of the whole genome but also sufficiently complete annotation data for drug targets and resistance mechanisms of closely related strains or organisms. Since *E. coli* and related bacteria have been widely studied before in this regard, we show here that one can computationally identify antibiotic drugs that, potentially, can effectively target a new pathogen with available genome, such as the enterohemorrhagic O104:H4 *E. coli* strain. The steps to achieve this are essentially routine bioinformatics work but typically not easily accessible to clinicians.

First, the available genome sequences (http://www.ncbi.nlm.nih.gov/Traces/wgs/?val=AFOB01) were searched with BLASTX [56] for close to identical sequence matches against a database of known drug targets from DrugBank [57]. Requiring at least 97% sequence identity of the *E. coli* sequences to the proteins known to be drug targets ensures that also their structure will be highly similar and hence should represent the same drug binding properties. Second, we repeat the sequence search but this time against a database of known drug resistance factors from ARDB [58] requiring a lower threshold of at least 60% identity to conservatively pick up also more remote similarities to possible resistance factors. Third, we use a Perl script to parse the hits from the BLAST outputs as well as the drug target and resistance annotation data from the two databases and finally identify the list of drugs for which a known target gene was found in the genome but no respective associated resistance factor.

TABLE 1: Predicted potentially effective drugs against enterohemorrhagic *E. coli*

Antibiotic	Exp.	Comp.
Piperacillin/Tazobactam	R*	S
Cefoxitin	R	R
Ceftazidim	R	R
Cefpodoxim	R	R
Imipenem	S	S
Meropenem	S	S
Amikacin	S	S
Gentamicin	S	S
Kanamycin	S	S
Tobramycin	S	S
Streptomycin	R	R
Tetracyclin	R	R
Nitrofurantoin	S	S
Trimethoprim/Sulfamethoxazol	R	R
Fosfomycin	S	R

*Experimentally measured (Exp.) versus computationally predicted (Comp.) antibiotics susceptibility profile. R ... resistant; S... sensitive; * ... defined as resistant (AES VITEK). Prediction and experimentally determined results coincide except for two cases (Piperacillin/Tazobactam and Fosfomycin) which are discussed in the text in detail.*

In order to validate the results, we compared our computational antibiotic susceptibility profile with the experimental results. To our positive surprise, 15 out of 25 experimentally tested antibiotics were also covered by the existing databases and could, hence, be assessed through our computational workflow. The identity thresholds for the two sequence searches described above have been selected to produce the best possible match with the experimental data. Table 1 shows that the in silico approach correctly assigns resistance or sensitivity for 13 of the 15 antibiotics. In detail, the new bacterial strain was correctly predicted to be sensitive to 7 antibiotics and resistant to 6 drugs from the list. The only two cases of a mismatch from the prediction with the clinical experimental result are interesting and discussed below.

The first case is the combination drug Piperacillin/Tazobactam which we flag as sensitive but the Robert Koch Institute as resistant. Sequence searches identified a TEM-1 metallo beta-lactamase in O104:H4 *E. coli* which causes resistance to penicillins (including Piperacillin) by degrading them but we also find that there exists a specific inhibitor against TEM-1 metallo beta-lactamases, Tazobactam, which is given in combination with Piperacillin to inhibit the beta-lactamase and, therefore, increase efficacy of penicillins to which this strain should otherwise be resistant. In theory, this means that the computational prediction that Piperacillin/Tazobactam is effective should be correct. However, it turns out that, in clinical practice, this drug is recommended to be avoided due to possible inoculum effects. Hence, the resistant flag from the clinical judgement according to the used VITEK AES experimental classification system.

The second case is Fosfomycin, to which the new strain was experimentally found to be sensitive while the computational approach assumed resistance due to the identification of a multidrugefflux pump protein annotated to also export Fosfomycin. This means that either the annotation is inaccurate or it would be interesting to further look into the detail of the few sequence differences between the new and the previously known transporter (99% identity) to find determinants of activity and substrate specificity which could be considered in a future more comprehensive approach.

Overall, this crude workflow utilizing available databases shows that a computational antibiotics susceptibility profile can be derived with some accuracy by combining next generation genome sequencing with further computational analysis, but it definitely still needs a critical experienced doctor who further scrutinizes and selects the most suitable treatment according to the circumstances of the infected patient as well as includes any new clinical findings on drug responses of the respective strain.

8.3.3 BACTERIAL COMMUNICATION AND COOPERATION IN HEALTH AND DISEASE

The analysis of human microbiomes and small bacterial communities causing multi-bacterial diseases are among the most challenging and intriguing tasks of medical genome research today [59-61] also including the field of plant diseases [62]. The discovery of chemical communication among bacteria in the 1990s has fundamentally changed the traditional view that pictures bacteria as single-celled organisms living in isolation [63-66]. In the last fifteen years, it has become increasingly evident that bacteria have the potential to establish highly complex communities. Many microbes live in large, multispecies communities in which the participants jointly exploit the resources. Multispecies microbial consortia constitute a major form of life that is found in environments ranging from high-altitude mountains (more than 8 km above sea level) to more than 10 km below the surface of the oceans, and have always been among the most important members and maintainers of the planet's ecosystem. The medical importance of this phenomenon is sweeping. Opportunistic pathogenes, such as *Pseudonomas* and *Burkholderia* species abound in hospital environments, ready to attack patients weakened by disease or injury. For instance, *Pseudomonas aeruginosa* usually does not harm a healthy human organism, but can be lethal in the lung of cystic fibrosis (CF) patients, or in burn wounds [67].

Many prokaryotes possess inter-cellular signaling systems which allow species to colonise new habitats, to invade hosts and to spread over surfaces [63-66]. A typical example is quorum sensing (QS) which enables bacteria to switch from low activity to high activity regimes using signaling molecules as well as "public goods" (e.g. surfactants, enzymes, siderophores) that facilitate movement, nutrient uptake amongst other things [65,66]. We share the widespread opinion that the "change of bacterial lifestyle" is crucial for colonizing habitats and infecting susceptible hosts

– unfortunately the signalling systems that orchestrate the underlying communication and collaboration mechanisms are not accurately annotated in bacterial genomes. Therefore, a systematic characterization of QS systems in Gram negative bacteria was carried out [68,69] and a modelling effort to map out the theoretically possible consequences of communication and collaboration in bacterial populations was initiated [70-72]. Virulence and adaptability of many Gram-negative bacterial species are associated with an N-acylhomoserine lactone (AHL) gene regulation mechanism called quorum sensing (QS). The arrangement of quorum sensing genes is variable throughout bacterial genomes, although there are unifying themes that are common among the various topological arrangements. A bioinformatics survey of 1403 complete bacterial genomes revealed characteristic gene topologies in 152 genomes that could be classified into 16 topological groups [68,69]. A concise notation for the patterns was developed and it was shown that the sequences of *LuxR* regulators and *LuxI* autoinducer synthase proteins cluster according to the topological patterns.

The macroscopic behavior of bacterial communities is notoriously difficult to study, colony patterns, invasion/colonization events depend on a multitude of parameters many of which cannot be reproduced in lab cultures. Therefore, computational modeling, and particularly the use of simplified minimal models is a very important tool for studying the behavior of populations in rational terms. Agent-based models of communicating and collaborating bacteria have developed [70]. The bacterial cells are represented by agents randomly moving on a plain (such as an agar surface), while consuming nutrients, secreting signal molecules and "public goods". Nutrients, signals and public goods are diffusing on the surface, and their local concentration exceeds a threshold, the metabolism and movement of bacterial agent switches to a more intensive state. In this model signals are the means of communications, and public goods are the means of cooperation as can be observed in QS bacteria. Even though highly simplified, the model reflects the crucial behavior patterns of communicating/cooperating bacteria in an open, nutrient/limited environment. Namely, 1) isolated bacteria cannot survive; only bacteria reaching a critical population size ("quorum") have a chance for survival. 2) Bacteria self-organize into compact communities or "active zones" in which signals and public goods are present in sufficient amounts [70]. 3) Collaborating communities can collapse if non-cooperating mutants are present [71,72].

Modeling the mutants of QS mechanisms is highly relevant for disease prevention. There is a very vivid interest from the pharmaceutical and pesticide industries, analysts agree that interventions targeting quorum sensing are among the major trends of the future. Since many bacteria use quorum sensing for infection, it is plausible to think about jamming strategies. According to one such scenario, one can saturate the surface of a plant with a signal molecule that will call bacteria to attack. If a lonely pathogen lands on the surface, it will immediately start to attack, but at the wrong time and place. Since it is alone, it will perish. Or, we can put a gene into the plant that produces an enzyme capable of destroying the signal molecule of the pathogenic bacteria, so that those will never wage an attack. But both strategies can strike back since they can also destroy the signaling of the beneficial bacteria that are essential to the host. According to a third scenario one may prevent the growth of an infecting pathogen by a greedy but antibiotic sensitive mutant of the same species, and then we eliminate the mutant by an antibiotic that specifically acts on that mutant. This is very appealing, but what do we do if the mutant created to heal gets some harmful genes or looses its antibiotic susceptibility? Many similar questions can be studied using computational models [73].

8.4 IMPACT OF BIOIMAGE INFORMATICS ON HEALTHCARE

Most likely, the penetration of automated evaluation tools for the analysis of clinically relevant histological images in diagnostic contexts is one of the areas that will experience great changes in the near future. The process of biomedical imaging involves little or no discomfort to the patients, while providing an effective tool for diagnosis. However, successful usage of images requires a high level of human intelligence, making automated image analysis by machines a challenging task. Currently, the gold standard for diagnosis through imaging is by experienced clinicians, typically radiologists or pathologists. It takes many years to train proficient clinicians to analyze images manually and, despite that, this gold standard is not perfect and suffers from subjective variations between different clinicians.

Advances in image processing, pattern recognition and computer vision in the past decades have boosted the possibilities for the application of computing technology. Currently, the focus is on computer aided diagnosis rather than to achieve a fully automated approach. Software that can support decision making and reduce the workload of clinicians, especially in routine operations, is extremely useful and valuable. Besides the direct derivation of clinically relevant conclusions from the images, such systems call also for the integration with databases of medical ontologies, the patients' medical records, etc.

Computational image analysis methods can be broadly categorized into those used for assessment, diagnosis and surgery. This section attempts to cover several exemplary areas of imaging and image analysis in healthcare. Because of the large extent of research work ongoing in academic bioimage informatics and medical image analysis and the growing engagement of the industry, this section cannot be comprehensive but rather we seek to cover a broad spectrum.

8.4.1 DIGITAL PATHOLOGY

Advances in computer vision and microscopy instrumentation have made digital pathology an important emerging field. The objective is to aid the pathologist in the analysis of high resolution cellular images obtained through biopsy. For example, highlighting regions of interest or reducing diagnostic variation can generate a big impact. Histological images from various organs such as prostate [74], breast [75] and liver have been the object of algorithm development.

Here, we shall focus our discussion on prostate digital pathology. Prostate cancer has a high prevalence rate worldwide. For example, it is the most common non-cutaneous male cancer in the United States [76] and it is the 3rd most common male cancer in Singapore [16]. The American Cancer Society report in 2009 estimates 192,280 new prostate cancer cases with 27,360 prostate cancer specific death [76]. The severity of prostate cancer diagnostics is compounded by disagreements between individual pathologists with regard to grading using the Gleason classification [77].

This agreement between different pathologist can be as low as 70% [78] and up to 29% of Gleason gradings were different between pre- and post-operative prostate cancer specimen [79]. Hence, having objective computer algorithms to aid in prostate pathology assessment is essential to improve diagnosis.

Most computational methods are developed to analyze microscopy images on the standard hematoxylin/eosin stain. The goals are gland segmentation since the architecture of glands is critical for Gleason grading and the identification and segmentation of nuclei since this is useful for detecting nuclei signatures specific to cancerous cells. Common computer vision techniques used are level sets [80], fractal analysis [81] and machine learning [80,82-86]. These techniques are used to segment glands [80,85] and nuclei [82,84] or to identify regions of malignancy directly [83].

8.4.2 COMPUTER VISION IN DERMATOLOGY

Assessment of skin condition and health is both important for clinical medicine as well as for the cosmetics industry. At present, assessment of the skin typically involves a trained dermatologist who will examine features such as textures and landmarks. While training of dermatologists takes many years, the subsequent diagnosis suffers from subjective interpretation differing among dermatologists. Hence, a more objective approach is in demand.

Considerable effort is ongoing to analyze skin surfaces through the use of objective computational methods. Protocols to ensure objective and consistent imaging of human skin (for example, in a well-controlled lighting environment) are vital for reliable diagnosis by computer algorithms [87-89]. Image acquisition is followed by the application of task-dependent image processing and computer vision methods. Liu et al. [90] use texture analysis to create an objective way of evaluating the effectiveness of treatment. A neural network framework has been developed to analyze the human skin conditions such as color, roughness, glossiness or tension [91,92]. Skin images have also been studied with data mining methods [88,93] and via modeling/reconstructing the skin surface [89,94].

8.4.3 COMPUTER VISION IN EYE DISEASES

Imaging methods for eye diseases are unique among bioimaging techniques because images of the eyes are easily accessible using conventional light cameras. There is no need for expensive and sophisticated machines such as a computer tomograph or magnet resonance imager. A common imaging modality is the optical coherence tomography; other imaging methods such as fundus photography, ultrasound and infra-red imaging are also used. Although image analysis has been used in the assessment of many eye diseases, we will focus our discussion on glaucoma and dry eye disease in this paper.

8.4.3.1 ANGLE CLOSURE GLAUCOMA

According to a world health organization report [95], glaucoma is a major global cause of blindness (approximately 5.2 million cases and about 15% of all cases of blindness). The impact of glaucoma on public health will increase with an aging population. However, the lack of a comprehensive measure of glaucoma compounded with its ability to cause sudden blindness makes it hard for treatment planning. Surprisingly, about 50-90% of potential patients in the world are unaware that they have glaucoma [96,97].

Glaucoma is classified into angle closure and open angle glaucoma according to the drainage angle, the angle between the cornea and iris. Primary angle closure glaucoma is the major form of glaucoma in Asia, in particular, among the Chinese population. It was suggested that angle closure glaucoma causes more blindness than open angle glaucoma in relative terms [98].

A common way for assessment of angle closure glaucoma is through gonioscopy in which the doctor uses an optical instrument to look at the anterior chamber to decide if the drainage angle is open or close.

Ultrasound [99] and optical coherence tomography (OCT) [100] images are also used for assessment. Computer vision techniques are used for analyzing eye images derived from the different modalities. As it takes much effort to master the technique of gonioscopy, Cheng et al. [101] developed a computational technique for RetCam images. A machine-learning based method aids glaucoma diagnosis by analyzing the cup-to-disc ratio measured on fundus images [102]. OCT images provide high resolution and a 3D view of the anterior chamber. Image analysis software has been developed to make precise measurements of important geometric information such as anterior chamber area, anterior chamber width, iris thickness, etc. on OCT images [103]. These data can then be correlated to generate new clinical knowledge [104,105].

8.4.3.2 IMAGE ANALYSIS IN ASSESSING THE DRY EYE CONDITION

The disease of dry eye has no clear definition; generally, it is a condition in which there is an unstable tear film during the open eye state. The dry eye condition has a prevalence rate of 10-20% in Sweden, Japan, Australia and several other countries. The most common treatment of dry eye is application of eye drops [106].

One cause of dry eye disease is meibomian gland dysfunction. The meibomian glad is located at the inside of the tarsel plate that supplies meibum, an oily substance, which forms a protective layer to the tear film. Dysfunction of meibomian glands causes lack of meibum and, often, resulted in degeneration of meibomian glands.

The morphology of meibomian glands can be imaged using an infrared camera mounted on a conventional slit lamp camera [106]. This imaging technology has enabled the application of advanced computer vision techniques for better diagnosis and patient management. Images from healthy meibomian glands shows a strip like pattern in gland morphology; with the strips being relatively straight, parallel and equally spaced. Images of highly degenerated glands show no strip like patterns at all, but only small isolated regions of remnant glands. Morphology for early stage disease shows twisting, non-parallel and unequally spaced strip like patterns [106].

While the process of imaging is simple and relatively cheap, the analysis of the morphology of meibomian glands and other clinical examinations that eventually lead to diagnosis and treatment require trained ophthalmologists with experience in handling dry eye patients. Unfortunately, there is no clear objective criteria for grading meibomian glands morphology degeneration, although some schemes have been suggested [106]. Inter-individual variation will also cause problems. Hence, large population screens on meibomian glands morphology does not directly lead to overall increase in better management of the disease.

An effective way to circumvent the problem of cost and inter-individual variation is to develop advanced computer vision techniques to process and grade images of meibomian glands. A team from Singapore has developed an image analysis software that can enhance infra-red images of meibomian glands, segment the strip-like patterns and extract important features for classifying the images [107].

8.4.5 IMAGE ANALYSIS FOR ASSISTED SURGERY

Pre-planning is an important component to the success of surgery, so that surgical operations can be performed systematically, completely and swiftly. Usually, planning involves studies of 3D images of the part of the patient's body where the operation will be performed. Image assisted surgery is available or being developed for almost all parts of the human body, for example for brain, liver, heart, gastrointestinal tract and for hand reconstruction surgery. The digital 3D image is enhanced by advanced computer graphics, visualization and various forms of accurate geometrical measurements done by the computer. This enhancement is very important because the human mind cannot decipher 3D objects represented on a 2D computer screen effectively. We are also unable to make accurate geometrical measurements. In this case, the computer essentially provides the "ruler" to make measurements.

8.4.6 TUMOR SEGMENTATION

Accurate measurements are particularly important in the case of surgery aimed at removing tumors. The size of the tumor is an important prognostic factor for treatment. 1D and 2D measurements such as tumor length, the largest axis length or cross sectional area had been used as a measure of tumor sizes. However, studies have shown that tumor volume provides a more accurate estimate of the tumor size [108,109]. Accurate measurement of tumor sizes calls for effective segmentation of tumors. Once properly segmented, the tumor size can be calculated trivially. Tumors occur in many parts of the human body and different segmentation algorithms are developed for segmenting tumors in different organs. The literature in this area is vast. In the following, we focus on liver tumors. Liver cancer accounts for about one million deaths per year [110]. Segmentation is usually done on computer tomography images. Many techniques have been developed to segment liver and its tumor including region growing [109,111], statistical techniques [109], machine learning [108,109], active contours [112], fuzzy c-means [113] and watershed [114].

Surgery planning also needs careful consideration of the vasculature structure around the tumor and their relationship with the tumor. Hence, segmentation of the vasculature structure can aid the surgeon to visualize the structure and location of vessels [115].

8.5 CONCLUDING REMARKS

The development and implementation of analytical and computational tools provided from the side of bioinformatics and bioimaging analysis provide opportunities for quality interaction among biotechnology, fundamental life science research and clinical studies. Bioinformatics findings can be translated into innovations that are adopted by the healthcare system and biomedical industry in form of diagnostic kits, analysis programs, etc. after the validation in both bench and clinical studies. In this article, we present several examples of how clinically relevant conclusions can be drawn from sequencing, expression profiling or histopathological bioimaging data with computational biology algorithms.

Unfortunately, considerable basic research is still necessary to make full use of the potential opportunities that are associated with the increasing availability of high-throughput technologies such as genome sequencing, mainly since most of the genome's hidden functional information is not known; the understanding of biomolecular mechanisms that translate genotype into phenotype is limited. But the progress in this field is uneven; pathogen sequencing can already provide important insights in contrast, for example, to sequencing of cancer samples.

Since an efficient healthcare system must be aligned to social, economic and political infrastructure of the country and focus on evidence-based prophylactic, prevention, diagnosis, prognosis, prediction and treatments that are proven to provide quality service and clinical outcome in a cost-effective manner, genomics, proteomics and other new technologies will first have to demonstrate in a research hospital setting that they can have a dramatic effect in improving health care, also cost-wise in addition to providing better quality of life, before the approaches will penetrate the routine healthcare systems. Nevertheless, it is very clear that major advances in diagnostics and treatments for infections as well as cancers, circulatory and metabolic diseases that are critical for improving most healthcare systems will arise from these developments in a medium to longer time frame.

As we have seen above, genome information of pathogens linked with the geographic origin allows tracing the spread of infections and parasites. Similarly, analyzing the geographic, even better spatio-temporal distribution of disease occurrences can provide hints for environmental influences [116,117]. Generally, going beyond the patient-centric approach and the linking of biomolecular and clinical data of populations with geographic information, data on food and environment, etc. will be an important source for improving public health, for stopping epidemics, for finding sources of food or environmental poisoning and for improving life styles.

REFERENCES

1.   Eisenhaber F: A decade after the first full human genome sequencing: When will we understand our own genome? J Bioinformatics Comp Biol 2012, 10:12710.

2. The Human Genome Project Completion: Frequently Asked Questions3-26-2012 http://www.genome.gov/11006943

3. Sequencing competition heats up3-28-2012 http://rna-seqblog.com/news/sequencing-competition-heats-up/

4. Eisenhaber F: Bioinformatics: Mystery, Astrology or Service Technology. Preface. In Discovering Biomolecular Mechanisms with Computational Biology. 1st edition. Edited by Eisenhaber F. Georgetown: Landes Biosciences and Eurekah.com; 2006::1-10.

5. Bio-IT World 10th Anniversary 2002-20122-8-2012 http://www.bio-itworld.com/10th-Anniversary/

6. Collins FS, McKusick VA: Implications of the Human Genome Project for medical science. JAMA 2001, 285:540-544.

7. Lander ES: Initial impact of the sequencing of the human genome. Nature 2011, 470:187-197.

8. Dethoff EA, Chugh J, Mustoe AM, Al-Hashimi HM: Functional complexity and regulation through RNA dynamics. Nature 2012, 482:322-330.

9. Guttman M, Rinn JL: Modular regulatory principles of large non-coding RNAs. Nature 2012, 482:339-346.

10. DNA-sequencing penetrates Norway's healthcare system3-2-2012 http://www.fiercemedicaldevices.com/story/dna-sequencing-penetrates-norways-healthcare-system/2012-02-03

11. Lupski JR, Reid JG, Gonzaga-Jauregui C, Rio Deiros D, Chen DC, Nazareth L, Bainbridge M, Dinh H, Jing C, Wheeler DA, et al.: Whole-genome sequencing in a patient with Charcot-Marie-Tooth neuropathy. N Engl J Med 2010, 362:1181-1191.

12. Peterson J, Garges S, Giovanni M, McInnes P, Wang L, Schloss JA, Bonazzi V, McEwen JE, Wetterstrand KA, Deal C, et al.: The NIH Human Microbiome Project. Genome Res 2009, 19:2317-2323.

13. Hajibabaei M, Singer GA, Hebert PD, Hickey DA: DNA barcoding: how it complements taxonomy, molecular phylogenetics and population genetics. Trends Genet 2007, 23:167-172.

14. Kozarewa I, Turner DJ: 96-plex molecular barcoding for the Illumina Genome Analyzer. Methods Mol Biol 2011, 733:279-298.

15. Stratton MR, Campbell PJ, Futreal PA: The cancer genome. Nature 2009, 458:719-724.

16. Singapore Cancer Registry: Trends in cancer incidence in Singapore 2001-2005. Singapore Cancer Registry interim report. 2008.

17. Trusheim MR, Berndt ER, Douglas FL: Stratified medicine: strategic and economic implications of combining drugs and clinical biomarkers. Nat Rev Drug Discov 2007, 6:287-293.

18. Valencia A, Hidalgo M: Getting personalized cancer genome analysis into the clinic: the challenges in bioinformatics. Genome Med 2012, 4:61.

19. Kris NG, Meropol NJ, Winer EP: ASCO's Blueprint for Transforming Clinical and Translational Cancer Research, November 2011. Alexandria, VA: American Society of Clinical Oncology; 2011.

20. Sequist LV, Heist RS, Shaw AT, Fidias P, Rosovsky R, Temel JS, Lennes IT, Digumarthy S, Waltman BA, Bast E, et al.: Implementing multiplexed genotyping of non-small-cell lung cancers into routine clinical practice. Ann Oncol 2011, 22:2616-2624.

21. Ion AmpliSeq™ Cancer Panel8-10-2012 http://www.iontorrent.com/lib/images/PDFs/ampliseq%20appnote.pdf

22. Schiffer CA: BCR-ABL tyrosine kinase inhibitors for chronic myelogenous leukemia. N Engl J Med 2007, 357:258-265.

23. Weisberg E, Manley PW, Cowan-Jacob SW, Hochhaus A, Griffin JD: Second generation inhibitors of BCR-ABL for the treatment of imatinib-resistant chronic myeloid leukaemia. Nat Rev Cancer 2007, 7:345-356.

24. Chapman PB, Hauschild A, Robert C, Haanen JB, Ascierto P, Larkin J, Dummer R, Garbe C, Testori A, Maio M, et al.: Improved survival with vemurafenib in melanoma with BRAF V600E mutation. N Engl J Med 2011, 364:2507-2516.

25. Poulikakos PI, Persaud Y, Janakiraman M, Kong X, Ng C, Moriceau G, Shi H, Atefi M, Titz B, Gabay MT, et al.: RAF inhibitor resistance is mediated by dimerization of aberrantly spliced BRAF(V600E). Nature 2011, 480:387-390.

26. Saintigny P, Burger JA: Recent advances in non-small cell lung cancer biology and clinical management. Discov Med 2012, 13:287-297.

27. Toh SH, Prathipati P, Motakis E, Kwoh CK, Yenamandra SP, Kuznetsov VA: A robust tool for discriminative analysis and feature selection in paired samples impacts the identification of the genes essential for reprogramming lung tissue to adenocarcinoma. BMC Genomics 2011, 12(Suppl 3):S24.

28. Motakis E, Ivshina AV, Kuznetsov VA: Data-driven approach to predict survival of cancer patients: estimation of microarray genes' prediction significance by Cox proportional hazard regression model. IEEE Eng Med Biol Mag 2009, 28:58-66.

29. Grinchuk OV, Motakis E, Kuznetsov VA: Complex sense-antisense architecture of TNFAIP1/POLDIP2 on 17q11.2 represents a novel transcriptional structural-functional gene module involved in breast cancer progression. BMC Genomics 2010, 11(Suppl 1):S9.

30. Ivshina AV, George J, Senko O, Mow B, Putti TC, Smeds J, Lindahl T, Pawitan Y, Hall P, Nordgren H, et al.: Genetic reclassification of histologic grade delineates new clinical subtypes of breast cancer. Cancer Res 2006, 66:10292-10301.

31. Bard-Chapeau EA, Jeyakani J, Kok CH, Muller J, Chua BQ, Gunaratne J, Batagov A, Jenjaroenpun P, Kuznetsov VA, Wei CL, et al.: Ecotopic viral integration site 1 (EVI1) regulates multiple cellular processes important for cancer and is a synergistic partner for FOS protein in invasive tumors. Proc Natl Acad Sci U S A 2012, 109:2168-2173.

32. Batagov AO, Kuznetsov VA, Kurochkin IV: Identification of nucleotide patterns enriched in secreted RNAs as putative cis-acting elements targeting them to exosome nano-vesicles. BMC Genomics 2011, 12(Suppl 3):S18.

33. Sirota FL, Batagov A, Schneider G, Eisenhaber B, Eisenhaber F, Maurer-Stroh S: Beware of moving targets: reference proteome content fluctuates substantially over the years. J Bioinform Comput Biol 2012, 10:1250020.

34. Vogel M, Schwarze-Zander C, Wasmuth JC, Spengler U, Sauerbruch T, Rockstroh JK: The treatment of patients with HIV. Dtsch Arztebl Int 2010, 107:507-515.

35. Wilson LE, Gallant JE: HIV/AIDS: the management of treatment-experienced HIV-infected patients: new drugs and drug combinations. Clin Infect Dis 2009, 48:214-221.
36. Eberle J, Gurtler L: The evolution of drug resistance interpretation algorithms: ANRS, REGA and extension of resistance analysis to HIV-1 group O and HIV-2. Intervirology 2012, 55:128-133.
37. Martinez-Cajas JL, Wainberg MA: Antiretroviral therapy: optimal sequencing of therapy to avoid resistance. Drugs 2008, 68:43-72.
38. Gianella S, Richman DD: Minority variants of drug-resistant HIV. J Infect Dis 2010, 202:657-666.
39. Deurenberg RH, Stobberingh EE: The evolution of Staphylococcus aureus. Infect Genet Evol 2008, 8:747-763.
40. Garten RJ, Davis CT, Russell CA, Shu B, Lindstrom S, Balish A, Sessions WM, Xu X, Skepner E, Deyde V, et al.: Antigenic and genetic characteristics of swine-origin 2009 A(H1N1) influenza viruses circulating in humans. Science 2009, 325:197-201.
41. Maurer-Stroh S, Ma J, Lee RT, Sirota FL, Eisenhaber F: Mapping the sequence mutations of the 2009 H1N1 influenza A virus neuraminidase relative to drug and antibody binding sites. Biol Direct 2009, 4:18.
42. Smith GJ, Vijaykrishna D, Bahl J, Lycett SJ, Worobey M, Pybus OG, Ma SK, Cheung CL, Raghwani J, Bhatt S, et al.: Origins and evolutionary genomics of the 2009 swine-origin H1N1 influenza A epidemic. Nature 2009, 459:1122-1125.
43. Kilander A, Rykkvin R, Dudman SG, Hungnes O: Observed association between the HA1 mutation D222G in the 2009 pandemic influenza A(H1N1) virus and severe clinical outcome, Norway 2009-2010. Euro Surveill 2010, 15:19498.
44. Stevens J, Blixt O, Tumpey TM, Taubenberger JK, Paulson JC, Wilson IA: Structure and receptor specificity of the hemagglutinin from an H5N1 influenza virus. Science 2006, 312:404-410.
45. Shinya K, Ebina M, Yamada S, Ono M, Kasai N, Kawaoka Y: Avian flu: influenza virus receptors in the human airway. Nature 2006, 440:435-436.
46. Watanabe T, Shinya K, Watanabe S, Imai M, Hatta M, Li C, Wolter BF, Neumann G, Hanson A, Ozawa M, et al.: Avian-type receptor-binding ability can increase influenza virus pathogenicity in macaques. J Virol 2011, 85:13195-13203.
47. Maurer-Stroh S, Lee RT, Eisenhaber F, Cui L, Phuah SP, Lin RT: A new common mutation in the hemagglutinin of the 2009 (H1N1) influenza A virus. PLoS Curr 2010, 2:RRN1162.
48. Barr IG, Cui L, Komadina N, Lee RT, Lin RT, Deng Y, Caldwell N, Shaw R, Maurer-Stroh S: A new pandemic influenza A(H1N1) genetic variant predominated in the winter 2010 influenza season in Australia, New Zealand and Singapore. Euro Surveill 2010, 15:19692.
49. Inoue M, Barkham T, Leo YS, Chan KP, Chow A, Wong CW, Tze Chuen LR, Maurer-Stroh S, Lin R, Lin C: Emergence of oseltamivir-resistant pandemic (H1N1) 2009 virus within 48 hours. Emerg Infect Dis 2010, 16:1633-1636.
50. Hurt AC, Lee RT, Leang SK, Cui L, Deng YM, Phuah SP, Caldwell N, Freeman K, Komadina N, Smith D, et al.: Increased detection in Australia and Singapore of a novel influenza A(H1N1)2009 variant with reduced oseltamivir and zanamivir sensitivity due to a S247N neuraminidase mutation. Euro Surveill 2011, 16:19884.

51. Nguyen HT, Trujillo AA, Sheu TG, Levine M, Mishin VP, Shaw M, Ades EW, Klimov AI, Fry AM, Gubareva LV: Analysis of influenza viruses from patients clinically suspected of infection with an oseltamivir resistant virus during the 2009 pandemic in the United States. Antiviral Res 2012, 93:381-386.

52. Van der Vries E, Veldhuis Kroeze EJ, Stittelaar KJ, Linster M, der LA V, Schrauwen EJ, Leijten LM, Van AG, Schutten M, Kuiken T, et al.: Multidrug resistant 2009 A/H1N1 influenza clinical isolate with a neuraminidase I223R mutation retains its virulence and transmissibility in ferrets. PLoS Pathog 2011, 7:e1002276.

53. Frank C, Faber MS, Askar M, Bernard H, Fruth A, Gilsdorf A, Hohle M, Karch H, Krause G, Prager R, et al.: Large and ongoing outbreak of haemolytic uraemic syndrome, Germany, May 2011. Euro Surveill 2011., 16

54. Mellmann A, Harmsen D, Cummings CA, Zentz EB, Leopold SR, Rico A, Prior K, Szczepanowski R, Ji Y, Zhang W, et al.: Prospective genomic characterization of the German enterohemorrhagic *Escherichia coli* O104:H4 outbreak by rapid next generation sequencing technology. PLoS One 2011, 6:e22751.

55. Characterization of EHEC O104:H46-30-2011 http://www.rki.de/cln_178/nn_217400/EN/Home/EHECO104

56. Altschul SF, Madden TL, Schaffer AA, Zhang J, Zhang Z, Miller W, Lipman DJ: Gapped BLAST and PSI-BLAST: a new generation of protein database search programs. Nucleic Acids Res 1997, 25:3389-3402.

57. Knox C, Law V, Jewison T, Liu P, Ly S, Frolkis A, Pon A, Banco K, Mak C, Neveu V, et al.: DrugBank 3.0: a comprehensive resource for 'omics' research on drugs. Nucleic Acids Res 2011, 39:D1035-D1041.

58. Liu B, Pop M: ARDB–Antibiotic Resistance Genes Database. Nucleic Acids Res 2009, 37:D443-D447.

59. Sibley CD, Rabin H, Surette MG: Cystic fibrosis: a polymicrobial infectious disease. Future Microbiol 2006, 1:53-61.

60. Sibley CD, Duan K, Fischer C, Parkins MD, Storey DG, Rabin HR, Surette MG: Discerning the complexity of community interactions using a Drosophila model of polymicrobial infections. PLoS Pathog 2008, 4:e1000184.

61. Sibley CD, Parkins MD, Rabin HR, Duan K, Norgaard JC, Surette MG: A polymicrobial perspective of pulmonary infections exposes an enigmatic pathogen in cystic fibrosis patients. Proc Natl Acad Sci U S A 2008, 105:15070-15075.

62. Hosni T, Moretti C, Devescovi G, Suarez-Moreno ZR, Fatmi MB, Guarnaccia C, Pongor S, Onofri A, Buonaurio R, Venturi V: Sharing of quorum-sensing signals and role of interspecies communities in a bacterial plant disease. ISME J 2011, 5:1857-1870.

63. Bassler BL: Small talk. Cell-to-cell communication in bacteria. Cell 2002, 109:421-424.

64. Camilli A, Bassler BL: Bacterial small-molecule signaling pathways. Science 2006, 311:1113-1116.

65. Fuqua C, Parsek MR, Greenberg EP: Regulation of gene expression by cell-to-cell communication: acyl-homoserine lactone quorum sensing. Annu Rev Genet 2001, 35:439-468.

66. Fuqua C, Greenberg EP: Listening in on bacteria: acyl-homoserine lactone signalling. Nat Rev Mol Cell Biol 2002, 3:685-695.

67. Collier DN, Anderson L, McKnight SL, Noah TL, Knowles M, Boucher R, Schwab U, Gilligan P, Pesci EC: A bacterial cell to cell signal in the lungs of cystic fibrosis patients. FEMS Microbiol Lett 2002, 215:41-46.

68. Gelencsér Z, Choudhary KS, Coutinho BG, Hudaiberdiev S, Galbáts B, Venturi V, Pongor S: Classifying the topology of AHL-driven quorum sensing circuits in proteobacterial genomes. Sensors 2012, 12:5432-5444.

69. Gelencsér Z, Galbáts B, Gonzalez JF, Choudhary KS, Hudaiberdiev S, Venturi V, Pongor S: Chromosomal arrangement of AHL.driven quorum sensing circuits in Pseudomonas. ISRN Microbiology 2012, 2012:484176.

70. Netotea S, Bertani I, Steindler L, Kerenyi A, Venturi V, Pongor S: A simple model for the early events of quorum sensing in Pseudomonas aeruginosa: modeling bacterial swarming as the movement of an "activation zone". Biol Direct 2009, 4:6.

71. Venturi V, Bertani I, Kerenyi A, Netotea S, Pongor S: Co-swarming and local collapse: quorum sensing conveys resilience to bacterial communities by localizing cheater mutants in Pseudomonas aeruginosa. PLoS One 2010, 5:e9998.

72. Venturi V, Kerenyi A, Reiz B, Bihary D, Pongor S: Locality versus globality in bacterial signalling: can local communication stabilize bacterial communities? Biol Direct 2010, 5:30.

73. Kerényi A, Suárez-Moreno ZR, Venturi V, Pongor S: Multispecies microbial communities. Part II: Principles of molecular communications. Medical Mycology 2010, 17:113-116.

74. Demir C, Yener B: Automated cancer diagnosis based on histopathological images: a systematic survey. Technical report TR-05-09. 2005. [Computer Science Department of Rensselaer Polytechnic Institute]

75. Huang CH, Veillard A, Roux L, Lomenie N, Racoceanu D: Time-efficient sparse analysis of histopathological whole slide images. Comput Med Imaging Graph 2011, 35:579-591.

76. American Cancer Society, Inc: Surveillance Research, Updated March 2010. 2010.

77. Gleason DF: Histologic grading of prostate cancer: a perspective. Hum Pathol 1992, 23:273-279.

78. Allsbrook WC Jr, Mangold KA, Johnson MH, Lane RB, Lane CG, Epstein JI: Interobserver reproducibility of Gleason grading of prostatic carcinoma: general pathologist. Hum Pathol 2001, 32:81-88.

79. Grossfeld GD, Chang JJ, Broering JM, Li YP, Lubeck DP, Flanders SC, Carroll PR: Under staging and under grading in a contemporary series of patients undergoing radical prostatectomy: results from the Cancer of the Prostate Strategic Urologic Research Endeavor database. J Urol 2001, 165:851-856.

80. Naik S, Doyle S, Agner S, Madabhushi A, Feldman M, Tomaszewski J: Automated gland and nuclei segmentation for grading prostate and breast cancer histopathology. Proc IEEE Int Symp Biomed Imaging 2008, :284-287.

81. Huang P-W, Lee C-H: Automated classification for pathological prostate images based on fractal analysis. IEEE Trans Medical Imaging 2009, 28:1037-1050.

82. Arif M, Rajpot N: Classification of potential nuclei in prostate histology images using shape manifold learning. Proc. Int Conference in Machine Vision 2007, :113-118.

83. Doyle S, Rodriguez C, Madabhushi A, Tomaszeweski J, Feldman M: Detecting prostatic adenocarcinoma from digitized histology using a multi-scale hierarchical classification approach. Conf Proc IEEE Eng Med Biol Soc 2006, 1:4759-4762.

84. Hafiane A, Bunyak F, Palaniappan K: Level set-based histology image segmentation with region-based comparison.Microscopic Image Analysis with Applications in Biology Workshop 2008, :1-6.

85. Naik S, Doyle S, Feldman M, Tomaszewski J, Madabhushi A: Gland segmentation and computerized Gleason grading of prostate cancer histology by integrating low-, high-level and domain specific information. Microscopic Image Analysis with Applications in Biology Workshop 2007, :1-8.

86. Teverovskiy M, Kumar V, Ma J, Kotsianti A, Verbel D, Tabesh A, Pang H-Y, Vengrenyuk Y, Fogarasi S, Saidi O: Improved prediction of prostate cancer reoccurrence based on an automated tissue image analysis system. IEEE Intl.Symp.Biomedical Imaging 2004, :257-260.

87. Jacques SL, Ramella-Roman JC, Lee K: Imaging skin pathology with polarized light. J Biomed Opt 2002, 7:329-340.

88. Nakao D, Tsumura N, Miyake Y: Real-time multi-spectral image processing for mapping pigmentation in human skin. Proc 9th IS&T/SID Color Imaging Conference 1995, 9:80-84.

89. Yamada T, Saito H, Ozawa S: 3D shape inspection of skin surface from rotation of light source. Intl Conf Quality Control by Artificial Vision 1999, :245-251.

90. Liu J, Bowyer K, Goldgof D, Sarkar S: A comparative study of texture measures for human skin treatment. Intl Conf Information, Communications and Signal Processing 1997, :170-174.

91. Takemae Y, Morimaya T, Ozawa S: The correspondence between physical features and subjective evaluation on skin image. Proceedings of The 1999 IEICE General conference 1999, :268.

92. Takemae Y, Saito H, Ozawa S: The evaluating system of human skin surface condition by image processing. IEEE Intl Conf System, Man and Cybernetics 2000, :218-223.

93. Sparavigna A, Marazzato R: An image processing analysis of skin textures. Skin Res Technol 2010, 16:161-167.

94. Yamada T, Saito H, Ozawa S: 3D reconstruction of skin surface from image sequence. IAPR Workshop on Machine Vision Applications 1998, :742-745.

95. Thylefors B, Negrel AD: The global impact of glaucoma. Bull World Health Organ 1994, 72:323-326.

96. Foster PJ, Oen FT, Machin D, Ng TP, Devereux JG, Johnson GJ, Khaw PT, Seah SK: The prevalence of glaucoma in Chinese residents of Singapore: a cross-sectional population survey of the Tanjong Pagar district. Arch Ophthalmol 2000, 118:1105-1111.

97. Sathyamangalam RV, Paul PG, George R, Baskaran M, Hemamalini A, Madan RV, Augustian J, Prema R, Lingam V: Determinants of glaucoma awareness and knowledge in urban Chennai. Indian J Ophthalmol 2009, 57:355-360.

98. Quigley HA, Congdon NG, Friedman DS: Glaucoma in China (and worldwide): changes in established thinking will decrease preventable blindness. Br J Ophthalmol 2001, 85:1271-1272.

99. Amerasinghe N, Aung T: Angle-closure: risk factors, diagnosis and treatment. Prog Brain Res 2008, 173:31-45.

100. Nolan WP, See JL, Chew PT, Friedman DS, Smith SD, Radhakrishnan S, Zheng C, Foster PJ, Aung T: Detection of primary angle closure using anterior segment optical coherence tomography in Asian eyes. Ophthalmology 2007, 114:33-39.

101. Cheng J, Tao D, Liu J, Wong DW, Lee BH, Baskaran M, Wong TY, Aung T: Focal biologically inspired feature for glaucoma type classification. Med Image Comput Comput Assist Interv 2011, 14:91-98.

102. Ahmed IIK, MacKeen LD: A new approach to imaging the angle. Glaucoma Today 2007, 2007:28-30.

103. Console JW, Sakata LM, Aung T, Friedman DS, He M: Quantitative analysis of anterior segment optical coherence tomography images: the Zhongshan Angle Assessment Program. Br J Ophthalmol 2008, 92:1612-1616.

104. Wang B, Sakata LM, Friedman DS, Chan YH, He M, Lavanya R, Wong TY, Aung T: Quantitative iris parameters and association with narrow angles. Ophthalmology 2010, 117:11-17.

105. Wang BS, Narayanaswamy A, Amerasinghe N, Zheng C, He M, Chan YH, Nongpiur ME, Friedman DS, Aung T: Increased iris thickness and association with primary angle closure glaucoma. Br J Ophthalmol 2011, 95:46-50.

106. Arita R, Itoh K, Inoue K, Amano S: Noncontact infrared meibography to document age-related changes of the meibomian glands in a normal population. Ophthalmology 2008, 115:911-915.

107. Koh YW, Celik T, Lee HK, Petznick A, Tong L: Detection of meibomian glands and classification of meibigraphy images. J Biomed Optics 2012, 17:086008.

108. Zhou J, Xiong W, Ding F, Qi T, Wang Z, Oo T, Venkatesh SK: Liver workbench: a tool suite for liver and liver tumor segmentation and modeling. Advances in Soft Computing 2012, 120:193-208.

109. Zhou JY, Wong DW, Ding F, Venkatesh SK, Tian Q, Qi YY, Xiong W, Liu JJ, Leow WK: Liver tumour segmentation using contrast-enhanced multi-detector CT data: performance benchmarking of three semiautomated methods. Eur Radiol 2010, 20:1738-1748.

110. Bosch FX, Ribes J, Borras J: Epidemiology of primary liver cancer. Semin Liver Dis 1999, 19:271-285.

111. Zhao B, Schwartz LH, Jiang L, Colville J, Moskowitz C, Wang L, Leftowitz R, Liu F, Kalaigian J: Shape-constraint region growing for delineation of hepatic metastases on contrast-enhanced computed tomograph scans. Invest Radiol 2006, 41:753-762.

112. Yim PJ, Foran DJ: Volumetry of hepatic metastases in computed tomography using watershed and active contour algorithms. IEEE Symp Computer-Based Medical Systems 2003, :329-335.

113. Yim PJ, Vora AV, Raghavan D, Prasad R, McAullife M, Ohman-Strickland P, Nosher JL: Volumetric analysis of liver metastases in computed tomography with the fuzzy C-means algorithm. J Comput Assist Tomogr 2006, 30:212-220.

114. Ray S, Hagge R, Gillen M, Cerejo M, Shakeri S, Beckett L, Greasby T, Badawi RD: Comparison of two-dimensional and three-dimensional iterative watershed segmentation methods in hepatic tumor volumetrics. Med Phys 2008, 35:5869-5881.

115. Chi Y, Liu J, Venkatesh S, Huang S, Zhou J, Tian Q, Nowinski W: Segmentation of Liver Vasculature from Contrast Enhanced CT Images using Context-based Voting. IEEE Trans Biomed Eng 2011, 58:2144-2153.

116. Bai H, Ge Y, Wang J-F, Liao YL: Using rough set theory to identify villages affected by birth defects: the example of Heshun, Shanxi, China. Int J Geographical Information Science 2010, 24:559-576.
117. Kolovos A, Angulo J, Modis K, Papantonopoulos G, Wang JF, Christakos G: Model-driven development of covariances for spatiotemporal environmental health assessment. Environ Monit Assess 2012, 2012:1-17.
118. Wikimedia commons: patient body and organ8-15-2012 http://commons.wikimedia.org/wiki/File:Symptoms_of_swine_flu.svg
119. Wikimedia commons: infected cell8-15-2012 http://commons.wikimedia.org/wiki/File:Virus_Replication.svg

CHAPTER 9

APPLICATION OF "OMICS" TO PRION BIOMARKER DISCOVERY

RHIANNON L. C. H. HUZAREWICH, CHRISTINE G. SIEMENS, AND STEPHANIE A. BOOTH

9.1 INTRODUCTION

Prion diseases, or Transmissible Spongiform Encephalopathies (TSEs), are invariably fatal neurodegenerative diseases associated with the conversion of the normal host cellular prion protein (PrP^C) into the abnormal protease-resistant isoform (PrP^{Sc}) [1]. They occur in a wide range of host species including humans, the most common of which is sporadic CJD (sCJD), occurring at a rate of approximately 1 case per million a year worldwide and accounts for greater than 80% of CJD cases [2]. Amino acid changes, which include point or insertional mutations in the normal (cellular) prion protein (PrP^C) encoded by the *PRNP* gene, are linked to genetic prion diseases such as Gerstmann-Strausler-Sheinker (GSS) disease, fatal familial insomnia (FFI), and genetically associated Creutzfeldt-Jakob disease (CJD). Acquired forms of disease are caused by ingestion

This chapter was originally published under the Creative Commons Attribution License. Huzarewich RLCH, Siemens CG, and Booth SA. Application of "omics" to Prion Biomarker Discovery. Journal of Biomedicine and Biotechnology **2010** (2010). http://dx.doi.org/10.1155/2010/613504.

of, or exposure to, contaminated biological material via food or during medical procedures. Kuru, found amongst the Fore tribe in Papua-New Guinea, was the first known human transmissible spongiform encephalopathy and resulted from exposure to infected material during ritualistic cannibalism. More recently a new human prion disease has emerged, variant CJD (vCJD), which is associated with exposure to the BSE agent in beef. Cases of iatrogenic transmission have also occurred through the use of improperly sterilized surgical instruments, the use of human growth hormone derived from cadaveric pituitaries, and transplantation of corneas and dura mater from infected patients [3]. Recently, human-to-human transmission of vCJD has been reported through blood transfusion [4]; human-adapted prions are more readily transmitted from human to human via this route than via ingestion of BSE prions from contaminated meat products [5].

Animals affected by TSEs include sheep (Scrapie), cattle (BSE) and mule, deer, elk (CWD). The impact of animal TSEs is twofold; firstly, there is a risk of transmission to humans, and secondly, the economic impact on animal production has been substantial. Although scrapie has been endemic for hundreds of years in many parts of the world its transmission to humans has never been reported. However, when vCJD in humans was determined to be associated with consumption of contaminated food there was concern as to what extent the population has been exposed. In the recent years, the incidence of CWD has increased markedly within North America and although it has not been linked to CJD either epidemiologically, or by laboratory confirmation, there is concern about the possibility for cross-species transmission [6, 7]. TSEs in animals have caused huge economic loses. Since the BSE epidemic began in 1986, millions of cattle have been slaughtered and bans on the importation of beef have affected many countries and cost billions of dollars.

The threat posed to public health by dietary and medical exposure to prions has driven tremendous efforts to develop sensitive methods of detection of prions to control the spread of human and animal TSEs. All the commercially available diagnostic tests for TSEs rely on the direct detection of the proteinase K resistant, misfolded form (PrP^{Sc}) of cellular prion protein in the central nervous system (CNS). Although methodologies are sensitive and specific for postmortem diagnosis, the use of PrP^{Sc} as a preclinical or general biomarker for surveillance is difficult, due to the fact that it is present in extremely small amounts in accessible tissues or body fluids such as blood, urine, saliva, and cerebrospinal fluid (CSF). Recently, amplification techniques have been developed which have enabled increased sensitivity. These are based on the ability of the disease-related abnormal isoform, PrP^{Sc}, to convert a pool of normal PrP^C to a proteinase K resistant form thus "amplifying" the original infectious seed. Amplification can be increased by breaking down the resulting aggregated seeds of PrP^{Sc} to smaller units which in turn act as seeds for further replication until levels of PrP^{Sc} detectable by Western blot or ELISA are produced. These developments may provide the sensitivity necessary for a blood or food screening test useful for some of the transmissible TSEs. However, it has recently been reported that proteinase K sensitive, pathological isoforms of PrP may have a significant role in the pathogenesis of some prion diseases [8]. Novel PrP^{Sc} isoforms with unique biochemical properties may be generated in sporadic or acquired disease that exhibit increased sensitivity to PK digestion. Therefore, conventional tests may show significant discordance between the amounts of PrP^{Sc} detected and the infectivity observed. Accordingly, the development of new diagnostic tests that do not rely on PK digestion is desirable. Another challenge for diagnosis and surveillance is that hosts can incubate infectious prion agents for many months or years, during which time they exhibit no overt symptoms. Incubation periods for some human prion diseases can be as long as 40 years and given the recent cases of vCJD transmitted by blood transfusion the need for development of a test for screening blood has increased. Furthermore, a noninvasive test to identify the early stages of CJD would be valuable in the development of treatment strategies for TSEs.

A biomarker is defined as a discriminative feature that can be measured objectively and used as an indicator of biological processes such as normal health, pathogenic processes, or pharmacological responses to a therapeutic treatment. Biomarkers include physical traits such as temperature or blood pressure, imaging of pathological features such as amyloid deposition or ventricular volume changes in the brain, and the presence of biological molecules in tissues and body fluids such as blood or urine. One aim of biomarker discovery is the detection of molecular correlates of disease that can be used as early diagnostic tools. However, this type of marker has a crucial requirement for high sensitivity and specificity. Few markers of this type have emerged from omics studies, not only for prion diseases but also for the multitude of other diseases that have been investigated. Biomarkers with broad specificity for neurodegeneration (not only in TSEs) may also be useful as general indicators of disease pathology; identification of biomarkers to follow the progression of disease would significantly impact the time and cost required to evaluate the efficacy of therapeutic interventions.

The search for biomarkers (other than PrP^{Sc}) as tools for diagnosis of prion diseases has a long history; in fact there are several protein markers in cerebrospinal fluid (CSF) that are useful for diagnosis of human prion diseases. In 1980, two proteins were identified by 2D electrophoresis in the CSF of sporadic CJD patients. One of these, a 30 kDa polypeptide, was identified as a member of the 14-3-3 family of proteins, a normal neuronal protein that is released into the CSF after neuronal insult. The CSF detection of 14-3-3 protein by Western blot is widely used as diagnostic evidence of CJD, in conjunction with clinical indicators of prion disease [9]. The detection of 14-3-3 in CSF is a highly sensitive marker for sCJD, iCJD, and the genetic form of CJD, however, is much less sensitive for diagnosis of vCJD, GSS, and FFI [10]. A number of other proteins are also increased in the CSF of CJD patients including Tau and phospho-Tau, S-100β, and neuron-specific enolase (NSE). Levels of the tau protein are raised in patients affected by all forms of CJD including vCJD. A recent study determined that the detection of tau in CSF has a sensitivity of 80% and specificity of 94% for vCJD, higher than any of the other markers tested. Testing for the presence of multiple markers, 14-3-3 protein in CSF plus tau, results in the highest sensitivity for the use of these biomarkers in diagnosis of human TSEs [11]. These CSF biomarkers have proven to be extremely useful in confirmatory diagnostics of CJD cases, and their widespread use illustrates an important role for surrogate marker detection in prion disease. They do not, however, have comprehensive value for surveillance of transmitted TSEs and are useful only when the disease is already at an advanced state. Ideally biomarkers able to detect all TSEs even at preclinical stages of infection are desirable. This paper will focus on recent efforts to harness the plethora of omic technologies to identify not only potentially diagnostic biomarkers, but also markers to follow disease progression or which have risk determining potential.

9.2 TOOLS FOR BIOMARKER IDENTIFICATION IN TSES

In the last few years technologies to study all the genes and proteins expressed in an organism or cells simultaneously have become accessible for most laboratories, and these provide a platform for biomarker discovery. Experimental strategies to detect biomarkers generally involve comparisons of mRNA, protein, peptide, and metabolite abundances between samples collected from infected versus control tissues. The most commonly used technologies are described here, followed by discussion of approaches that are being used in the identification of useful biomarkers in relation to prion disease, as well as potential directions for future research.

9.2.1 DIFFERENTIALLY EXPRESSED MRNA BIOMARKERS

High-throughput genomic techniques, most commonly DNA microarrays and subtractive hybridization approaches, are the most frequently reported methodologies used for the identification of deregulated genes in tissues and cells [12]. These expression profiles, or "signatures," can themselves be used grossly as biomarkers, are relatively easy to generate, and the techniques can be readily adapted to high throughput. Signatures can be compared across multiple time points, tissues, or experimental populations to look for molecules predictive of disease. With the advent of ultra-high-throughput sequencing technologies, researchers are increasingly turning to deep sequencing for gene expression studies [13–15]. Advantages over microarray approaches are that different variants or "isoforms" of mRNA generated by differential splicing, alternative termination, and alternative transcriptional start sites are all identified. Additionally, these methodologies are also well suited for the identification and profiling of small RNAs, such as miRNAs, which are increasingly thought to play important roles in neurodegenerative diseases and may be useful biomarkers [16]. One recent innovation that has been applied to high-throughput sequencing is the purification of RNA from ribosome complexes prior to deep sequencing to capture those templates actively undergoing translation. This method has been found to be more reflective of protein abundance than are traditional microarray or sequenced mRNA profiles and may improve the ability to infer protein biomarkers from RNA profiles [17].

TABLE 1: Genes with differential abundances in prion disease and other neurodegenerative disorders.

Gene	Description	Reference for Prion Disease	Other Neurodegenerative Disorder*
ABCA1	ATP-binding cassette, subfamily A (ABC1), member 1	[20, 22, 23]	AD
APLP1	Amyloid beta (A4) precursor-like protein 1	[20]	AD
APOD	Apolipoprotein D	[20, 20, 23, 25, 28]	AD, NPC
APOE	Apolipoprotein E	[20, 22, 25]	AD, PD, and MTS
B2M	Beta-2-microglobulin	[20, 22, 25, 29]	AD, Tay-Sachs, Sandoff disease, and MTS
CD9	CD9 molecule	[20, 22, 23, 25]	SSPE, CMT
CLU	Clusterin	[20, 25, 30]	AD, PD
CST3	Cystatin C (amyloid angiopathy and cerebral hemorrhage)	[20, 25. 30]	AD, MTS
CTSB	Cathepsin B	[20, 22, 25]	AD, seizures, Tay-Sachs, and Sandhoff disease
CTSS	Cathepsin S	[20, 22, 23, 25, 28]	AD
GFAP	Glial fibrillary acidic protein	[20, 22, 23, 25, 28]	Tay-Saches, Sandhoff disease, MTS, and AD
SPARC	Secreted protein, acidic, cysteine-rich (osteonectin)	[20]	Tay-Sachs, Sandhoff disease, and MTS
SPP1	Secreted phosphoprotein 1 (ostepontin, bone sialoprotein I, early T-lymphocyte activation 1)	[20, 22]	PD

AD: Alzheimer's Disease; PD: Parkinson's Disease; NPCL Niemann Pick type C; MTSL Mesial temporal sclerosis; SSPE: Subacute sclerosing panencephalitis; CMT: Charcot-Marie Tooth disease

Changes in mRNA profiles in brain tissue from CJD patients are infrequently studied due to the rarity of cases. Only one report has been published using tissues isolated from the postmortem brain samples of sporadic CJD patients; transcriptional changes pointed to alterations in neuronal dysfunction pathways including the cell cycle, cell death, and the stress response [18]. A number of genomic analyses of brain tissue from rodent adapted models of prion diseases including CJD, scrapie and BSE have been performed, as well as investigation of samples from larger animals, sheep infected with scrapie and cattle infected with BSE [19–27]. These studies have revealed widespread alterations of multiple cellular pathways correlating with the onset of pathological disease including cholesterol homeostasis, ion homeostasis, and regulation of apoptosis, stress response, and metal ion homeostasis. The most consistent finding between experimental models, relates to the onset of neuroinflammation, a process common to many neurodegenerative diseases that is likely induced by damage and death of neurons. Accordingly, many of these transcriptional changes have been consistently identified in multiple neurodegenerative diseases and a selection of these is listed in Table 1. Although differential expression of these genes may not be specific to prion diseases, neuroinflammation-related gene expression may be an excellent choice as candidate biomarkers to track the stage of development of the neurodegenerative process and to predict the response to therapy.

Studies to correlate the temporal changes in neuronal health during disease have not yet been reported in prion-infected neurons; however, hippocampal neurons from Alzheimer's disease (AD) patients have revealed a transcriptional response comprising thousands of genes that significantly correlates with AD markers [31, 32]. It is possible that at least in part, these biological processes may be common to degenerating neurons in multiple degenerative conditions. In this case these biomarkers may well be broadly applicable to track the progression of neurodegeneration. As the vast majority of human samples are collected postmortem, animal models may be the only practical way of assessing early markers of neuronal status prior to obvious clinical symptoms. A number of studies have attempted to use genomics to determine transcriptional changes at preclinical stages of disease; however, due to the cellular complexity of brain tissue only modest fold changes are revealed. Subtle alterations occurring in a small number of neurons at the onset of disease are likely masked. To get the best results from this type of study it is essential to use large sample numbers for statistical significance. Neuron specific expression changes may be masked even at late stages of disease by the extensive astrocytosis and gliosis that accompany neuronal degeneration. Neurons, and therefore their genetic material, are outnumbered 10 or 20 to 1. Laser capture microdissection to excise specific cell populations is set to overcome many of the limitations of whole tissue analysis and will undoubtedly provide a new "layer" of information regarding specific cellular responses to prion replication.

9.2.2 DIFFERENTIALLY EXPRESSED PROTEIN BIOMARKERS

Protein biomarkers are particularly well suited for measuring and detecting phenotypic characteristics of disease processes. Proteomic technologies enable the exhaustive analysis of the protein content of a tissue or bodily fluid sample. Only in the recent years has technological advances facilitated the differential measurement of protein abundance levels between multiple conditions at a given time, and just as importantly, provided sufficient "through-put" to attach statistical significance to protein biomarker detection. Proteomics of prion infected tissues also suffer from some of the same drawbacks of genomics studies; samples are often very heterogeneous due to cellular complexity and the stage of disease. One of the major caveats of proteomics for prion disease discovery is that the commonly used rodent models provide very small sample volumes, especially in terms of bodily fluids such as blood, from which only the most abundant proteins can be identified. The laser capture microdissection techniques hold much promise for genomic studies to reduce cell heterogeneity as the small amounts of nucleic acid can be amplified by polymerase chain reaction (PCR). However, proteins cannot be amplified and ultra-sensitive techniques must be developed in order to perform similar proteomic studies. New techniques for labelling small amounts of protein such as fluorescence saturation labelling may be one step in the right direction to overcome these limitations. This technique has enabled proteomic analysis of hippocampal CA1 neurons in an Alzheimer's mouse model; however identification of such small amounts of protein requires increased sensitivity of mass spectrometry techniques [33]. Given these problems and the scarcity of samples from large animal models and human TSE cases, it is not surprising that only a very small number of prion-related proteomic studies have been reported. Table 2 provides a general

summary of these, and some examples from similar studies of neurodegenerative diseases with similar aetiologies as proof of principle.

TABLE 2: Potential Biomarkers of Neurodegenerative Diseases Identified by Mass
spectrometry (MS) and 2D-Gel Electrophoresis.

Marker	Fluid	Disease*	Reference
10 kDa subunit of vitronectin	Serum	AD	[34, 35]
alpha 1-acid glycoprotein	Serum	AD	[34, 35]
alpha 1-antichymotrypsin	Urine	CJD	[36]
Apolipoprotein B100	Serum	AD	[34, 35]
Apolipoprotein E	Serum and CSF	AD and PD	[29, 34, 35]
Cathelicidin antimicrobial peptide (Bos taurus)	Urine	BSE	[37]
Clusterin	Urine, CSF, blood, and plasma	BSE and AD	[37–39]
Complement C3 component C3dg	Serum	ALS and PD	[40]
Complement C3 components of C3c family	Serum	ALS, PD, and AD	[34, 35, 40]
Complement C4	Serum	AD	[34, 35]
Complement Factor H	Serum and Plasma	ALS, PD, and AD	[34, 35, 40, 41]
Fragment Bb of Complement Factor B	Serum	PD	[40]
Haptoglobin α-2 chain	Serum and CSF	AD and PD	[29, 34, 35]
Heart-type fatty acid binding protein (H-FABP)	Plasma and CSF	CJD and AD	[42]
Hemoglobin α-2 chain	Serum	AD	[34, 35]
Histidine-rich glycoprotein	Serum	AD	[34, 35]
Ig Gamma-2 chain C region (Bos taurus)	Urine	BSE	[37]
Transthyretin	Serume and CSF	CJD, AD, and PD	[29, 34, 35, 43]
Uroguanylin	Urine	BSE	[37]
Vitronectin precursor	Serum	AD	[34, 35]
α-2-macroglobulin (α-$_2$M)	Plasm and Serum	AD	[34, 35, 41]

AD: Alzheimer's Disease' PD: Parkinson's Disease; ALS: Amyotrophic lateral sclerosis

Current methodologies for proteomics follow two principal steps. Firstly proteins are separated to provide a sample with decreased complexity and then mass spectrometry (MS) is used for protein identification. Separation techniques include surface-enhanced laser desorption ionizing time of flight (SELDI-TOF), two-dimensional gel electrophoresis (2D-GE) and the recently developed two-dimensional differential gel electrophoresis (2D-DIGE) for differential protein analysis, and liquid chromatography (LC). All of these methodologies are not inherently quantitative but have been adapted to allow the user to identify qualitative changes between samples, for example, differential abundances of proteins between diseased and control tissues.

Surface-enhanced laser desorption ionizing time of flight (SELDI-TOF) is a mass spectrometry technique that is based on a combination of techniques, chromatography, and matrix-assisted laser desorption ionization time of flight (MALDI-TOF) [44]. In the first step of SELDI-TOF MS proteins are captured using different platform chemistries such as absorption, electrostatic interaction, and affinity chromatography which reduce the complexity of the original sample. This happens on a small "protein-chip" surface which enables multiple separations to be performed on very small sample volumes at high throughput. Following this, the bound proteins are cocrystallized with an energy absorbing matrix (EAM) which is then vaporized propelling the ionized proteins down the flight tube through an electric field based upon the particles mass/charge (m/z) ratio. Contact of complexes with the detector at the end of the flight tube results in a unique peak resolved by m/z ratio of the original protein or protein isoform. The advantages of this method are the ability to perform high-throughput analysis of hundreds or thousands of samples, resolving power of the captured proteins, and the ease of analysis using dedicated user friendly software. However, protein peaks of interest must be experimentally isolated for identification using peptide mass fingerprinting along with protein purification techniques. This step requires much larger sample volumes than used in the initial discovery stage and can be arduous and time-consuming.

In 2D gel electrophoresis separation of proteins is performed in a polyacrylamide gel according to their size and their charge thus enabling resolution of multiple isoforms of an individual protein. The spot intensities can be used to calculate differences in protein abundance between different samples and individual protein spots can then be excised from the gel and identified using MS. Practical issues such as variations in sample preparation make it very difficult to get consistently reproducible gels making this methodology labour intensive. Recent adaptations to this methodology have improved the situation somewhat, with the most significant innovation being labelling of samples using fluorescence dyes (2D-DIGE). In this way it is possible to include three samples per gel, control and infected samples plus an internal standard (pooled samples). This creates a standard for each protein in the analysis resulting in the user being able to make comparisons across different gels with a high degree of confidence [45].

In LC the sample components interact to a varying extent with a chromatographic packing material in a column (stationary phase). A pump moves the mobile phase through the column, and the sample is separated based on a retention time through the column which varies depending on the interactions between the stationary phase, the molecules being analyzed, and the solvent used. Generally protein samples are enzymatically digested prior to loading on the column. This differs from both SELDI and 2D-GE in which it is generally intact proteins that are separated. In one dimension the peptide mixtures are generally too complex to separate; however, in combination with chromatography or 2D gel electrophoresis the methodology provides a means to perform large-scale proteomic analysis with good dynamic range. Labelling of peptides with isotopes enables this methodology to be used to identify the differential abundance of peptides between samples. To do this a stable isotope is used to label peptides. The labelled peptide is chemically identical to its native counterpart, so it behaves in an identical fashion during chromatographic separation, however it is distinguishable by MS, therefore variation in abundance between a tagged and untagged sample can be determined. A number of approaches using this methodology have been described including Isotope Coded Affinity Tagging (ICAT) and the recently described Isobaric Tagging for Relative and Absolute Protein Quantification (iTRAQ) quantitative proteomic approach [46]. iTRAQ is ideally suited for biomarker applications, as it provides both quantification and allows some degree of multiplexing in a single reagent. The isotopic tag can be incorporated either during sample labelling or in vivo (stable isotope labelling, SILAC), further increasing its scope [47].

MS to identify the mass/charge ratio of the peptide/protein of interest or to determine the primary sequence of the peptide is the final step in all proteomic approaches. This step involves ionization of the sample; MALDI and electrospray ionization (ESI) are the most commonly used technologies for this. In MALDI the sample is mixed with a matrix, applied to a target surface and inserted into a vacuum chamber and a laser is used to activate sample ionization. In ESI the sample is dissolved in a solvent and pumped through a narrow, stainless steel capillary. A high voltage is applied to the tip of the capillary, which is situated within the ionisation source of the mass spectrometer, and as a consequence of this strong electric field, the sample emerging from the tip is dispersed into an aerosol of highly charged droplets. A gas, usually nitrogen, helps this process and directs the spray emerging from the capillary tip towards the mass spectrometer. The ionized sample is generally resolved based on the m/z ratio in a time-of-flight (TOF) analyser. However, a number of choices for ion sources are available and these can be combined with different spectrometers. One promising innovation is fourier transform mass spectrometry (FTMS), a popular tool for discovery due to its high resolving power, mass measurement accuracy, multistage MS/MS potential, and extended dynamic range [48]. When accompanied by 1D or 2D electrophoresis, FTMS has demonstrated an excellent ability to deal with sample complexity for biomarker discovery [49].

One further innovative method sometimes used in biomarker discovery is the antibody microarray, the proteomic equivalent of gene microarrays. In these arrays specific antibodies are spotted onto glass or membranes, or bound to beads in fluidic arrays. Target proteins are then captured from samples of plasma or disrupted tissue and detected using an ELISA type approach using labelled secondary antibodies. The use of these arrays has not been described in prion disease; however a study to screen the abundance of 120 signalling proteins in plasma from Alzheimer's patients was recently reported. A total of 259 samples were analysed with the antibody panel, and 18 proteins were identified as potential biomarkers. These proteins were used to classify blinded samples from Alzheimer's and control subjects with close to 90% accuracy as well as identifying patients with mild cognitive impairment that progressed to Alzheimer's disease 2–6 years later. The 18 proteins are involved in biological processes known to be disrupted in neurodegeneration including deregulation of haematopoiesis, immune responses, apoptosis and neuronal support [50].

9.3 SEARCHING THE BODY FOR PRION RELATED BIOMARKERS

The complexity of prion-induced neurodegenerative diseases along with their unique molecular mechanisms poses huge challenges to understand their biology and to identify antemortem biomarkers. In addition the diseases are aetiologically heterogeneous. Prion diseases are unique in that they can occur in one of three ways, spontaneously, via genetic changes, or acquired through oral or iatrogenic transmission of the infectious agent. Spontaneous or genetic forms of the disease arise and progress solely in the brain, with minimal to no agent replication in the periphery. Only in the case of a TSE transmitted by digestion or blood transfusion, such as in vCJD, does the initial replication of the agent take place in the periphery. Implications of these aetiologies are that diagnostic biomarkers, especially for preclinical stages of disease, will unlikely encompass all forms of prion disease.

Prions acquired from different sources, strains, or different genetic origins present with differing symptoms, incubation periods, and pathobiological features will result in ambiguities in biomarker detection. Accordingly, the tissues and bodily fluids chosen for biomarker selection need to be tailored to the TSE under study and the specific aim of biomarker selection. For example, blood or lymphoid tissue may be the sample of choice for selection of a preclinical marker of vCJD infection, brain tissue for indicators of prognosis, and blood or urine to follow disease progression or perhaps to identify individuals more susceptible to disease or particular treatments.

While specificity to prion disease would be a requirement for identifying preclinical cases or screening donated blood, for example, progression of disease could be followed using markers of broader specificity such as indicators of CNS damage and neuronal death. To further complicate the selection of biomarkers in prion disease the long incubation period prior to development of clinical symptoms, from many months to many years, may well result in temporal differences in marker expression. Therefore disease stage, as well as target tissue, needs to be taken into account when deciding on a sampling strategy and evaluating biomarkers. The greatest public health risk accompanies those TSEs that can be transmitted in food, medical products, and blood such as vCJD or any future novel outbreaks. A closer look at prion pathogenesis in these instances may lead to the identification of appropriate tissues and body fluids for early detection of prion diseases. A summary of these potential tissues and body fluids is given in Figure 1(a). In Figure 1(b) a schema illustrating the incubation period of a typical TSE indicating disease stages optimal for identification of biomarkers for different purposes is provided. In the next section we describe some of these tissues and bodily fluids that are potential reservoirs for biomarkers in more detail, and review related biomarker studies.

9.3.1 BIOMARKERS IN LYMPHOID TISSUES

Following ingestion of contaminated foodstuffs, PrP^{Sc} must be transported from the gut to the brain. Current data suggests that PrP^{Sc} crosses the gut epithelium, possibly through M cells, and rapidly accumulates in the gut-associated lymphoid tissues (GALTs), mainly in the mesenteric lymph nodes, and then in the spleen early in the preclinical phase prior to neuroinvasion. Two studies of gene expression changes have been described using tissue from infected and control Peyers patches. In the first samples from cattle orally infected with BSE revealed 90 genes and 16 ESTs to be differentially expressed. Of these genes, five were found to be related to immune function. These were major histocompatibility complex (MHC) class II, MHC class II DQ alpha, L-RAP, and two hypothetical proteins. Other differentially expressed genes identified related to cellular and metabolic processes including the development and maturation of cells [51]. In the second study the mRNA level of a pancreatitis-associated protein (PAP)-like protein was found to be elevated in the ileal Peyer's patch of lambs during the early phase of scrapie infection [52]. Although the first study tissue analysed was 12 months following oral inoculation Peyers Patches may be sources of very early disease specific markers.

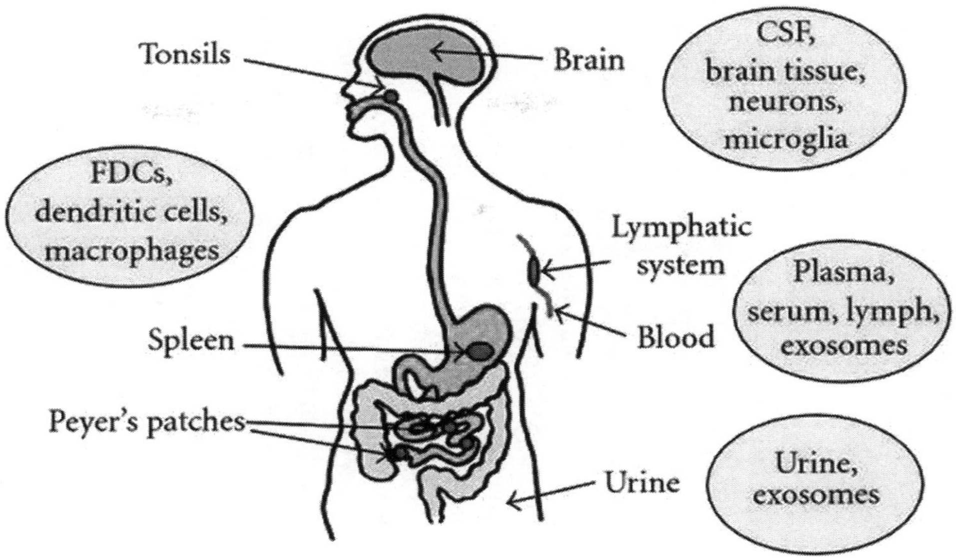

FIGURE 1: Summary of the tissues, cell populations, and bodily fluids that provide a source of discovery for biomarkers of prion infection (a). Schema to illustrate the stage-specific diagnostic and therapeutic windows for biomarker identification for diagnosis, disease progression, and monitoring pharmacological interventions (b).

In B cell follicles, PrP^{Sc} is detected on follicular dendritic cells (FDCs) networks and macrophages within germinal centres (GCs) [53–55] of the spleen and lymph nodes. Although prion infection still occurs in the absence of FDCs, the infection is severely delayed suggesting that FDCs are significant cellular sites of peripheral replication [53, 56, 57]. The lympho-invasion step is an appealing point in the disease process for the identification of biomarkers as it occurs prior to neuroinvasion in those TSEs with ingestion aetiologies and is thus the optimal time point for early diagnosis or therapeutic intervention [58]. In addition lymphoid tissues are also more accessible than brain tissue for sampling purposes. Despite this, very few studies describing gene or protein biomarkers associated with prion replication in spleen or lymph nodes have been published. A recent study has identified changes in the expression of *St6gal1, St3gal5, Man2a1, Hexb, Pigq*, glycosylation-related genes, in the spleens of scrapie infected mice [59]. The authors suggest that this indicates modification of the splenic metabolism of glycosphingolipids associated with prion disease.

FDCs themselves express high levels of PrP^{C} and are able to retain antigens for relatively long periods of time, including replicating PrP^{Sc}, making them good cell candidates for the identification of biomarkers. However, FDCs make up less than 1% of the total cells within the spleen or lymph node which likely means that most disease associated expression changes are masked when looking at whole tissues. They are also tightly associated with other cell types, especially B cells, which along with the gap in knowledge regarding their lineage and molecular characteristics make them difficult to isolate for independent analysis. One study, however, has shown an increase in clusterin expression in association with abnormal PrP accumulation expression on FDCs during TSE disease, particularly human vCJD cases [60].

Lymph fluid passes through lymph nodes and contains a mixture of proteins and antigens picked up from the interstitial tissues which it drains; it therefore reflects changes associated with any immunoinflammatory response within the node itself. The protein composition therefore emulates that of blood as well as being highly reflective of the host response to mucosal challenge. Given the route of transmission of acquired prion infections following ingestion and the subsequent preclinical replication in lymph nodes, lymph fluid may be an excellent, as yet untested source, from which biomarkers that accompany preclinical prion disease progression can be identified. In the human genetic and sporadic forms of prion diseases, the disease occurs spontaneously within brain tissue and there is no preliminary involvement of peripheral tissues, and so FDCs and macrophages do not play a role in disease pathogenesis. Indeed PrP^{Sc} is most often absent in the lymphoid tissues, although, secondary infection of lymph nodes can occur in some instances, as recent studies show that PrP^{Sc} can be detected in spleens of patients with sCJD [61]. Biomarkers specifically expressed in lymphoid tissue or FDCs would therefore be useful for the detection of TSEs acquired specifically by peripheral exposure.

9.3.2 BIOMARKERS IN CSF

Following neuroinvasion and establishment of prion disease in the brain of transmitted TSEs and following the onset of sporadic or genetic forms of disease, the CSF has been the tissue of choice for diagnosis and biomarker identification, due to its obvious association with the CNS and the fact that it is somewhat more accessible than CNS tissue itself. CSF is ideal for use in protein and gene expression profiling techniques to identify biomarkers both to track progression of neurodegeneration, as well as having the potential to contain biomarkers specific to prion replication.

Studies using 2D gel electrophoresis to profile proteins in CSF have the longest history in biomarker identification for CJD. A number of studies have identified cystatin C, transferrin, ubiquitin, Apoliprotein J, lactate dehydrogenase, 14-3-3 proteins plus other as yet unidentified polypeptides as potential biomarkers [43, 62, 63]. In addition a study employing SELDI-TOF analysis of CSF revealed a 13.4 KDa protein. Further analysis using cationic exchange chromatography, sodium dodecyl sulfate-polyacrylamide gel electrophoresis (SDS-PAGE), and liquid chromatography-tandem mass spectrometry (LC-MS/MS) revealed this protein to be cystatin C. Immunoblotting confirmed the significantly increased abundance of cystatin C in all eight CJD-affected patients included in this preliminary study [64]. Interestingly many, if not all, of the genes mentioned above have also been identified as differentially expressed in brain tissue of rodents infected with prions. A study by Brown et al. also links the levels of differentially expressed genes to protein levels in the CSF, an observation that suggests that candidate gene panels identified from animal studies could be used for prediction of disease-associated CSF biomarkers [30, 65–67]. The detection of biomarkers in the CSF appears promising; however, all of the above-mentioned proteins have been observed to increase in abundance in the brains and/or CSF in other neurodegenerative conditions such as Alzheimer's disease or in traumatic brain injury. This finding suggests that the proteins identified to date are not specific markers for prion disease but general biomarkers of neurodegeneration disease or trauma. However, the use patterns of expression of panels of these markers may confer specificity, or alternatively, these markers could well be used as useful indicators of disease progression.

Recently, a proof of principle study revealed that the combination of MALDI-FTMS, in addition to machine learning for the classification of mass spectral features, is able to identify preclinical protein signatures from the CSF of prion infected animals with reasonable predicative accuracy [68]. In this study CSF was isolated from 21 infected and 22 control hamsters at a time-point when approximately 80% of the expected incubation period had been completed. CSF was isolated and subjected to trypsin digestion without further fractionation and subjected to MALDI-FTMS, a methodology described earlier in the chapter. Peptide profiles were identified and the peaks compared using IonSpec peak picking software; a number of peptide peaks exhibiting differential abundances were identified. It was reported that these peaks were amongst the least abundant peptides detected in the study, highlighting the need for improved methodologies to target low abundance proteins and peptides in biomarker studies. A linear support vector machines (SVM) and 10-fold leave-one-out cross validation was used to evaluate the predictive accuracy of the peptide peaks showing the greatest differences in abundance between infected and non-infected hamsters. The predictive accuracy was determined to be 72%; true positive rate of 73% and false positive rate of 27% using a 10-fold leave-one-out cross validation demonstrated a potential for the use of proteomic profiling of CSF for the identification of multiple biomarkers with diagnostic value. However, the identity of these peptides was not resolved in this study. Although specificity was fairly low, as mentioned by Herbst et al. a disease-specific protein signature clearly exists in the CSF. This type of approach combined with a prefractionation step to improve the accuracy of biomarker detection in the range of low abundance proteins could well result in identification of a panel of markers with diagnostic potential. In this case the small size of hamsters and small volumes of CSF precludes this approach so larger animal models or human samples would be required. The comparison of protein profiles with other forms of neurodegenerative disorders would be the next step in increasing specificity to prion diseases.

9.3.3 BIOMARKERS IN BLOOD

Blood is the ideal reservoir for markers indicative of the progression of disease processes in the body, samples are easy to obtain and noninvasive to the patient, and as it circulates throughout the whole body, is a repository for biomarkers of general health and disease. Although no secondary transmission of the sporadic or genetic forms of CJD has been reported, secondary transmission of vCJD from "human-to-human" can occur via blood transfusion. The infectious agent itself is present in the blood in this instance probably following replication in peripheral lymphoid tissue. An in vitro test to detect vCJD prion contamination in human blood or blood products is therefore one of the priorities for the development of sensitive and specific tests. Diagnostic signatures of BSE have been identified in serum of infected cattle by multivariate analysis of infrared spectra, at a sensitivity of 85% and a specificity of 90%, strongly supporting the hypothesis that infection with prion agents leads to specific changes in the molecular content of serum [69]. However, no predictive tests for CJD or other prion disease have yet been validated in blood, including detection of the prion specific biomarker PrP^{Sc}.

A number of studies have identified differential abundances of a handful of proteins in the blood of patients with CJD. One study has shown an increase of the S-100β protein and another, an increase in cystatin C [70]. Another recent study found elevated levels of heart fatty acid binding protein (H-FABP) in the serum of patients with CJD [42]. Fatty acid binding proteins are located within the cell and are responsible for the shuttling of fatty acids in the cytosol and are released from the cell in response to cell damage [42]. However, again, high levels of this protein have also been observed in acute myocardial infarctions, in stroke patients, and Alzheimer's patients, implying that this is not a specific prion disease biomarker.

An early genomics study using differential display reverse-transcriptase PCR (DDRT-PCR) to determine differentially expressed genes in blood identified lower levels of the erythroid differentiation-related factor (ERAF) in the spleen, bone marrow and blood of scrapie infected mice [71]. The same group also observed differential expression of other erythroid-related genes (KEL, GYPA) in the spleens of infected mice as a common feature of murine scrapie [72]. However, these genes were found to be expressed at highly variable levels between individuals, thus precluding their usefulness as accurate markers for diagnosis.

Blood (serum and plasma) is one of the most difficult tissues to analyse using omic technologies. Blood is a highly complex tissue that displays a huge dynamic range of protein abundances challenging the identification of the less abundant species; these rare proteins and peptides likely include the majority of disease specific biomarkers. Prefractionation steps are absolutely necessary to deplete the most abundant proteins [73]. The most common of these are immunodepletion, used extensively for the specific removal of high abundance proteins, based on the action of specific antibodies. More recently, saturation protein binding to a random peptide library has been proposed as an alternative method [38]. Not surprisingly, given these difficulties and the scarcity of samples from CJD patients and large animal models, no proteomic screens for biomarkers of prion disease have been reported to date. Protein profiling of plasma has been reported in a number of studies of Alzheimer's disease and these show some success as evidenced by cross-study reproducibility and validation (albeit fairly low sensitivity and specificity) in patient cohorts. In one study, mass spectrometric analysis of the changes observed in two-dimensional electrophoresis from the plasma of 50 Alzheimers patients and 50 matched controls identified a number of proteins previously implicated in the disease pathology. These included complement factor H (CFH) precursor and alpha-2-macroglobulin (alpha-2M). Although the specificity and sensitivity was fairly low, elevation of CFH and alpha-2M was shown to be specific for Alzheimer's disease and to correlate with disease severity [38]. Two other studies also identified these proteins as upregulated in the plasma of

Alzhiemers patients; given that prion diseases have similar aetiologies it is likely that plasma may well be a rich source for biomarkers to monitor disease progression, and potentially for use in diagnosis [34, 35, 74].

Blood contains a number of circulating cells such as lymphocytes, macrophages, dendritic cells and platelets. Another approach in the search for biomarkers is to isolate specific cell populations and use these as a basis for gene or protein profiling studies. Targeting cells that may be involved in prion replication may increase the chance of picking up disease specific changes; however, no such studies have yet been done. Circulating immune cells such as macrophages and dendritic cells can carry infectious PrP^{Sc} and may therefore traffic infectious prion agent around the body. Macrophages have been found to contain PrP^{Sc} even in the absence of FDCs, thus leading to the speculation that they might serve as alternative sites of prion accumulation and replication when there are no functional FDCs [56]. Dendritic cells are also mobile cells that can retain endocytosed particles without degradation for long periods of time, therefore ideal candidates for propagating prion proteins throughout the body [75].

Activation of the innate immune system in the brain is a general response during neurodegeneration, including that induced by prions. Studies have identified inflammatory genes that are significantly induced or suppressed in microglia isolated from CJD infected brains and these may be a source for potential candidate markers. In one study the CJD expression profile obtained contrasted with that of uninfected microglia exposed to prototypic inflammatory stimuli such as lipopolysaccharide and IFN-gamma, as well as PrP amyloid. Transcript profiles unique for microglia and other myeloid cells involved in neurodegeneration provide opportunities for the discovery of disease specfic biomarkers [19]. A second study also describes the expression of a number of potentially neuroprotecive genes in macrophages/microglia from CJD infected patients [76]. The serum levels of immunomarkers may reflect the inflammatory process in the brain so that monitoring the levels of a panel of these in the serum of infected individuals may track the progress of the neurodegenerative process in patients. Whether or not the inflammatory process is reflected in the serum in prion diseases has not been investigated and is an area for further work. However, a number of studies in other diseases support this possibility including the identification of upregulated neuroinflammatory markers in the blood of Parkinson's affected individuals [34, 35] and a study by Ray et al to identify plasma biomarkers of Alzheimer's disease using antibody arrays [48]. These arrays were used to identify 18 plasma protein biomarkers that one able to discriminate Alzheimer's disease with 90% accuracy, the majority of which were immune related cytokines and growth factors.

9.3.4 BIOMARKERS IN URINE

Urine is commonly used for diagnostic testing in many different conditions and being somewhat less complex than serum is amenable to exploratory biomarker analysis. Two recent studies have applied proteomics for the identification of prion-induced biomarkers. In the first study, the urine of infected cattle over the time course of disease was examined using a combination of 2D-DIGE and mass spectrometry analysis [37]. Four classifier proteins were identified, two of these proteins, immunoglobulin Gamma-2 chain C region and clusterin significantly increased in abundance over time. Increase in the abundance of immunoglobulins has also been reported previously in the urine of scrapie-infected hamsters [77]. Levels of an isoform of clusterin were found to predict with 100% accuracy during infection with BSE, however, the study size was extremely small and limited to a single sample group and so requires validation. Clusterin is a multifunctional glycoprotein found ubiquitously expressed throughout the body and is abundant in astrocytes, CSF, and blood plasma [39, 78]. Its expression has been found to increase in the brains of prion infected mice as well as in other neurodegenerative diseases, and on insult to the brain. A recently reported identified clusterin as a blood borne biomarker following plasma profiling in Huntington's disease patients and additionally saw its upregulation in the CSF of affected individuals [79]. These studies suggest clusterin could have general utility as an inflammatory associated marker for multiple conditions including neurodegeneration.

In the second study, urinary alpha1-antichymotrypsin was found to be dramatically increased in urine of patients suffering from sporadic Creutzfeldt-Jakob disease and a number of other animal models of prion disease [36]. Alpha1-antichymotrypsin, like clusterin, has been identified as a potential disease marker in many disparate diseases including the response to renal and other injuries, and deregulation of expression in many cancers and is therefore not specific to TSEs. It is likely that as both clusterin and alpha1-antichymotrypsin levels are highly responsive to multiple diseases and trauma that the levels in a normal population would preclude utility as a diagnostic marker. However, as levels of both clusterin and alpha1-antichymotrypsin were reported to increase incrementally during the course of disease, this type of marker could potentially be used to monitor the progress of degeneration in individuals during treatment.

9.3.5 EXOSOMES

Exosomes have been investigated for their value as "repositories" for biomarker detection. Exosomes are small (50–90 nm) microvesicles that originate in the cell and following release are thought to be able to migrate and interact with or on other cells [80]. They are often released by cells undergoing stress or other stimuli and may therefore act as carriers of potential biomarkers. Additionally, exosomes are easily isolated from multiple biological fluids and have a much less complex protein component than whole blood, serum, CSF or urine; during the formation of intraluminal vesicles in exosomes extensive sorting of proteins and lipids occurs at the membrane of endosomes which results in them containing a specific group of proteins [80]. Recently the presence of both PrP^C [81] and PrP^{Sc} [82] on exosomes has been demonstrated. In prion infection, exosomes may be ideal candidates for biomarker discovery as they have been reported to be released from several cell types that are involved in prion infection, including intestinal epithelial cells, neurons, neuroglial cells, and DCs.

PrP^{Sc} is associated with exosomes from neuroglial and epithelial cells and these may provide a means of cell-to-cell transfer of infectious prions [82]. FDCs which are actively involved in peripheral prion replication release significant numbers of exosomes on stimulation and it is possible that these are involved in the extracellular transport of PrP^{Sc} to nerve endings, although the mechanism by which prions travel from FDCs to the nervous system is presently unknown. It has, however, been shown that the topographical location between FDCs and nerve endings plays a key role in determining the efficiency of neuroinvasion; the process is accelerated when FDCs are in closer contact with the nerve endings [83]. As exosomes in human plasma may have a multitude of cellular origins including release from human platelets, epithelial cells and hemopoetic cells such as mast cells we believe they could be both a source of biomarkers for early detection of PrP^{Sc} in peripheral infection or for the identification of biomarkers specific to prion replication. Alternatively, they may be reservoirs of inflammation- or stress-related biomarkers that could be detected in plasma, CSF or urine. Although these avenues have yet to be explored in prion diseases, other studies to identify exosomal biomarkers provide proof-of-principle, such as the identification of Fetuin-A as a potential biomarker from urine in patients with acute kidney injuries and glioblastoma [84, 85].

9.4 FUTURE PERSPECTIVES IN PRION BIOMARKER RESEARCH

Techniques for protein profiling are rapidly evolving as are techniques for rapid genome scale sequencing for gene expression profiling. Other novel methodologies can be applied to the fractionation and isolation of pertinent cell types from which to isolate more specific markers of prion disease and neurodegenerative processes. One technique that shows promise in this regard is laser-capture microdissection (LCM) capable of isolating individual cells from cut tissue sections, thus allowing identification of RNA and protein changes specifically in prion-replicating cells. These biomarkers may well be too scarce to pick up on examination of whole tissues or body fluids. It is a useful tool for either markers of peripheral infection in cells from the spleen, gut mucosa, and lymphoid tissue such as tonsils, or to identify prion-replication associated markers, or neuronal health related markers in brain tissues which may well translate to markers in CSF or blood. So far LCM biomarker research on prion diseases and other neurodegenerative disorders is in its infancy but is a promising area for future research.

A burgeoning area for biomarker research is the identification of dysregulated small noncoding RNAs, especially the recently identified family of microRNAs (miRNAs) which are involved in post-transcriptional regulation of gene expression in both plants and animals [86]. These short RNAs have been determined to have regulatory roles that are vital to many cellular processes and appear to be particularly active in controlling complex functions in the nervous system such as neurodevelopment and neuronal function. Recently, compelling evidence for the involvement of microRNAs (miRNAs) in neurodegenerative diseases including Alzheimer's, Parkinson's and prion diseases, has been published [87–89]. Indeed two miRNAs exhibiting increased expression in the brains of rodent models of scrapie were similarly upregulated in the brains of BSE infected macaques illustrating the potential for consistency across species [90]. The potential of miRNAs as biomarkers for diagnosis and prognosis has also been endorsed by studies showing that expression of miRNAs in various cancers can be highly specific and discriminatory profiles between diseased and non-diseased tissues can be readily identified [91].

9.5 CONCLUSION

Significant advances in recent years in technologies for high-throughput sequencing and proteomics mean that the future is bright for biomarker discovery in relation to prion diseases. Of particular note are the ability to obtain transcriptional profiles from homogeneous cell populations at different stages of disease, advances in prefractionation methods for proteomic studies, and the possibility of high-throughput proteomics to identify ever increasing numbers of individual proteins from a single sample. However, a number of unique hurdles and pitfalls remain in relation to prion diseases; these include the very small number of clinical cases for validatory studies, the long incubation period, and the variability of pathogenesis between strains and routes of infection. It is this heterogeneity among prion disease phenotypes that requires careful choice of tissues and time points to use as starting materials for biomarker discovery. Given these factors it may well be impossible to, for example, identify a single preclinical biomarker congruent to the diagnosis of all prion diseases. Conversely, similarities between molecular mechanisms leading to damage and death of neurons in multiple degenerative conditions may allow the broad utility of biomarkers to track disease progression or to predict the onset of disease between prion and other neurodegenerative conditions. Another factor of note that contributes to the relatively slow progress of research in this area relates to the physical properties of the agent itself. The resistance of prions to conventional chemical and physical procedures designed to inactivate viruses and bacteria means that infected tissues must be analysed under biocontainment conditions. Analysis equipment must in many instances be dedicated to TSE biomarker discovery following contamination with potentially infectious prions. These issues often preclude the use of the most up-to-date techniques that rely on expensive, often core-facility-based, apparatus.

REFERENCES

1. S. B. Prusiner, "Prions," Proceedings of the National Academy of Sciences of the United States of America, vol. 95, no. 23, pp. 13363–13383, 1998.
2. A. Ladogana, M. Puopolo, E. A. Croes, et al., "Mortality from Creutzfeldt-Jakob disease and related disorders in Europe, Australia, and Canada," Neurology, vol. 64, no. 9, pp. 1586–1591, 2005.
3. P. Brown, J.-P. Brandel, M. Preese, and T. Sato, "Iatrogenic Creutzfeldt-Jakob disease: the waning of an era," Neurology, vol. 67, no. 3, pp. 389–393, 2006.
4. C. A. Llewelyn, P. E. Hewitt, R. S. G. Knight, et al., "Possible transmission of variant Creutzfeldt-Jakob disease by blood transfusion," The Lancet, vol. 363, no. 9407, pp. 417–421, 2004.
5. M. Bishop, P. Hart, L. Aitchison, et al., "Predicting susceptibility and incubation time of human-to-human transmission of vCJD," The Lancet Neurology, vol. 5, no. 5, pp. 393–398, 2006.
6. E. D. Belay, R. A. Maddox, E. S. Williams, M. W. Miller, P. Gambetti, and L. B. Schonberger, "Chronic wasting disease and potential transmission to humans," Emerging Infectious Diseases, vol. 10, no. 6, pp. 977–984, 2004.
7. G. J. Raymond, A. Bossers, L. D. Raymond, et al., "Evidence of a molecular barrier limiting susceptibility of humans, cattle and sheep to chronic wasting disease," The EMBO Journal, vol. 19, no. 17, pp. 4425–4430, 2000.
8. P. Gambetti, Z. Dong, J. Yuan, et al., "A novel human disease with abnormal prion protein sensitive to protease," Annals of Neurology, vol. 63, no. 6, pp. 697–708, 2008.
9. K. Blennow, A. Johansson, and H. Zetterberg, "Diagnostic value of 14-3-3beta immunoblot and T-tau/P-tau ratio in clinically suspected Creutzfeldt-Jakob disease," International Journal of Molecular Medicine, vol. 16, no. 6, pp. 1147–1149, 2005.
10. P. Sanchez-Juan, A. Green, A. Ladogana, et al., "CSF tests in the differential diagnosis of Creutzfeldt-Jakob disease," Neurology, vol. 67, no. 4, pp. 637–643, 2006.

11. A. J. E. Green, E. J. Thompson, G. E. Stewart, et al., "Use of 14-3-3 and other brain-specific proteins in CSF in the diagnosis of variant Creutzfeldt-Jakob disease," Journal of Neurology Neurosurgery and Psychiatry, vol. 70, no. 6, pp. 744–748, 2001.

12. S. A. Ness, "Microarray analysis: basic strategies for successful experiments," Molecular Biotechnology, vol. 36, no. 3, pp. 205–219, 2007.

13. A. Mortazavi, B. A. Williams, K. McCue, L. Schaeffer, and B. Wold, "Mapping and quantifying mammalian transcriptomes by RNA-Seq," Nature Methods, vol. 5, no. 7, pp. 621–628, 2008.

14. B. T. Wilhelm, S. Marguerat, S. Watt, et al., "Dynamic repertoire of a eukaryotic transcriptome surveyed at single-nucleotide resolution," Nature, vol. 453, no. 7199, pp. 1239–1243, 2008.

15. R. Rosenkranz, T. Borodina, H. Lehrach, and H. Himmelbauer, "Characterizing the mouse ES cell transcriptome with Illumina sequencing," Genomics, vol. 92, no. 4, pp. 187–194, 2008.

16. N. Bushati and S. M. Cohen, "MicroRNAs in neurodegeneration," Current Opinion in Neurobiology, vol. 18, no. 3, pp. 292–296, 2008.

17. N. T. Ingolia, S. Ghaemmaghami, J. R. S. Newman, and J. S. Weissman, "Genome-wide analysis in vivo of translation with nucleotide resolution using ribosome profiling," Science, vol. 324, no. 5924, pp. 218–223, 2009.

18. W. Xiang, O. Windl, I. M. Westner, et al., "Cerebral gene expression profiles in sporadic Creutzfeldt-Jakob disease," Annals of Neurology, vol. 58, no. 2, pp. 242–257, 2005.

19. C. A. Baker and L. Manuelidis, "Unique inflammatory RNA profiles of microglia in Creutzfeldt-Jakob disease," Proceedings of the National Academy of Sciences of the United States of America, vol. 100, no. 2, pp. 675–679, 2003.

20. G. Sorensen, S. Medina, D. Parchaliuk, C. Phillipson, C. Robertson, and S. A. Booth, "Comprehensive transcriptional profiling of prion infection in mouse models reveals networks of responsive genes," BMC Genomics, vol. 9, article 114, 2008.

21. S. Booth, C. Bowman, R. Baumgartner, et al., "Identification of central nervous system genes involved in the host response to the scrapie agent during preclinical and clinical infection," Journal of General Virology, vol. 85, no. 11, pp. 3459–3471, 2004.

22. W. Xiang, O. Windl, G. Wünsch, et al., "Identification of differentially expressed genes in scrapie-infected mouse brains by using global gene expression technology," Journal of Virology, vol. 78, no. 20, pp. 11051–11060, 2004.

23. C. Riemer, S. Neidhold, M. Burwinkel, et al., "Gene expression profiling of scrapie-infected brain tissue," Biochemical and Biophysical Research Communications, vol. 323, no. 2, pp. 556–564, 2004.

24. G. P. Sawiris, K. G. Becker, E. J. Elliott, R. Moulden, and R. G. Rohwer, "Molecular analysis of bovine spongiform encephalopathy infection by cDNA arrays," Journal of General Virology, vol. 88, no. 4, pp. 1356–1362, 2007.

25. P. J. Skinner, H. Abbassi, B. Chesebro, R. E. Race, C. Reilly, and A. T. Haase, "Gene expression alterations in brains of mice infected with three strains of scrapie," BMC Genomics, vol. 7, article 114, 2006.

26. Y. Tang, W. Xiang, S. A. C. Hawkins, H. A. Kretzschmar, and O. Windl, "Transcriptional changes in the brains of cattle orally infected with the bovine spongiform

encephalopathy agent precede detection of infectivity," Journal of Virology, vol. 83, no. 18, pp. 9464–9473, 2009.

27. G. M. Cosseddu, O. Andréoletti, C. Maestrale, et al., "Gene expression profiling on sheep brain reveals differential transcripts in scrapie-affected/not-affected animals," Brain Research, vol. 1142, no. 1, pp. 217–222, 2007.

28. F. Dandoy-Dron, F. Guillo, L. Benboudjema, et al., "Gene expression of scrapie: cloning of a new scrapie-responsive gene and the identification of increased levels of seven other mRNA transcripts," Journal of Biological Chemistry, vol. 273, no. 13, pp. 7691–7697, 1998.

29. I. Rite, S. Argüelles, J. L. Venero, et al., "Proteomic identification of biomarkers in the cerebrospinal fluid in a rat model of nigrostriatal dopaminergic degeneration," Journal of Neuroscience Research, vol. 85, no. 16, pp. 3607–3618, 2007.

30. A. R. Brown, J. Webb, S. Rebus, A. Williams, and J. K. Fazakerley, "Identification of up-regulated genes by array analysis in scrapie-infected mouse brains," Neuropathology and Applied Neurobiology, vol. 30, no. 5, pp. 555–567, 2004.

31. E. M. Blalock, J. W. Geddes, K. C. Chen, N. M. Porter, W. R. Markesbery, and P. W. Landfield, "Incipient Alzheimer's disease: microarray correlation analyses reveal major transcriptional and tumor suppressor responses," Proceedings of the National Academy of Sciences of the United States of America, vol. 101, no. 7, pp. 2173–2178, 2004.

32. Z.-L. Wu, J. R. Ciallella, D. G. Flood, T. M. O'Kane, D. Bozyczko-Coyne, and M. J. Savage, "Comparative analysis of cortical gene expression in mouse models of Alzheimer's disease," Neurobiology of Aging, vol. 27, no. 3, pp. 377–386, 2006.

33. K. E. Wilson, R. Marouga, J. E. Prime, et al., "Comparative proteomic analysis using samples obtained with laser microdissection and saturation dye labelling," Proteomics, vol. 5, no. 15, pp. 3851–3858, 2005.

34. R. Zhang, L. Barker, D. Pinchev, et al., "Mining biomarkers in human sera using proteomic tools," Proteomics, vol. 4, no. 1, pp. 244–256, 2004.

35. M. Dufek, M. Hamanová, J. Lokaj, et al., "Serum inflammatory biomarkers in Parkinson's disease," Parkinsonism and Related Disorders, vol. 15, no. 4, pp. 318–320, 2009.

36. G. Miele, H. Seeger, D. Marino, et al., "Urinary α1-antichymotrypsin: a biomarker of prion infection," PLoS ONE, vol. 3, no. 12, article e3870, 2008.

37. S. L. R. Simon, L. Lamoureux, M. Plews, et al., "The identification of disease-induced biomarkers in the urine of BSE infected cattle," Proteome Science, vol. 6, article 23, 2008.

38. C. Sihlbom, I. Kanmert, H. von Bahr, and P. Davidsson, "Evaluation of the combination of bead technology with SELDI-TOF-MS and 2-D DIGE for detection of plasma proteins," Journal of Proteome Research, vol. 7, no. 9, pp. 4191–4198, 2008.

39. K. Sasaki, K. Doh-ura, J. W. Ironside, and T. Iwaki, "Increased clusterin (apolipo-protein J) expression in human and mouse brains infected with transmissible spongiform encephalopathies," Acta Neuropathologica, vol. 103, no. 3, pp. 199–208, 2002.

40. I. L. Goldknopf, E. A. Sheta, J. Bryson, et al., "Complement C3c and related protein biomarkers in amyotrophic lateral sclerosis and Parkinson's disease," Biochemical and Biophysical Research Communications, vol. 342, no. 4, pp. 1034–1039, 2006.

41. A. Hye, S. Lynham, M. Thambisetty, et al., "Proteome-based plasma biomarkers for Alzheimer's disease," Brain, vol. 129, no. 11, pp. 3042–3050, 2006.

42. E. Guillaume, C. Zimmermann, P. R. Burkhard, D. F. Hochstrasser, and J.-C. Sanchez, "A potential cerebrospinal fluid and plasmatic marker for the diagnosis of Creutzfeldt-Jakob disease," Proteomics, vol. 3, no. 8, pp. 1495–1499, 2003.

43. P. Brechlin, O. Jahn, P. Steinacker, et al., "Cerebrospinal fluid-optimized two-dimensional difference gel electrophoresis (2-D DIGE) facilitates the differential diagnosis of Creutzfeldt-Jakob disease," Proteomics, vol. 8, no. 20, pp. 4357–4366, 2008.

44. T. C. W. Poon, "Opportunities and limitations of SELDI-TOF-MS in biomedical research: practical advices," Expert Review of Proteomics, vol. 4, no. 1, pp. 51–65, 2007.

45. W. Winkler, M. Zellner, M. Diestinger, et al., "Biological variation of the platelet proteome in the elderly population and its implication for biomarker research," Molecular and Cellular Proteomics, vol. 7, no. 1, pp. 193–203, 2008.

46. P. L. Ross, Y. N. Huang, J. N. Marchese, et al., "Multiplexed protein quantitation in Saccharomyces cerevisiae using amine-reactive isobaric tagging reagents," Molecular and Cellular Proteomics, vol. 3, no. 12, pp. 1154–1169, 2004.

47. S. E. Ong, B. Blagoev, I. Kratchmarova, et al., "Stable isotope labeling by amino acids in cell culture, SILAC, as a simple and accurate approach to expression proteomics," Molecular and Cellular Proteomics, vol. 1, no. 5, pp. 376–386, 2002.

48. M. T. Cancilla, S. G. Penn, and C. B. Lebrilla, "Alkaline degradation of oligosaccharides coupled with matrix-assisted laser desorption/ionization Fourier transform mass spectrometry: a method for sequencing oligosaccharides," Analytical Chemistry, vol. 70, no. 4, pp. 663–672, 1998.

49. M. J. Chalmers, C. L. Mackay, C. L. Hendrickson, et al., "Combined top-down and bottom-up mass spectrometric approach to characterization of biomarkers for renal disease," Analytical Chemistry, vol. 77, no. 22, pp. 7163–7171, 2005.

50. S. Ray, M. Britschgi, C. Herbert, et al., "Classification and prediction of clinical Alzheimer's diagnosis based on plasma signaling proteins," Nature Medicine, vol. 13, no. 11, pp. 1359–1362, 2007.

51. B. Khaniya, L. Almeida, U. Basu, et al., "Microarray analysis of differentially expressed genes from Peyer's patches of cattle orally challenged with bovine spongiform encephalopathy," Journal of Toxicology and Environmental Health. Part A, vol. 72, no. 17-18, pp. 1008–1013, 2009.

52. L. Austbø, A. Kampmann, U. Müller-Ladner, E. Neumann, I. Olsaker, and G. Skretting, "Identification of differentially expressed genes in ileal Peyer's patch of scrapie-infected sheep using RNA arbitrarily primed PCR," BMC Veterinary Research, vol. 4, article 12, 2008.

53. K. L. Brown, K. Stewart, D. L. Ritchie, et al., "Scrapie replication in lymphoid tissues depends on prion protein- expressing follicular dendritic cells," Nature Medicine, vol. 5, no. 11, pp. 1308–1312, 1999.

54. M. Lötscher, M. Recher, L. Hunziker, and M. A. Klein, "Immunologically induced, complement-dependent up-regulation of the prion protein in the mouse spleen: follicular dendritic cells versus capsule and trabeculae," Journal of Immunology, vol. 170, no. 12, pp. 6040–6047, 2003.

55. N. A. Mabbott and M. E. Bruce, "The immunobiology of TSE diseases," Journal of General Virology, vol. 82, no. 10, pp. 2307–2318, 2001.

56. N. A. Mabbott, F. Mackay, F. Minns, and M. E. Bruce, "Temporary inactivation of follicular dendritic cells delays neuroinvasion of scrapie," Nature Medicine, vol. 6, no. 7, pp. 719–720, 2000.

57. N. A. Mabbott, J. Young, I. McConnell, and M. E. Bruce, "Follicular dendritic cell dedifferentiation by treatment with an inhibitor of the lymphotoxin pathway dramatically reduces scrapie susceptibility," Journal of Virology, vol. 77, no. 12, pp. 6845–6854, 2003.

58. P. Aucouturier and C. Carnaud, "The immune system and prion diseases: a relationship of complicity and blindness," Journal of Leukocyte Biology, vol. 72, no. 6, pp. 1075–1083, 2002.

59. F. Guillerme-Bosselut, L. Forestier, C. Jayat-Vignoles, et al., "Glycosylation-related gene expression profiling in the brain and spleen of scrapie-affected mouse," Glycobiology, vol. 19, no. 8, pp. 879–889, 2009.

60. K. Sasaki, K. Doh-ura, J. W. Ironside, N. Mabbott, and T. Iwaki, "Clusterin expression in follicular dendritic cells associated with prion protein accumulation," Journal of Pathology, vol. 209, no. 4, pp. 484–491, 2006.

61. M. Glatzel, E. Abela, M. Maissen, and A. Aguzzi, "Extraneural pathologic prion protein in sporadic Creutzfeldt-Jakob disease," The New England Journal of Medicine, vol. 349, no. 19, pp. 1812–1820, 2003.

62. C. Piubelli, M. Fiorini, G. Zanusso, et al., "Searching for markers of Creutzfeldt-Jakob disease in cerebrospinal fluid by two-dimensional mapping," Proteomics, vol. 6, supplement 1, pp. S256–S261, 2006.

63. L. Cepek, P. Brechlin, P. Steinacker, et al., "Proteomic analysis of the cerebrospinal fluid of patients with Creutzfeldt-Jakob disease," Dementia and Geriatric Cognitive Disorders, vol. 23, no. 1, pp. 22–28, 2006.

64. J.-C. Sanchez, E. Guillaume, P. Lescuyer, et al., "Cystatin C as a potential cerebrospinal fluid marker for the diagnosis of Creutzfeldt-Jakob disease," Proteomics, vol. 4, no. 8, pp. 2229–2233, 2004.

65. O. Carrette, I. Demalte, A. Scherl, et al., "A panel of cerebrospinal fluid potential biomarkers for the diagnosis of Alzheimer's disease," Proteomics, vol. 3, no. 8, pp. 1486–1494, 2003.

66. W.-M. Gao, M. S. Chadha, R. P. Berger, et al., "A gel-based proteomic comparison of human cerebrospinal fluid between inflicted and non-inflicted pediatric traumatic brain injury," Journal of Neurotrauma, vol. 24, no. 1, pp. 43–53, 2007.

67. L. Pucci, S. Triscornia, D. Lucchesi, et al., "Cystatin C and estimates of renal function: searching for a better measure of kidney function in diabetic patients," Clinical Chemistry, vol. 53, no. 3, pp. 480–488, 2007.

68. A. Herbst, S. McIlwain, J. J. Schmidt, J. M. Aiken, C. D. Page, and L. Li, "Prion disease diagnosis by proteomic profiling," Journal of Proteome Research, vol. 8, no. 2, pp. 1030–1036, 2009.

69. T. C. Martin, J. Moecks, A. Belooussov, et al., "Classification of signatures of Bovine Spongiform Encephalopathy in serum using infrared spectroscopy," Analyst, vol. 129, no. 10, pp. 897–901, 2004.

70. M. Otto, J. Wiltfang, E. Schütz, et al., "Diagnosis of Creutzfeldt-Jakob disease by measurement of S100 protein in serum: prospective case-control study," British Medical Journal, vol. 316, no. 7131, pp. 577–582, 1998.

71. G. Miele, J. Manson, and M. Clinton, "A novel erythroid-specific marker of transmissible spongiform encephalopathies," Nature Medicine, vol. 7, no. 3, pp. 361–363, 2001.

72. A. R. Brown, A. R. A. Blanco, G. Miele, et al., "Differential expression of erythroid genes in prion disease," Biochemical and Biophysical Research Communications, vol. 364, no. 2, pp. 366–371, 2007.

73. X. Li, Y. Gong, Y. Wang, et al., "Comparison of alternative analytical techniques for the characterisation of the human serum proteome in HUPO Plasma Proteome Project," Proteomics, vol. 5, no. 13, pp. 3423–3441, 2005.

74. I. Ueno, T. Sakai, M. Yamaoka, R. Yoshida, and A. Tsugita, "Analysis of blood plasma proteins in patients with Alzheimer's disease by two-dimensional electrophoresis, sequence homology and immunodetection," Electrophoresis, vol. 21, no. 9, pp. 1832–1845, 2000.

75. P. Aucouturier, F. Geissmann, D. Damotte, et al., "Infected splenic dendritic cells are sufficient for prion transmission to the CNS in mouse scrapie," Journal of Clinical Investigation, vol. 108, no. 5, pp. 703–708, 2001.

76. F. Chrétien, G. Le Pavec, A.-V. Vallat-Decouvelaere, et al., "Expression of excitatory amino acid transporter-1 (EAAT-1) in brain macrophages and microglia of patients with prion diseases," Journal of Neuropathology and Experimental Neurology, vol. 63, no. 10, pp. 1058–1071, 2004.

77. A. Serban, G. Legname, K. Hansen, N. Kovaleva, and S. B. Prusiner, "Immunoglobulins in urine of hamsters with scrapie," Journal of Biological Chemistry, vol. 279, no. 47, pp. 48817–48820, 2004.

78. S. McHattie, G. A. H. Wells, J. Bee, and N. Edington, "Clusterin in bovine spongiform encephalopathy (BSE)," Journal of Comparative Pathology, vol. 121, no. 2, pp. 159–171, 1999.

79. A. Dalrymple, E. J. Wild, R. Joubert, et al., "Proteomic profiling of plasma in Huntington's disease reveals neuroinflammatory activation and biomarker candidates," Journal of Proteome Research, vol. 6, no. 7, pp. 2833–2840, 2007.

80. G. van Niel, I. Porto-Carreiro, S. Simoes, and G. Raposo, "Exosomes: a common pathway for a specialized function," Journal of Biochemistry, vol. 140, no. 1, pp. 13–21, 2006.

81. C. Robertson, S. A. Booth, D. R. Beniac, M. B. Coulthart, T. F. Booth, and A. McNicol, "Cellular prion protein is released on exosomes from activated platelets," Blood, vol. 107, no. 10, pp. 3907–3911, 2006.

82. B. Fevrier, D. Vilette, F. Archer, et al., "Cells release prions in association with exosomes," Proceedings of the National Academy of Sciences of the United States of America, vol. 101, no. 26, pp. 9683–9688, 2004.

83. M. Prinz, M. Helkenwalder, T. Junt, et al., "Positioning of follicular dendritic cells within the spleen controls prion neuroinvasion," Nature, vol. 425, no. 6961, pp. 957–962, 2003.

84. H. Zhou, T. Pisitkun, A. Aponte, et al., "Exosomal Fetuin-A identified by proteomics: a novel urinary biomarker for detecting acute kidney injury," Kidney International, vol. 70, no. 10, pp. 1847–1857, 2006.

85. J. Skog, T. Würdinger, S. van Rijn, et al., "Glioblastoma microvesicles transport RNA and proteins that promote tumour growth and provide diagnostic biomarkers," Nature Cell Biology, vol. 10, no. 12, pp. 1470–1476, 2008.

86. D. P. Bartel, "MicroRNAs: genomics, biogenesis, mechanism, and function," Cell, vol. 116, no. 2, pp. 281–297, 2004.

87. R. Saba, C. D. Goodman, R. L. C. H. Huzarewich, C. Robertson, and S. A. Booth, "A miRNA signature of prion induced neurodegeneration," PLoS ONE, vol. 3, no. 11, article e3652, 2008.

88. W. J. Lukiw and A. I. Pogue, "Induction of specific micro RNA (miRNA) species by ROS-generating metal sulfates in primary human brain cells," Journal of Inorganic Biochemistry, vol. 101, no. 9, pp. 1265–1269, 2007.

89. J. Kim, K. Inoue, J. Ishii, et al., "A microRNA feedback circuit in midbrain dopamine neurons," Science, vol. 317, no. 5842, pp. 1220–1224, 2007.

90. J. Montag, R. Hitt, L. Opitz, W. J. Schulz-Schaeffer, G. Hunsmann, and D. Motzkus, "Upregulation of miRNA hsa-miR-342-3p in experimental and idiopathic prion disease," Molecular Neurodegeneration, vol. 4, no. 1, article 36, 2009.

91. M. Osaki, F. Takeshita, and T. Ochiya, "MicroRNAs as biomarkers and therapeutic drugs in human cancer," Biomarkers, vol. 13, no. 7-8, pp. 658–670, 2008.

PART IV

OMICS AND PATHOGENS

CHAPTER 10

INSIGHTS FROM GENOMICS INTO BACTERIAL PATHOGEN POPULATIONS

DANIEL J. WILSON

10.1 INTRODUCTION

Bacteria are the most abundant group of organisms, and a major source of human disease and mortality. Bacterial cells account for most of the earth's biomass [1], and the 100 trillion microbial residents of the human body outnumber human cells 10 to 1 [2]. Bacteria that cause pneumonia, diarrhea, and tuberculosis are leading causes of death worldwide [3], [4]. In countries with a low overall burden of infectious disease such as the United States, bacteria are nevertheless responsible for more than 60% of the deaths attributable to communicable disease, with hospital-associated infections, HIV-associated infections, and tuberculosis most prominent (Table 1).

This chapter was originally published under the Creative Commons Attribution License. Wilson DJ. Insights from Genomics into Bacterial Pathogen Populations. PLoS Pathogens 8,9 (2012). doi:10.1371/journal.ppat.1002874.

Since the introduction of the earliest antibiotics, bacteria have evolved resistance [5]. Treatment options continue to be eroded by the spread of antibiotic resistance [6], not only in countries with advanced health care infrastructure, but globally [7], [8]. However, advances in DNA sequencing capacity offer hope in the fight against pathogenic bacteria because the ability to sequence populations of bacterial genomes is illuminating our understanding of bacterial evolution and virulence. Ultimately these insights will underpin translational research into improved medical practice, drug and vaccine targets, and public health policy.

High-throughput whole genome sequencing (Figure 1) represents a genuine step change for the study of bacterial populations because current approaches are based on the analysis of gene fragments amounting to just a few thousandths the total length of the genome [9], [10]. Population genomics offers unprecedented sensitivity for the detection of rare genetic variants, vastly improved resolution for population studies, and direct sequencing of functionally relevant loci. As a result, it is driving new understanding of within-host evolution, transmission, and population structure. Moreover, the advent of rapid benchtop sequencing is changing the way that microbiology is conducted, signaling a new era of real-time genomics and disseminated collaborative analysis in response to rapidly changing public health emergencies.

TABLE 1: Major bacterial causes of death: World and United States.

Cause of Death	Total Deaths (Thosands)	% Communi- cable Disease Deaths	Key Bacterial Species
Global (2008 estimates) [3, 83]			
Lower respiratory infections	3,742	30.6	*Streptococcus pneumoniae, Haemophilus influenzae*
Tuberculosis	1,833	15.0	*Mycobacterium tuberculosis*
Directly attributable	1,250	10.2	
HIV-associated[a]	583	4.8	
Diarrhoeal disease	1,687	13.8	*Vibrio cholerae, Escherichia coli, Salmonella typhi*

TABLE 1: *Cont.*

Cause of Death	Total Deaths (Thosands)	% Communi- cable Disease Deaths	Key Bacterial Species
Meningitis	270	2.2	*Neisseria meningitidis*
Pertussis	194	1.6	*Bordella pertussis*
Tetanus	128	1.0	*Clostridium tetani*
Syphilis	81	0.7	*Treponema pallidum*
Upper repiratory infections	69	0.6	*Streptococcus pyogenes*
Chlamydia	7	0.1	*Chlamydia trachomatis*
Other communicable disease[b]	4,231	34.5	
United States of America (1999–2007) [84]			
Sepsis[b]	280.3	48.17	
Clostridium difficile infection	30.2	5.19	*Clostridium difficile*
Staphylococcal infection	16.6	2.86	*Staphylococcus aureus*
HIV-associated[b]	9.7	1.66	
Tuberculosis	8.8	1.50	*Mycobacterium tuberculosis*
Directly attributable	7.4	1.26	
HIV-assoicated	1.4	0.24	
Streptococcal infection	6.4	1.09	*Streptococcus pneumoniae*
Meningococcal disease	1.4	0.24	*Neisserie meningitidis*
Legionnaires' disease	0.7	0.12	*Legionella pneumophila*
Other bacterial disease[b]	17.6	4.57	
Other communicable disease[b]	210.1	36.1	

The total number of deaths attributable to communicable diseases is shown for the world (2008 estimates) and United States (1999–2007), with key bacterial species highlighted. At the global level, the WHO classifications for causes of death are broad and usually encompass multiple etiological agents, not only bacterial species. The United States and some other countries classify deaths based on detailed ICD-10 four-digit codes that frequently specify the bacterial species responsible.
[a]Estimated from the total number of HIV deaths assuming 26% are associated with tuberculosis [85].
[b]Excluding other causes of death mentioned explicitly.

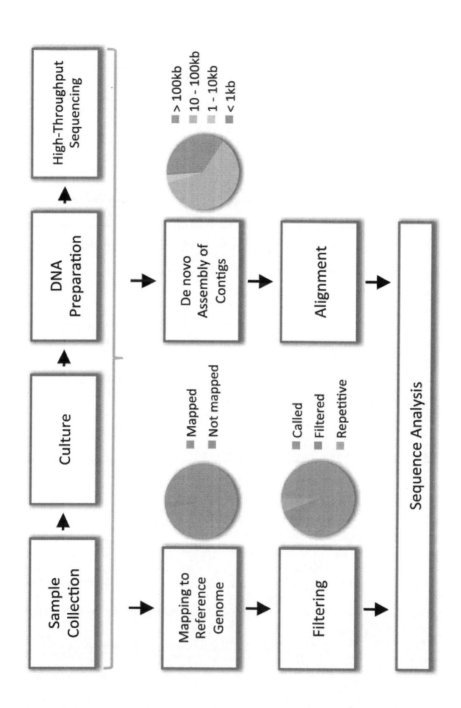

FIGURE 1: An example workflow for high-throughput whole genome sequencing in bacteria. Sample collection. A biological sample (e.g., blood) is collected. Culture. Bacterial colonies are isolated from the sample by culturing on appropriate media. DNA Preparation. DNA is extracted from the colonies and a DNA library is prepared ready for sequencing. High-Throughput Sequencing. Millions of short sequence reads are yielded, typically several hundred nucleotides long or less. To reconstruct the genome, one of two approaches is generally adopted. Mapping to Reference Genome. In reference-based mapping, the short sequences are mapped (i.e., aligned) to a reference genome using an algorithm (e.g., [73], [74]). Preferably the reference genome is high quality, complete, and closely related. The pie chart illustrates that not all reads necessarily map to the reference genome (e.g., because of novel regions not present in the reference). Filtering. Short reads cannot be mapped reliably to repetitive regions of the reference genome, so these are identified and filtered out. Sites that are problematic for other reasons (e.g., because too few reads have mapped or because the consensus nucleotide is ambiguous) are also filtered out. The pie chart illustrates that some portion of the reference genome does not get called due to filtering. In the mapped genome, these positions will receive an ambiguity code (i.e., N rather than A, C, G, or T). De novo Assembly of Contigs. An alternative to mapping is de novo assembly, in which no reference genome is used. An algorithm (e.g., [75], [76]) is used to assemble short reads into longer sequences known as contigs. The number and length of contigs will depend on general factors such as the length of sequence reads and the total amount of DNA sequence produced, as well as local factors such as the presence of repetitive regions. The pie chart shows an example of the proportion of all reads that assemble into contigs of a given length. Alignment. For further analysis, it is necessary to align local regions (e.g., genes) or whole genomes using appropriate algorithms (e.g., [77]–[79]). There is a trade-off in computational terms between the length of region and the number of sequences that can be aligned. Sequence Analysis. The two approaches produce sequence alignments that represent pairwise alignments against a reference (mapping) or multiple alignments one to another (de novo assembly). These alignments can be analyzed directly, or processed further to detect variants such as single nucleotide polymorphisms, insertions, and deletions. The pie charts are meant to be illustrative only, and were produced from data in [27].

10.2 WITHIN-HOST EVOLUTION

Successful colonization of a host is essential to the lifecycle of the pathogen, and the dynamics of the host-pathogen interaction determine the outcome of the interaction, including the severity of disease. DNA/RNA sequencing has greatly advanced the understanding of viral dynamics during infection [11], [12], including the ability to predict disease progression [13], [14], but progress in bacteria has lagged behind, owing to much larger genomes and sparser genetic variation [15]. However, whole genome sequencing in populations of bacteria colonizing individual hosts is shedding new light on the host-pathogen interaction, and the dynamics of bacterial evolution within the host.

At the whole genome scale, genetic variation has been discovered in singly infected hosts colonized by species as disparate as *Mycobacterium tuberculosis* [16], *Salmonella enterica* [17], and *Staphylococcus aureus* [18]. The absolute number of variable sites detected in singly infected hosts is small, frequently fewer than 10 single nucleotide polymorphisms (SNPs), although this varies by species and depends on the number of genomes sequenced and the time elapsed between sampling. Other forms of genetic variation observable at the species level [19] have also been detected, including short insertions and deletions (indels), and variation in the presence or absence of mobile elements such as prophages [16]–[18].

The real-time mutation rate is a key factor in determining the potential for bacterial pathogens to adapt to the host immune system or drug intervention. Traditional estimates of bacterial substitution rates over geological timescales predict fewer than 0.01 mutations per megabase (Mb) per year [20], [21]. Yet laboratory estimates and limited sequencing of longitudinal samples suggest rates 100 or 1,000 times faster [22]–[26]. These estimates have been put to the test by whole genome sequencing, yielding within-host mutation rates ranging from 0.1/Mb/year in *Mycobacterium tuberculosis* [16] through 2.7/Mb/year in *Staphylococcus aureus* [27] to 19/Mb/year in *Helicobacter pylori* [28]. This supports the conclusion that short-term substitution rates in bacteria are several orders of magnitude faster than long-term rates [23], [26], a finding that may be explained by the delayed action of purifying selection [29], [30]. In other words, over longer evolutionary periods the substitution rate depends on selection as well as mutation. It also demonstrates the potential for bacteria to adapt within the host. For example, the discovery that the genome-wide mutation rate in latent tuberculosis infection is similar to that in active disease may explain reports that found treating even latent infections with the antibiotic isoniazid was a risk factor for the emergence of isoniazid resistance [16].

Many bacterial pathogens are common constituents of the body's natural flora [31]. Evolution during colonization may trigger a transition from healthy carriage to invasive disease. For example, 27% of adults carry

Staphylococcus aureus asymptomatically in the nose [32], a known risk factor for disease [33]. In a study of one long-term carrier who developed a bloodstream infection, the genomes of invasive bacteria were found to possess an excess of mutations that truncated proteins, including a transcriptional regulator implicated in pathogenicity [27], [34]. Although further work would be needed to establish causality, this demonstrates the potential for loss-of-function mutations to induce radical functional change during colonization.

Unusual patterns of mutation in the genome during colonization may signal adaptive change and reveal mechanisms of virulence or immune evasion. A study of a 16-year outbreak of chronic *Burkholderia dolosa* infection among cystic fibrosis patients revealed evidence for parallel adaptive evolution [35]. Seventeen genes accrued three or more mutations across the 14 patients, of which a significant excess altered the encoded protein. Some of these mutations affected important phenotypes, including oxygen-dependent gene regulation—which may be pertinent to lung infection—antibiotic resistance, and outer membrane synthesis. Mutations not previously implicated in pathogenesis present novel therapeutic targets.

In hosts colonized multiple times by distinct genotypes, whole genome sequencing affords an opportunity to investigate recombination in vivo. Horizontal gene transfer, also known as recombination, is a fundamental process that generates diversity and facilitates the spread of advantageous genes [36], [37]. A longitudinal study of the highly promiscuous gut pathogen *Helicobacter pylori* identified recombination events from the clustering of SNP differences introduced by the import of DNA from one strain to another [28]. Surprisingly, multiple fragments of around 400 bases appeared to be simultaneously imported in a span up to 20 kilobases long. This pattern of integration was implied by the results of a similar study in *Streptococcus pneumoniae* [38], demonstrating the power of whole genome sequencing to illuminate molecular mechanisms.

10.3 DETECTION OF TRANSMISSION EVENTS

Whole genome sequencing offers unprecedented resolution to distinguish degrees of relatedness among bacterial isolates, and this is a powerful tool for microbial forensics [39]. Genome sequencing complements existing epidemiological tools by providing a means to reconstruct recent chains of transmission, identify sequential acquisition of strains by persistent carriers, and identify cryptic outbreaks that might otherwise go unnoticed.

The superiority of genomics over traditional approaches to molecular epidemiology was demonstrated in a study of *Staphylococcus aureus* ST-239 [40], a widely disseminated multi-drug resistant clonal lineage dominant in much of Asia. Traditional typing methods offer little discriminatory power for subtyping ST-239, but 5,842 SNPs were discovered by whole genome sequencing, revealing detailed geographical structure within the lineage. Against this backdrop of geographical differentiation, examples of recent intercontinental transmission were evident from the clustering of two isolates from England and Denmark among the Thai group. Moreover, a cluster of five isolates sampled over 11 weeks from adjacent blocks of a Thai hospital differed by just 14 SNPs, providing evidence of recent hospital transmission.

Population genomics offers complementary tools to routine outbreak investigation. Following the discovery of two index cases, an outbreak of *Mycobacterium tuberculosis* was uncovered in British Columbia using contact tracing and social network questionnaires [41]. The transitivity of the social network and inability to distinguish isolates by traditional genotyping prevented identification of the source. Whole genome sequencing revealed two distinct lineages, ruling out transmission between social contacts infected with discordant types. Further epidemiological investigation intimated a complex scenario in which an increase in crack cocaine usage triggered simultaneous outbreaks that were sustained by key members of a high-risk social network.

In some cases, the direction of transmission may be discernible from patterns of relatedness and associated epidemiological information. In their study of chronic *Burkholderia dolosa* infection among cystic fibrosis patients [35], the authors used the chronological accumulation of mutations to discriminate donors from recipients in the transmission network. By the same method, they were even able to infer repeated transmission from the airways to the bloodstream within patients (Figure 2). In a persistent *Escherichia coli* infection of members of a household over three years [42], whole genome sequencing revealed at least six transmission events between family members including at least two zoonotic transmissions to the family dog.

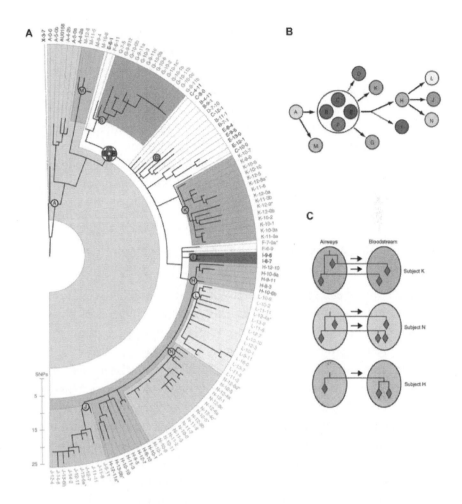

FIGURE 2: Whole genome sequencing reveals within-host evolution and recent transmission between patients. Lieberman, Michel, and colleagues [35] sequenced the genomes of 112 isolates of *Burkholderia dolosa* from 14 cystic fibrosis patients involved in an outbreak in Boston, Massachusetts in the 1990s. (A) The maximum likelihood tree relating the bacterial genomes, color-coded by patient, is broadly consistent with a single founding infection for each patient. (B) The date of sampling and the chronological accumulation of mutations implied a network of transmission events. (C) Interesting patterns emerged when comparing bacteria isolated from different sites in the same patient. For two patients (subjects K and N), multiple genotypes appeared to have been transmitted from the airways to the bloodstream during septicemia, either concurrently or over the course of the infection. By contrast, a single genotype appeared to have been transmitted from the airways to the bloodstream in subject H. Reproduced from [35] appearing in Nature Genetics (Volume 43, 2011).

Multiple transmission events resulting in serial acquisition by a single host can be distinguished from persistent or relapsing infection using whole genome sequencing. This is useful in infections such as invasive nontyphoidal *Salmonella*, a common cause of severe and recurring bloodstream infections among HIV-infected adults in Africa. A study of invasive nontyphoidal *Salmonella* in 14 Malawian patients discriminated recrudescent (i.e., relapsing) infection from multiple infection on the basis of relatedness inferred from genome-wide SNP differences [17]. Recrudescence accounted for 78% of recurring infections, although recrudescence and multiple infection in the same patient was not uncommon.

10.4 HISTORICAL PATTERNS OF TRANSMISSION

In addition to revealing fine-grained genetic structure that is informative about recent transmission, genomics offers unrivalled precision for reconstructing historical patterns of spread. With comprehensive sampling, we can identify the geographical and temporal origin of pandemics and the dominant transmission routes responsible for global dissemination. For example, a study of *Yersinia pestis* used genome sequencing to assist in the discovery of 933 SNPs subsequently typed in 286 global isolates [43]. Based on the diversity and juxtaposition of isolates close to the root of the tree, the authors concluded that the origin of plague more than 2,600 years ago occurred in or near China.

Understanding the circumstances under which epidemics emerge and take hold may help to manage contemporary threats and prevent future outbreaks. The history of the seventh and current pandemic of *Vibrio cholerae* was pieced together using population genomics [44]. An analysis of global isolates revealed three partially overlapping waves of pandemic cholera originating in the Bay of Bengal during the 1950s and leading to a succession of geographically restricted epidemics. Each wave was characterized by a particular armory of genetic elements including distinct forms of the cholera toxin and, from the second wave onwards, SXT/R391 integrative and conjugative elements that confer antibiotic resistance.

Sequencing ancient bacteria is a particularly powerful tool for investigating historical transmission. To reconstruct the history of leprosy, SNPs discovered by whole genome sequencing were typed in over 400 isolates of *Mycobacterium leprae*, including bacteria isolated from skeletal remains recovered from leprosy graveyards in and around Europe [45]. The paleomicrobiological samples resembled modern European isolates, supporting a model in which leprosy arose in East Africa before dispersing east and west by traders along the Silk Road [46]. The provenance of *Yersinia pestis* was investigated by sequencing bacteria isolated from teeth disinterred from the East Smithfield burial ground for Black Death victims in London [47]. The reconstructed genome closely resembled the most recent common ancestor of modern plague in humans, suggesting that the Black Death was the main historical event antecedent to contemporary plague worldwide (Figure 3).

Zoonosis is a major source of emerging infectious disease, with wildlife the most frequent origin [48]. In the United States, leprosy is rare and most infected individuals have a history of foreign residence. Yet a third of cases in Texas and Louisiana had no such explanation [49]. Genome sequencing and SNP typing revealed that a distinctive strain of leprosy was present in 33 wild armadillos and 26 of 29 unexplained human cases, strongly suggesting zoonotic transmission of *Mycobacterium leprae* from wild armadillos.

10.5 POPULATION STRUCTURE, CARRIAGE, AND DISEASE

Many bacterial pathogens cause diseases of varying severity, and some cause no disease at all most of the time (e.g., [32], [33]), constituting a normal part of the body flora [2], [31]. Such observations raise the question of why bacteria cause disease and have led to the notion of the accidental pathogen [50], [51]. Virulence may be accounted for by differences between strains and by the expression of genes encoding toxins, adhesins, and drug resistance, often carried by mobile elements (e.g., [52]–[56]). With whole genome sequencing, populations of virulent and avirulent bacteria can be compared to help explain disease from mechanistic and phylogenetic standpoints. Koch's postulates might be revised [57] to cover the discovery of associations between disease and individual genes or alleles, a process that is likely to accelerate rapidly in the 21st century. A fuller

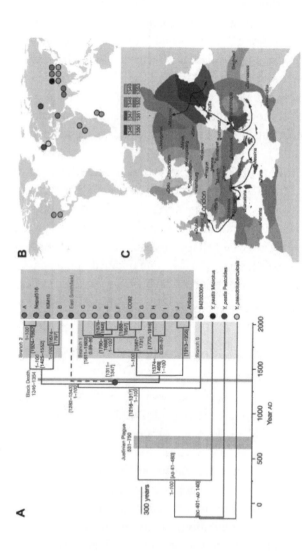

FIGURE 3: Patterns of historical transmission reconstructed by whole genome sequencing. Bos, Schuenemann, and colleagues [47] combined ancient DNA techniques with whole genome sequencing to reconstruct a draft genome of *Yersinia pestis*, the bacterium responsible for the Black Death, from five teeth recovered from a 660-year-old burial ground. (A) Genealogical reconstruction reveals that the bacteria responsible for the Black Death are positioned ancestral to modern Branch 1 *Yersinia pestis*, close to the most recent common ancestor of all modern *Yersinia pestis* pathogenic to humans. No derived mutations were observed in the ancient genome, suggesting that modern Branch 1 bacteria are essentially equivalent, and that differences in modern and 14th century epidemiology probably do not result from genetic changes in the bacteria. (B) Geographical origin of the bacterial isolates. (C) Inferred geographical spread of the Black Death through Europe [80]. Reproduced from [47] appearing in *Nature* (Volume 478, 2011).

understanding of bacterial population structure may also help predict the effects of interventions such as vaccination.

Bacteria of the same species isolated from patients with different clinical presentations can be compared directly by whole genome sequencing. *Streptococcus pyogenes*, also known as Group A *Streptococcus*, causes benign pharyngitis and invasive disease including scarlet fever. A comparison of around 300 isolates from Ontario indicated that invasive bacteria do not form genetically distinct populations. Rather, closely related bacteria may be invasive or pharyngitis-associated, demonstrating that the ability to cause invasive disease is not restricted to specific strains [58], [59]. Evidence for adaptation in genes involved in capsule synthesis and virulence regulation supported a model in which mutation in vivo plays an important role in progression to invasive disease.

Reconstructing the relationships between strains with different clinical manifestations can reveal the evolutionary origins of disease. A study of disparate *Clostridium difficile* isolates including representative members of the hypervirulent lineages denoted ribotypes 017, 027, and 078 found they were descended from multiple ancestors in the species tree, consistent with a scenario in which virulence evolved several times during evolution [60]. Whole genome sequencing afforded improved clarity for reconstructing the genealogy of *Chlamydia trachomatis* and appeared to show that strains causing trachoma, an eye infection, were evolutionarily descended from an ancestor causing urogenital disease [61].

The iatrogenic effect of public health intervention on bacterial pathogen populations, such as the introduction of a novel vaccine or the withdrawal of an antibiotic from agricultural use, is of major importance. Vaccine escape in *Streptococcus pneumoniae* has been a concern since the introduction of the heptavalent conjugate polysaccharide vaccine PCV7 in 2000 as it protects against many, but not all, serotypes. Two genomic studies found evidence for capsular switching, in which hybrid strains normally covered by the vaccine but expressing non-vaccine serotypes arise through recombination [62], [63]. One such strain has quickly established in the United States, spreading westwards from New England [63].

10.6 REAL-TIME PATHOGEN GENOMICS

The relentless demand for higher throughput, lower cost DNA sequencing has spurred dramatic advances in the capacity and rapidity of whole genome sequencing. Benchtop sequencers permit real-time applications of genomics by sequencing small batches of bacteria in a matter of hours. The outbreak of cholera on the island of Haiti in October 2010 provided an early example of the potential for real-time genomics [64]. In the wake of the devastating earthquakes of January 2010 that killed 230,000 people, two million residents were displaced from their homes. Cases were first reported on October 19. By July 2011, 419,511 cases and 5,968 deaths had been reported [65]. Initial investigation found many patients had drunk untreated river water. By November 1, 2010 culturing and PFGE confirmed the pathogen as *Vibrio cholerae* of probable South Asia origin [66]. First-round whole genome sequencing of Haitian isolates began on November 10 and completed within 2 days [64]. Genomic analysis showed they were essentially identical, but distinct from other cholera circulating in Latin America, instead resembling widely circulating Asia strains [64], a finding that was consistent with possible introduction by United Nations Peace-keeping troops dispatched from Nepal following the earthquakes [65].

Real-time genomics may prove particularly valuable in outbreaks involving newly emerged strains. However, processing genomic data in real time poses considerable analytical challenges. In the May 2011 outbreak of *Escherichia coli* in Germany, a novel crowd-sourcing experiment was trialed that foretells of the potential of real-time genomics to radically alter the way outbreaks are investigated [67]. The large outbreak was unusual in several aspects: high incidence in adults, greatly increased incidence of hemolytic-uremic syndrome, a preponderance of female patients, and a rare, Shiga toxin-producing serotype not previously linked to outbreaks [67], [68]. A first draft of the genome of an isolate sampled on May 17 was completed within 3 days, then released into the public domain, eliciting curiosity-driven analysis by scientists on four continents [67]. Within a week two dozen reports had been filed on a dedicated open-source wiki.

Analysis of this and other strains concluded that the outbreak was caused by the acquisition of a Shiga toxin-encoding prophage and a plasmid bearing an extended-spectrum beta-lactamase gene by an ancestral enteroaggregative strain [67], [68]. The striking virulence of the hybrid may be connected to the atypical presence of three SPATE genes, which are implicated in mucosal damage and intestinal colonization [68].

Two studies from hospitals in the United Kingdom have demonstrated the practical advantages of real-time whole genome sequencing as part of routine outbreak investigation and surveillance. Focusing on the most serious health-care-associated pathogens, *C. difficile* [69] and *S. aureus* [69], [70], bacterial samples were isolated from suspected outbreaks in four hospitals. The genomes were sequenced and analyzed within 5 working days of culture, confirming the suspected outbreaks of MRSA (methicillin-resistant *S. aureus*), but demonstrating that the epidemiologically linked cases of *C. difficile* infection were in fact genetically distinct. Characterization of the repertoire of resistance and toxin genes provided further information relevant to patient management.

10.7 SUMMARY AND PERSPECTIVES

In the future, population genomics will be central to an improved understanding of the epidemiology, etiology, and evolution of bacterial infectious diseases. However, there are obstacles yet to overcome. Pilot studies have demonstrated the potential genomics has for epidemiological investigation [40], [41], [64], [67]–[70], but creative solutions to the problem of integrating complex epidemiological and genomic data are now required. Currently, genome sequencing relies on culture to yield sufficient bacterial DNA and new technologies are needed to overcome this dependency. If the cost of DNA library preparation can be substantially reduced, genomics will come within reach of public health authorities as a tool for routine surveillance. These and other future challenges are discussed in Box 1.

BOX 1: FUTURE CHALLENGES FOR PATHOGEN WHOLE GENOME SEQUENCING

High-throughput whole genome sequencing has been demonstrated to be a practical tool for epidemiological and evolutionary investigation of bacterial pathogens, yet the current technology has certain limitations. The challenge for future advances in sequencing technology is to overcome these problems.

- **Culture**. Reliably sequencing the genomes of individual bacteria requires culture to obtain sufficient quantities of concentrated DNA. This takes time and effort, restricts the approach to culturable organisms, and may introduce artifacts such as in vitro mutation and laboratory cross-contamination. Direct sequencing without culture (e.g., [81]) may in the future relinquish this dependency on culture, but metagenomics approaches present additional challenges for bioinformatics and sequence analysis (see, e.g., [82]).

- **Library preparation**. Exponential increases in the capacity of high-throughput sequencers show no sign of abating. In principle, this should allow the cost of bacterial whole genome sequencing to continue to fall. However, the price per genome also depends on the cost of DNA library preparation, comprising both consumables and labor. Advances in automation and throughput will be required to prevent library preparation becoming a bottleneck, and to reduce the cost sufficiently that bacterial genome sequencing becomes affordable for routine surveillance.

- **Bioinformatics.** The development of bespoke bioinformatics pipelines for bacterial whole genome sequencing represents a considerable investment and a complex set of choices from among the many computational methods on offer. Some degree of normalization is required to ease the burden on users of whole genome sequencing, for example hospital microbiology laboratories, and to promote standardized and replicable workflows.

BOX 1: *CONT.*

- **Platforms.** High-throughput sequencing technologies yield large quantities of short read sequences but with substantially elevated error rates, compared to conventional capillary sequencing. The details of sequence length and error profile differ in important ways between platforms. In consequence, different results may be obtained when the same sample is sequenced on different platforms. Improved understanding of the error profiles of different architectures combined with efforts towards quantifying uncertainty in the DNA sequences generated will help minimize discrepancies of this kind.

- **Genome assembly.** De novo assembly is used to join together the short reads of DNA generated by the sequencing machines into longer genome fragments, known as contigs. The ultimate goal is to join all the fragments into a single contig representing the whole bacterial chromosome, known as a closed genome. However, variation in the number of reads sequenced from each part of the genome (the depth of coverage), and the existence of repetitive regions, conspire to prevent this. With longer reads, it should be possible to overcome these problems.

- **Public databases.** To accelerate the pace of discovery and assist collaboration between laboratories, well-organized publicly available databases are required from which bacterial genomes are readily downloaded in convenient formats. Raw data are currently available in short read archives (e.g., http://www.ncbi.nlm.nih.gov/sra and http://www.ebi.ac.uk/ena), but with standardization of bioinformatics processing it should become possible to provide pre-processed data which would dramatically reduce the workload for database users.

Improved understanding of disease etiology helps to direct research into therapies. Genomics is a promising tool for investigating the differences between invasive and non-invasive bacteria at the population and within-host levels [27], [59]–[61]. Tools from human genetics may help in this endeavor. Even so, investigations into bacterial population structure are required to assess the feasibility of genome-wide association studies [71]. Understanding the architecture of traits such as virulence would benefit from the development of high-throughput phenotyping assays. RNA sequencing is one such candidate [72], but differences in gene expression in culture and in vivo are a potential impediment to progress.

Population genomics also promises to improve our understanding of bacterial pathogen evolution. The resolution of whole genome sequencing allows precise calibration of evolutionary rates from longitudinal samples within populations and individual hosts [16]–[18], [28], [35], [40], [47]. This permits the origin of new species to be dated, but the discrepancy between short- and long-term rates requires further explanation [23], [30]. Investigating within-host dynamics will help identify the evolutionary mechanisms involved in disease progression [27], [35]. Sequencing populations of pathogens will reveal the prevalence of virulence factors and drug resistance, and the role of mobile elements in their spread [44]. Ultimately, however, we must pinpoint the evolutionary advantages that bacteria gain from inflicting illnesses if we are to fully understand the causes of bacterial disease.

REFERENCES

1. Whitman WB, Coleman DC, Wiebe WJ (1998) Prokaryotes: the unseen majority. Proc Natl Acad Sci U S A 95: 6578–6583. doi: 10.1073/pnas.95.12.6578
2. Qin J, Li R, Raes J, Arumugam M, Burgdorf KS, et al. (2010) A human gut microbial gene catalogue established by metagenomic sequencing. Nature 464: 59–65. doi: 10.1038/nature08821
3. World Health Organization (2008) The global burden of disease: 2004 update. Available: http://www.who.int/healthinfo/global_burden_disease. Accessed 10 August 2012.
4. World Health Organization (2012) Global invasive bacterial vaccine preventable diseases (IB-VPD) information and surveillance bulletin. Volume 5. Available: http://www.who.int/nuvi/surveillance/resources/en/index.html. Accessed 10 August 2012.

5. Davies J, Davies D (2010) Origins and evolution of antibiotic resistance. Microbiol Mol Biol Rev 74: 417–433. doi: 10.1128/MMBR.00016-10
6. Livermore DM (2012) Fourteen years in resistance. Int J Antimicrob Agents 39: 283–294. doi: 10.1016/j.ijantimicag.2011.12.012
7. Nordmann P, Poirel L, Walsh TR, Livermore DM (2011) The emerging NDM carbapenemases. Trends Microbiol 19: 588–595. doi: 10.1016/j.tim.2011.09.005
8. Schmieder R (2012) Insights into antibiotic resistance through metagenomics approaches. Future Microbiol 7: 73–89. doi: 10.2217/fmb.11.135
9. Maiden MC, Bygraves JA, Feil E, Morelli G, Russell JE, et al. (1998) Multilocus sequence typing: a portable approach to the identification of clones within populations of pathogenic microorganisms. Proc Natl Acad Sci U S A 95: 3140–3145. doi: 10.1073/pnas.95.6.3140
10. Urwin R, Maiden MC (2003) Multi-locus sequence typing: a tool for global epidemiology. Trends Microbiol 11: 479–487. doi: 10.1016/j.tim.2003.08.006
11. Grenfell BT, Pybus OG, Gog JR, Wood JL, Daly JM, et al. (2004) Unifying the epidemiological and evolutionary dynamics of pathogens. Science 303: 327–332. doi: 10.1126/science.1090727
12. Pybus OG, Rambaut A (2009) Evolutionary analysis of the dynamics of viral infectious disease. Nat Rev Genet 10: 540–550. doi: 10.1038/nrg2583
13. Connor RI, Sheridan KE, Ceradini D, Choe S, Landau NR (1997) Change in coreceptor use correlates with disease progression in HIV-1–infected individuals. J Exp Med 185: 621–628. doi: 10.1084/jem.185.4.621
14. Lemey P, Kosakovsky Pond SL, Drummond AJ, Pybus OG, Shapiro B, et al. (2007) Synonymous substitution rates predict HIV disease progression as a result of underlying replication dynamics. PLoS Comput Biol 3: e29 doi:10.1371/journal.pcbi.0030029.
15. Drake JW, Charlesworth B, Charlesworth D, Crow JF (1998) Rates of spontaneous mutation. Genetics 148: 1667–1686.
16. Ford CB, Lin PL, Chase MR, Shah RR, Iartchouk O, et al. (2011) Use of whole genome sequencing to estimate the mutation rate of *Mycobacterium tuberculosis* during latent infection. Nat Genet 43: 482–486. doi: 10.1038/ng.811
17. Okoro CK, Kingsley RA, Quail MA, Kankwatira AM, Feasey NA, et al. (2012) High-resolution single nucleotide polymorphism analysis distinguishes recrudescence and reinfection in recurrent invasive nontyphoidal *Salmonella* typhimurium disease. Clin Infect Dis 54: 955–963. doi: 10.1093/cid/cir1032
18. McAdam PR, Holmes A, Templeton KE, Fitzgerald JR (2011) Adaptive evolution of *Staphylococcus aureus* during chronic endobronchial infection of a cystic fibrosis patient. PLoS ONE 6: e24301 doi:10.1371/journal.pone.0024301.
19. Touzain F, Denamur E, Médigue C, Barbe V, El Karoui M, et al. (2010) Small variable segments constitute a major type of diversity of bacterial genomes at the species level. Genome Biol 11: R45. doi: 10.1186/gb-2010-11-4-r45
20. Ochman H, Wilson AC (1987) Evolution in bacteria: evidence for a universal substitution rate in cellular genomes. J Mol Evol 26: 74–86. doi: 10.1007/BF02111283
21. Moran NA, Munson MA, Baumann P, Ishikawa H (1993) A molecular clock in endosymbiotic bacteria is calibrated using insect hosts. Proc R Soc Lond B 253: 167–171. doi: 10.1098/rspb.1993.0098

22. Lenski RE, Winkworth CL, Riley MA (2003) Rates of DNA sequence evolution in experimental populations of *Escherichia coli* during 20,000 generations. J Mol Evol 56: 498–508. doi: 10.1007/s00239-002-2423-0

23. Ochman H (2003) Neutral mutations and neutral substitutions in bacterial genomes. Mol Biol Evol 20: 2091–2096. doi: 10.1093/molbev/msg229

24. Falush D, Kraft C, Taylor NS, Correa P, Fox JG, et al. (2001) Recombination and mutation during longterm gastric colonization by *Helicobacter pylori*: estimates of clock rates, recombination size, and minimal age. Proc Natl Acad Sci U S A 98: 15056–15061. doi: 10.1073/pnas.251396098

25. Pérez-Losada M, Crandall KA, Zenilman J, Viscidi RP (2007) Temporal trends in gonococcal population genetics in a high prevalence urban community. Infect Genet Evol 7: 271–278. doi: 10.1016/j.meegid.2006.11.003

26. Wilson DJ, Gabriel E, Leatherbarrow AJH, Cheesbrough J, Gee S, et al. (2009) Rapid evolution and the importance of recombination to the gastroenteric pathogen Campylobacter jejuni. Mol Biol Evol 26: 385–397. doi: 10.1093/molbev/msn264

27. Young BC, Golubchik T, Batty EM, Fung R, Larner-Svennson H, et al. (2012) Evolutionary dynamics of *Staphylococcus aureus* during progression from carriage to disease. Proc Natl Acad Sci U S A 109: 4550–4555. doi: 10.1073/pnas.1113219109

28. Kennemann L, Didelot X, Aebischer T, Kuhn S, Drescher B, et al. (2011) *Helicobacter pylori* evolution during human infection. Proc Natl Acad Sci U S A 108: 5033–5038. doi: 10.1073/pnas.1018444108

29. McDonald JH, Kreitman M (1991) Adaptive protein evolution at the Adh locus in Drosophila. Nature 351: 652–654. doi: 10.1038/351652a0

30. Rocha EPC, Maynard Smith J, Hurst LD, Holden MTG, Cooper JE, et al. (2006) Comparisons of dN/dS are time dependent for closely related bacterial genomes. J Theor Biol 239: 226–235. doi: 10.1016/j.jtbi.2005.08.037

31. Turnbaugh PJ, Ley RE, Hamady M, Fraser-Liggett CM, Knight R, et al. (2007) The human microbiome project. Nature 449: 804–810. doi: 10.1038/nature06244

32. Wertheim HF, Melles DC, Vos MC, van Leeuwen W, van Belkum A, et al. (2005) The role of nasal carriage in *Staphylococcus aureus* infections. Lancet Infect Dis 5: 751–762. doi: 10.1016/S1473-3099(05)70295-4

33. von Eiff C, Becker K, Machka K, Stammer H, Peters G (2001) Nasal carriage as a source of *Staphylococcus aureus* bacteremia. N Engl J Med 344: 11–16. doi: 10.1056/NEJM200101043440102

34. Yang J, Tauschek M, Robins-Browne RM (2011) Control of bacterial virulence by AraC-like regulators that respond to chemical signals. Trends Microbiol 19: 128–135. doi: 10.1016/j.tim.2010.12.001

35. Lieberman TD, Michel JB, Aingaran M, Potter-Bynoe G, Roux D, et al. (2011) Parallel bacterial evolution within multiple patients identifies candidate pathogenicity genes. Nat Genet 43: 1275–1280. doi: 10.1038/ng.997

36. Smith JM, Smith NH, O'Rourke M, Spratt BG (1993) How clonal are bacteria? Proc Natl Acad Sci U S A 90: 4384–4388. doi: 10.1073/pnas.90.10.4384

37. Weidenbeck J, Cohan FM (2011) Origins of bacterial diversity through horizontal genetic transfer and adaptation to new ecological niches. FEMS Microbiol Rev 5: 957–976. doi: 10.1111/j.1574-6976.2011.00292.x

38. Hiller NL, Ahmed A, Powell E, Martin DP, Eutsey R, et al. (2010) Generation of genic diversity among *Streptococcus pneumoniae* strains via horizontal gene transfer during a chronic polyclonal pediatric infection. PLoS Pathog 6: e1001108 doi:10.1371/journal.ppat.1001108.

39. Budowle B, Schutzer SE, Breeze RG, Keim PS, Morse SA (2011) editors (2011) Microbial forensics, second edition. Elsevier/Academic Press doi: 10.1111/j.1574-6976.2011.00292.x

40. Harris SR, Feil EJ, Holden MTG, Quail MA, Nickerson EK, et al. (2010) Evolution of MRSA during hospital transmission and intercontinental spread. Science 327: 469–474. doi: 10.1126/science.1182395

41. Gardy JL, Johnston JC, Ho Sui SJ, Cook VJ, Shah L, et al. (2011) Whole-genome sequencing and social-network analysis of a tuberculosis outbreak. N Engl J Med 364: 730–739. doi: 10.1056/NEJMoa1003176

42. Reeves PR, Liu B, Zhou Z, Li D, Guo D, et al. (2011) Rates of mutation and host transmission for an *Escherichia coli* clone over 3 years. PLoS ONE 6: e26907 doi:10.1371/journal.pone.0026907.

43. Morelli G, Song Y, Mazzoni CJ, Eppinger M, Roumagnac P, et al. (2010) *Yersinia pestis* genome sequencing identifies patterns of global phylogenetic diversity. Nat Genet 42: 1140–1143. doi: 10.1038/ng.705

44. Mutreja A, Kim DW, Thomson NR, Connor TR, Lee JH, et al. (2011) Evidence for several waves of global transmission in the seventh cholera pandemic. Nature 477: 462–465. doi: 10.1038/nature10392

45. Monot M, Honoré N, Garnier T, Zidane N, Sherafi D, et al. (2009) Comparative genomic and phylogeographic analysis of *Mycobacterium leprae*. Nat Genet 41: 1282–1291. doi: 10.1038/ng.477

46. Monot M, Honoré N, Garnier T, Araoz R, Coppée JY, et al. (2005) On the origin of leprosy. Science 308: 1040–1042. doi: 10.1126/science.1109759

47. Bos KI, Schuenemann VJ, Golding GB, Burbano HA, Waglechner N, et al. (2011) A draft genome of *Yersinia pestis* from victims of the Black Death. Nature 478: 506–510. doi: 10.1038/nature10549

48. Jones KE, Patel NG, Levy MA, Storeygard A, Balk D, et al. (2008) Global trends in emerging infectious diseases. Nature 451: 990–993. doi: 10.1038/nature06536

49. Truman RW, Singh P, Sharma R, Busso P, Rougemont J, et al. (2011) Probable zoonotic leprosy in the Southern United States. N Engl J Med 364: 1626–1633. doi: 10.1056/NEJMoa1010536

50. Moxon ER, Jansen VAA (2005) Phage variation: understanding the behaviour of an accidental pathogen. Trends Microbiol 13: 563–565. doi: 10.1016/j.tim.2005.10.004

51. Nandi T, Ong C, Singh AP, Boddey J, Atkins T, et al. (2010) A genomic survey of positive selection in Burkholderia pseudomallei provides insights into the evolution of accidental virulence. PLoS Pathog 6: e1000845 doi:10.1371/journal.ppat.1000845.

52. Peacock SJ, Moore CE, Justice A, Kantzanou M, Story L, et al. (2002) Virulent combinations of adhesin and toxin genes in natural populations of Staphylococcus aureus. Infect Immun 70: 4987–4996. doi: 10.1128/IAI.70.9.4987-4996.2002

53. Melles DC, Gorkink RF, Boelens HA, Snijders SV, Peeters JK, et al. (2004) Natural population dynamics and expansion of pathogenic clones of Staphylococcus aureus. J Clin Invest 114: 1732–1740. doi: 10.1172/JCI23083

54. Diep BA, Carleton HA, Chang RF, Sensabaugh GF, Perdreau-Remington F (2006) Roles of 34 virulence genes in the evolution of hospital- and community-associated strains of methicillin-resistant Staphylococcus aureus. J Infect Dis 193: 1495–1503. doi: 10.1086/503777

55. Lindsay JA, Moore CE, Day NP, Peacock SJ, Witney AA, et al. (2006) Microarrays reveal that each of the ten dominant lineages of *Staphylococcus aureus* has a unique combination of surface-associated and regulatory genes. J Bacteriol 188: 669–676. doi: 10.1128/JB.188.2.669-676.2006

56. Malachowa N, DeLeo FR (2010) Mobile genetic elements of Staphylococcus aureus. Cell Mol Life Sci 67: 3057–3071. doi: 10.1007/s00018-010-0389-4

57. Fredricks DN, Relman DA (1996) Sequence-based identification of microbial pathogens: a reconsideration of Koch's postulates. Clin Microb Rev 9: 18–33.

58. Beres SB, Carroll RK, Shea PR, Sitkiewicz I, Martinez-Gutierrez JC, et al. (2010) Molecular complexity of successive bacterial epidemics deconvoluted by comparative pathogenomics. Proc Natl Acad Sci U S A 107: 4371–4376. doi: 10.1073/pnas.0911295107

59. Shea PR, Beres SB, Flores AR, Ewbank AL, Gonzalez-Lugo JH, et al. (2011) Distinct signatures of diversifying selection revealed by genome analysis of respiratory tract and invasive bacterial populations. Proc Natl Acad Sci U S A 108: 5039–5044. doi: 10.1073/pnas.1016282108

60. He M, Sebaihia M, Lawley TD, Stabler RA, Dawson LF, et al. (2010) Evolutionary dynamics of *Clostridium difficile* over short and long time scales. Proc Natl Acad Sci U S A 107: 7527–7532. doi: 10.1073/pnas.0914322107

61. Harris SR, Clarke IN, Seth-Smith HMB, Solomon AW, Cutcliffe LT, et al. (2012) Whole-genome analysis of diverse *Chlamydia trachomatis* strains identifies phylogenetic relationships masked by current clinical typing. Nat Genet 44: 413–420. doi: 10.1038/ng.2214

62. Croucher NJ, Harris SR, Fraser C, Quail MA, Burton J, et al. (2011) Rapid pneumococcal evolution in response to clinical interventions. Science 331: 430–434. doi: 10.1126/science.1198545

63. Golubchik T, Brueggemann AB, Street T, Gertz Jr RE, Spencer CCA, et al. (2012) Pneumococcal genome sequencing tracks a vaccine escape variant formed through a multi-fragment recombination event. Nat Genet 44: 352–356. doi: 10.1038/ng.1072

64. Chin CS, Sorenson J, Harris JB, Robins WP, Charles RC, et al. (2011) The origin of the Haitian cholera outbreak strain. N Engl J Med 364: 33–42. doi: 10.1056/NEJMoa1012928

65. Tappero JW, Tauxe RV (2011) Lessons learned during public health response to cholera epidemic in Haiti and the Dominican Republic. Emerg Infect Dis 17: 2087–2093. doi: 10.3201/eid1711.110827

66. Centers for Disease Control and Prevention. Press release: laboratory test results of cholera outbreak strain in Haiti announced (1st November 2010).

67. Rohde H, Qin J, Cui Y, Dongfang L, Loman NJ, et al. (2011) Open-source genomic analysis of shiga-toxin-producing E. coli O104:H4. N Engl J Med 365: 718–724. doi: 10.1056/NEJMoa1107643

68. Rasko DA, Webster DR, Sahl JW, Bashir A, Boisen N, et al. (2011) Origins of the E. coli strain causing an outbreak of hemolytic-uremic syndrome in Germany. N Engl J Med 365: 709–717. doi: 10.1056/NEJMoa1106920

69. Eyre DW, Golubchik T, Gordon NC, Bowden R, Piazza P, et al. (2012) A pilot study of rapid benchtop sequencing of *Staphylococcus aureus* and *Clostridium difficile* for outbreak detection and surveillance. BMJ Open 2: e001124. doi: 10.1136/bmjopen-2012-001124

70. Köser CU, Holden MTG, Ellington MJ, Cartwright EJP, Brown NM, et al. (2012) Rapid whole-genome sequencing for investigation of a neonatal MRSA outbreak. N Engl J Med 366: 2267–2275. doi: 10.1056/NEJMoa1109910

71. Spencer CCA, Su Z, Donnelly P, Marchini J (2009) Designing genome-wide association studies: sample size, power, imputation and the choice of genotyping chip. PLoS Genet 5: e1000477 doi:10.1371/journal.pgen.1000477.

72. Pickrell JK, Marioni JC, Pai AA, Degner JF, Engelhardt BE, et al. (2010) Understanding mechanisms underlying expression variation with RNA sequencing. Nature 464: 768–772. doi: 10.1038/nature08872

73. Li H, Durbin R (2009) Fast and accurate short read alignment with Burrows-Wheeler Transform. Bioinformatics 25: 1754–1760. doi: 10.1093/bioinformatics/btp324

74. Lunter G, Goodson M (2011) Stampy: a statistical algorithm for sensitive and fast mapping of Illumina sequence reads. Genome Res 21: 936–939. doi: 10.1101/gr.111120.110

75. Zerbino DR, Birney E (2008) Velvet: algorithms for de novo short read assembly using de Bruijn graphs. Genome Res 18: 821–829. doi: 10.1101/gr.074492.107

76. Iqbal Z, Caccamo M, Turner I, Flicek P, McVean G (2012) De novo assembly and genotyping of variants using colored de Bruijn graphs. Nat Genet 44: 226–232. doi: 10.1038/ng.1028

77. Thompson JD, Gibson TJ, Plewniak F, Jeanmougin F, Higgins DG (1997) The CLUSTAL_X windows interface: flexible strategies for multiple sequence alignment aided by quality analysis tools. Nucl Acids Res 25: 4876–4882. doi: 10.1093/nar/25.24.4876

78. Kurtz S, Phillipy AL, Delcher M, Smoot M, Shumway M, et al. (2004) Versatile and open source software for comparing large genomes. Genome Biol 5: R12. doi: 10.1186/gb-2004-5-2-r12

79. Darling AE, Mau B, Perna NT (2010) ProgressiveMauve: multiple genome alignment with gene gain, loss, and rearrangement. PLoS ONE 5: e11147. doi: 10.1371/journal.pone.0011147

80. Benedictow OJ (2004) The Black Death 1346–1353: the complete history. Boydell Press.

81. Denef VJ, Banfield JF (2012) In situ evolutionary rate measurements show ecological success of recently emerged bacterial hybrids. Science 336: 462–466. doi: 10.1126/science.1218389

82. Chen K, Pachter L (2005) Bioinformatics for whole-genome shotgun sequencing of microbial communities. PLoS Comput Biol 1: e24 doi:10.1371/journal.pcbi.0010024.
83. World Health Organization (2008) Projections of mortality and burden of disease 2004–2030, baseline scenario. Available: http://www.who.int/healthinfo/global_burden_disease/projections/en/index.html. Accessed 12 April 2012.
84. World Health Organization (2011) World Health Organization Mortality Database, ICD-10. 24 November 2011 update. Available: http://www.who.int/whosis/mort/download/en/index.html. Accessed 12 April 2012.
85. Getahun H, Gunneberg C, Granich R, Nunn P (2010) HIV infection-associated tuberculosis: the epidemiology and the response. Clin Infect Dis 50: Suppl 3 S201–S207. doi: 10.1086/651492

CHAPTER 11

HIGH THROUGHPUT SEQUENCING AND PROTEOMICS TO IDENTIFY IMMUNOGENIC PROTEINS OF A NEW PATHOGEN: THE DIRTY GENOME APPROACH

GILBERT GREUB, CAROLE KEBBI-BEGHDADI, CLAIRE BERTELLI, FRANÇOIS COLLYN, BEAT M. RIEDERER, CAMILLE YERSIN, ANTONY CROXATTO, AND DIDIER RAOULT

11.1 INTRODUCTION

The recent availability of new generation sequencing technologies [1], [2] has provided unprecedented sequencing capacity, enabling the acquisition of genome-scale sequences at an extraordinary fast rate. These innovative techniques provide amazing opportunities for high-throughput structural and functional genomic researches and have been applied to date to a variety of contexts such as whole-genome sequencing [3] and resequencing [4], targeted resequencing [5], non coding RNA [6] or DNA-binding of modified histones [7], [8]. These high-throughput sequencing methods avoid the need for in vivo cloning and achieve a high accuracy. Even homopolymer problems, i.e. the major drawback of 454 pyrosequencing,

This chapter was originally published under the Creative Commons Attribution License. Greub G, Kebbi-Beghdadi C, Bertelli C, Collyn F, Riederer BM, Yersin C, Croxatto A, and Raoult D. High Throughput Sequencing and Proteomics to Identify Immunogenic Proteins of a New Pathogen: The Dirty Genome Approach. PLoS ONE 4,12 (2009). doi:10.1371/journal.pone.0008423.

may be overcome by reaching high sequence coverage [1]. These new technologies greatly reduce the work, time and expenses of such projects.

However, the relative short read length makes genome assembly problematic and their use in bacterial genomics has been fairly restricted to new strains closely related to already sequenced organisms to identify for example virulence factors [9], antibiotic resistance genes [10], or epidemiological markers [11]. Although improved techniques can now achieve paired-read information and longer reads [12], genomes still need a costly and time-consuming gap closure step, especially when containing highly repetitive elements such as transposases and recombination hot spots.

Still, complete genomic information is not necessarily needed and incomplete genome data obtained using high-throughput sequencing methods may potentially be informative enough to derive the pursued information. Moreover, the low time to results of such approaches (about 15 weeks [9]) is especially useful when genomic information are readily needed for instance in case of outbreaks (i) to search for the presence of specific pathogenicity island or virulence genes, (ii) to identify specific or multicopy gene targets in order to rapidly develop a reliable molecular diagnosis test, and (iii) to identify immunogenic proteins to set up a diagnostic tool for sero-epidemiological investigations or to develop a vaccine.

This strategy is particularly interesting for obligate intracellular bacteria such as members of the *Chlamydiales* order that lack a genetic manipulation system and only replicate within eukaryotic cells of different origins including humans, animals and amoebae [13]. One of them, *Parachlamydia acanthamoebae* strain Hall's coccus, was initially isolated from the water of an humidifier at the origin of a fever outbreak [14], and since then some evidences have accumulated suggesting the role of this species as an emerging human respiratory pathogen [15]. An emerging pathogen refers here to an agent that has been recently identified as pathogenic. Indeed, several serological and molecular studies supported a role of *P. acanthamoebae* in patients with community-acquired and aspiration pneumonia [16], [17], [18]. *P. acanthamoebae* also appeared to possibly cause bronchiolitis in children [19]. Moreover, pneumonia has been reproduced in a murine model following intranasal and intratracheal inoculation with *P. acanthamoebae* [20], [21]. Finally, the ability of *Parachlamydia* to resist to human macrophages [22], [23] further supported its human pathogenicity. Besides, the role of *P. acanthamoebae* in bovine abortion has been clearly demonstrated since the bacteria was detected by PCR, immunohistochemistry and electron microscopy in the placenta of aborted bovines [24]. The pathogenic potential of *Parachlamydia* towards humans and animals still remains largely unexplored since this strict intracellular bacterium does not grow on media routinely used for the detection of pathogens. To date, there are only few strains of *Parachlamydia acanthamoebae* available worldwide. Moreover, little information is available about strains from cattle and other animals, since no *Parachlamydia* strains have been isolated from animal samples by cell culture. It is thus important to develop new diagnostic approaches for *P. acanthamoebae* to better understand its epidemiological and pathogenic potentials in various human and animal diseases.

In this work, we undertook a proof of principle project that investigated the feasibility of combining genomic and proteomic approaches to rapidly identify immunogenic proteins. We showed that, even with relatively short reads from Genome Sequencer 20 (GS20) and after homopolymer correction through Solexa, we can gather almost the whole genome sequence of an emerging pathogen, allowing to analyze the proteome and to elaborate the first steps of an ELISA test, thus enabling to further evaluate its pathogenic role.

11.2 RESULTS

11.2.1 GENOME SEQUENCING

The pyrosequencing of *P. acanthamoebae* genomic DNA by two runs of GS20 yielded 566'453 reads of an average length of 111 nucleotides. In order to correct eventual frameshifts due to homopolymer errors, genomic DNA was also sequenced with Solexa technology, which produced 1'655'941 short reads of 36 bp that could be assembled in 8616 contigs. The latters were assembled with GS20 reads in 95 contigs larger than 500 bp, with a N50 size of 101'998 bp. The coverage obtained with 454 reads was of 17x whereas that obtained with Solexa reads was of 12x. The 95 contigs represents approximately 97% of the total genome and as many as 99.99% of all the non-repeated regions, i.e. when excluding contigs exhibiting a sequence depth higher than 30x with 454. As indicated by the total length of the contigs, the complete genome of *P. acanthamoebae* Hall's coccus stands around 3 Mb and was predicted to contain 4798 open reading frames larger than 90 nucleotides. More than 91% of the large contigs were covered with Solexa. The 1037 differences between Solexa and 454 were manually inspected. As many as 405 differences could be attributed to the presence of homopolymers and were corrected according to Solexa whereas the remaining 632 differences were mainly due to inaccurate Solexa contig ends and were not corrected.

11.2.2 IDENTIFICATION OF IMMUNOREACTIVE PROTEINS

To identify immunoreactive proteins that could be used in a diagnostic test, total proteins of *P. acanthamoebae* elementary bodies were separated by 2D gel electrophoresis and either Coomassie blue-stained or transferred onto nitrocellulose membranes. Immunoblots were performed with sera of rabbits immunized with *P. acanthamoebae* and with human *P. acanthamoebae* positive sera (Fig. 1A,C). Spots corresponding to immunogenic proteins reacting with at least one rabbit anti-*Parachlamydia* serum were selected by computer-assisted matching of the Coomassie blue-stained gel and immunoblots, and further analyzed by mass spectrometry. Eighteen different proteins were identified (Fig. 2A), out of which 5 reacted only against sera from immunized rabbits and 13 reacted with both rabbit and human *Parachlamydia* positive sera. Some of these proteins, such as chaperonin GroEL (Hsp60), DnaK (Hsp70), elongation factor Tu and the ribosomal proteins S1 and L7/L12, were already known to be antigenic [25] (Fig. 2B and Table S1). Some classical *Chlamydiales* immunogenic proteins, such as 60 kDa cysteine-rich OMP, LcrE or CPAF protease [25], [26], [27] were not detected. Since the corresponding genes were found in our contig assembly, these proteins are likely poorly expressed in elementary bodies or when the bacteria are co-cultivated in amoebae. Membranes were also probed with control human sera, i.e. either completely negative

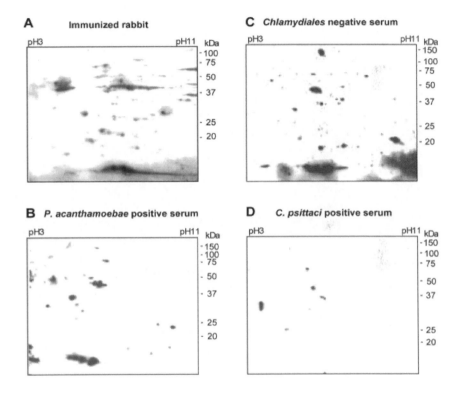

FIGURE 1: 2D patterns of the immunoreactive proteins of *P. acanthamoebae*. Proteins of *P. acanthamoebae* separated by 2D gel electrophoresis were probed with (A) serum from immunized rabbit #1, (B) a *Chlamydiales* negative human serum, (C) a *P. acanthamoebae* positive human serum, and, (D) a C. psittaci positive human serum. Five immunogenic proteins are numbered in reference to the following figures.

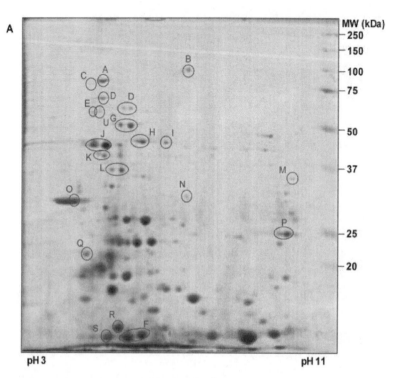

A

pH 3   pH 11

B

	Spot number	Best blast against *Chlamydiales* genomes
Good candidates	E	No significant similarity found
	N	No significant similarity found
Possible candidates	B	Superoxide dismutase
	M	NAD(P)H-dependent glycerol-3-phosphate dehydrogenase
	Q	Probable yciF protein
Poor candidates	A*°	Molecular chaperone DnaK
	C*°	30S ribosomal protein S1
	D/T/U*°	Chaperonin GroEL
	F*°	Probable 50S ribosomal protein L7/L12
	G°	Putative serine proteinase
	H*°	Elongation factor Tu
	I°	No significant similarity found
	J°	Putative hyperosmotically inducible periplasmic protein
	K°	Hypothetical protein pc1399
	L*°	Putative elongation factor Ts
	O°	Putative mip
	P°	Hypothetical protein pc0946
	R/S*	Co-chaperonin GroES

* High similarity with proteins of other species
° Cross reaction with *C. pneumoniae*, *C. psittaci* or negative controls

FIGURE 2: 2D map and identification of *P. acanthamoebae* immunogenic proteins. A. Proteins reacting with at least one rabbit anti-*Parachlamydia* serum were excised from gel and analysed by MALDI TOF mass spectrometry. Spots successfully identified are numbered. Molecular mass standards are indicated on the right side of the gel. B. The potential of 18 immunogenic proteins for use in a serological diagnostic test was evaluated based on their reactivity with control sera and on their sequence similarity in BLASTP results (see Table S2 for detailed analysis).

for any member of the *Chlamydiales* order (Fig. 1B), or positive only for *C. psittaci* (Fig. 1D) or *C. pneumoniae* (see Table S2). Based on 2D immunoblots with control sera and blast analysis of the MS identified proteins, the best candidates for a diagnostic assay of *P. acanthamoebae* infection were determined (see Table S2). Antigens displaying a high sequence homology with similar proteins in other species as well as proteins cross-reacting with non specific or negative sera were discarded. The two best candidate proteins were selected for evaluation in an ELISA test (Fig. 2B).

11.2.3 WESTERN-BLOT AND ELISA OF PROTEINS E AND N

The *Parachlamydial* protein E and N, which have no sequence homology with any known protein, were chosen to develop serological diagnostic tools. Recombinant proteins E (MW ~58 kDA) and N (MW ~30 kDA) were expressed in *E. coli* and purified thanks to a 6His tail fused to their N-terminal end. The purified recombinant proteins were detected by western blot with a rabbit anti-*Parachlamydia* serum or with a human *P. acanthamoebae* positive serum (Fig. 3A,C). Lower molecular weight bands also visible on these blots probably correspond to degradation products. Moreover, faint bands were detectable when protein E was probed with a *C. psittaci* positive serum indicating a low level of cross reaction with this organism. For both proteins, no signal was obtained when *Chlamydiales* negative or *C. pneumoniae* positive sera were tested.

When used as antigen in a direct ELISA assay, purified proteins E and N were detected by the sera of two rabbits immunized with *P. acanthamoebae* until a dilution value of 1/256 while only background reaction is observed with pre-immune sera at this and lower dilutions (Fig. 3B,D). For both proteins, a significant difference was observed in the level of reactivity of the two sera. However, both rabbit sera exhibited good antigen reactivity when tested by western-blot. Overall, these data demonstrate the potential of immunogenic proteins E and N for serological diagnostic tests that could be developed in the future.

11.2.4 COMPARISON BETWEEN COMBINED OR SEPARATED 454 AND SOLEXA APPROACHES TO IDENTIFY PROTEINS

In addition to identifying immunogenic proteins, the most abundantly expressed proteins of *P. acanthamoebae* elementary bodies were also analyzed. A total of 95 Coomassie blue-stained spots were analyzed by mass spectrometry and a reliable protein identification was obtained for 85 of them using the combined GS20 and Solexa sequences. Identification failed for 2 proteins due to the absence of signal by mass spectrometry and for 8 proteins due to the absence of hits in the genome-derived protein database. In many cases multiple spots on the gel corresponded to a single protein so that 61 different proteins were identified, including the immunoreactive proteins described above (see Fig. S1).

All 61 proteins identified using combined GS20 and Solexa sequences would also have been identified when using only the GS20 sequences-derived protein database. However, with uncorrected GS20 sequences, 4 ORFs presented a frameshift leading twice to a longer protein and twice to a premature end of the protein, i.e. splitting the ORF in two parts. Only 5 of the 61 proteins identified using combined GS20 and Solexa sequences were identical to the predicted ORFs using only Solexa sequences. The remaining 56 proteins were split between two to seven different small contigs preventing their accurate identification by Mascot. The limited performance of Solexa technology as compared to 454 is likely due to the short Solexa reads and the relative low sequence depth obtained.

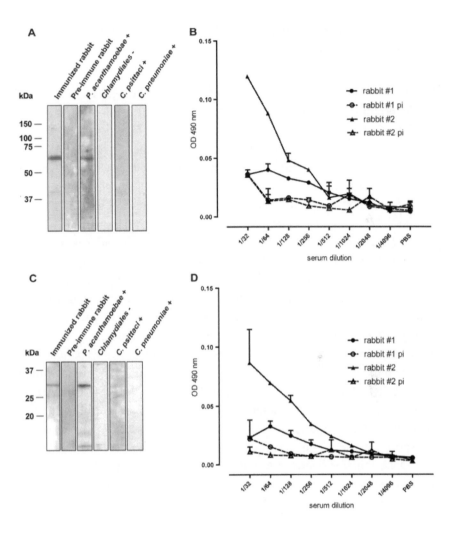

FIGURE 3: Western blot and ELISA with recombinant proteins E and N. Purified recombinant protein E (A) and N (C) were blotted on a nitrocellulose membrane and probed with a rabbit anti-*Parachlamydia* serum (rabbit #1), a rabbit pre-immune serum (rabbit #1), a *P. acanthamoebae* positive human serum, a *Chlamydiales* negative human serum, a *C. psittaci* positive human serum and a *C. pneumoniae* positive human serum. Proteins E (B) and N (D) were used as antigen in a direct ELISA. Sera from 2 rabbits immunized with *P. acanthamoebae* and pre-immune (pi) sera were tested in duplicates.

11.2.5 MOST ABUNDANTLY EXPRESSED PROTEINS AND VIRULENCE GENES

By BLAST against nr database, a function could be derived for half of the 61 proteins identified by mass spectrometry (for details, see Table S1 and Table S3), whereas one fourth have homologs of unknown function in *Protochlamydia amoebophila* genome [33] or in other organisms. Finally, the remaining proteins exhibit no significant homology with any known amino acid sequence.

Our assembly also enabled the identification of several virulence genes present in *P. acanthamoebae* genome. In addition to the previously mentioned LcrE and CPAF protease, *P. acanthamoebae* encodes a complete type three secretion system (T3SS), including components of the secretory apparatus, translocators, T3SS specific chaperones and effectors [28]. Like in other *Chlamydiales* species, T3SS genes are spread in distinct conserved genomic clusters (see Fig. S2). Interestingly, sctJ and sctC, two genes encoding components of the secretory apparatus, are duplicated in the genome. Moreover, four clear homologs to *Protochlamydia amoebophila* nucleotide transporters (*ntt_1, ntt_2, ntt_3* and *ntt_4*) which play a key feature in energy parasitism [29], [30] have been identified in *P. acanthamoebae*. A fifth putative ADP/ATP translocase candidate was also detected but homologies were not sufficient to establish the substrate transport specificity. Finally, four genes belonging to the F-like conjugative DNA transfer operon located on the genomic island of *Protochlamydia amoebophila* [31] were also detected in *P. acanthamoebae* (*traU, traN, traF* and *pc1435*).

11.3 DISCUSSION

In case of outbreak due to a new pathogen, diagnostic tests must be developed rapidly. Genome sequence is an important resource to develop various tools for molecular and serological diagnostic, specific monoclonal antibodies production or vaccine design. With the availability of high throughput sequencing strategies such as the widely used GS20/GSFLX [1], large sequence datasets are now obtained within very short time. However, the costly and time-consuming follow up necessary to close the gaps delays the release and accessibility of most genome sequences. As shown for *Francisella tularensis*, a rapid comparative genome analysis can be successfully applied on unfinished contigs enabling to uncover genomic rearrangements or gene mutations that could be involved in an increased strain virulence and resistance [9]. A similar approach was also proposed to study the role of *Helicobacter pylori* in chronic gastric infection by analyzing genetic changes in this species over time or between infected humans [32].

A rapid and public availability of raw genome data from an emerging pathogen at the origin of an outbreak is critical to permit the development of various diagnostic tools by medical microbiologists. The delay before genome release is especially crucial in case of new pathogenic agents with the absence of available closely related genomes, i.e. absence of scaffold that may be used to facilitate assembly and gap closure steps. This problem was faced here for *P. acanthamoebae* with the availability of a single published genome within the *Parachlamydia*caeae family, that of *Protochlamydia amoebophila* [33]. The presence of repeated elements in the genome significantly increases the number of contigs obtained, thus prolonging the gap closure. Although *Chlamydiaceae* genomes do not contain many transposases, the genome sequence of *Protochlamydia amoebophila* was much more invaded by such repeated components [33]. This suggests that sequence repetitions probably account for a large number of gaps in our own *Parachlamydia* genome project. Nevertheless, if these repeated elements can prevent an assembly in one unique contig, they do not hinder the availability of most coding sequences. Indeed, 90% of analyzed proteins could be identified, the remaining 10% being uncharacterized due either to the lack of mass spectrometry signal or to the lack of hits in the ORF database. Thus, although we could not determine the exact number of immunogenic proteins that have been missed due to the presence of the remaining gaps, we may estimate that only few (<10%) additional immunogenic proteins would have been identified if a complete genome sequence was available.

Our proteomic approach allowed us to detect 18 immunogenic proteins among which are several antigens already described as highly immunoreactive such as GroEL, DnaK or elongation factor Tu [25]. Five proteins represented good/possible candidates to develop a diagnostic test since exhibiting significant reactivity to sera taken from humans infected by *Parachlamydia* as well as from immunized rabbits and no cross-reactivity to sera from humans infected with *C. pneumoniae*, *C. psittaci* and negative controls. Then, we focused on only two of these five proteins, E and N, displaying no significant homology with any known amino acid sequence. Their potential to develop vaccine or diagnostic tools was suggested by western-blot and by preliminary ELISA tests despite the absence, in the heterologous protein expression system used, of post-translational modifications such as glycosylation or phosphorylation that might have resulted in poor serum recognition. Given its 96-well format, the ELISA test, once developed, would be very useful in large epidemiological studies to assess the precise seroprevalence of *Parachlamydia* antibodies in human population and to confirm the pathogenic role of this intracellular bacterium in human lower respiratory tract infections and in bovine abortion. Moreover, an ELISA based on a given immunogenic protein will be more specific than diagnostic microimmunofluorescence and western-blot assays based on whole bacterial proteins.

Interestingly, among the 85 ORFs identified using our dirty genome sequences, only 9 proteins could have been identified by Mascot versus protein sequences of the closest related bacteria available to date, *Protochlamydia amoebophila*, all of which are very conserved and, if immunogenic, would likely produce strong cross reactions when used in serological tests. In addition, among the five immunogenic proteins selected as good and possible candidates for the development of a diagnostic ELISA, none have been identified by Mascot versus SPTrembl database because of the differences between the peptides identified and the protein database. Moreover, the two proteins we considered as the best candidates have no homologs in other genomes and their sequences could not be derived from those of any related or unrelated bacteria.

This further emphasizes the need for a protein database directly derived from genome sequences of the studied emerging pathogen. Besides, rapid genome sequencing provides information useful not only for proteomics but also for comparative genomics, transcriptomics, cell biology and molecular biology. The availability of most genome sequences of a new emerging pathogen isolated during an outbreak may also be important to design molecular diagnostic tools, to define epidemiological marker as well as to identify virulence genetic traits and antibiotic resistance determinants.

The advantages of using mass spectrometry associated with an unfinished genome to identify immunogenic proteins, compared to other approaches such as phage display library [34], [35], comparative genomic [36] or systematic expression of all ORFs [37], resides mainly in the minimal necessary workload and in the rapidity of the method. Indeed, the whole process can take place in less than 4 months, with only 2 weeks necessary to obtain the contigs (Fig. 4). In addition, with the lowering of sequencing costs, the price of such an approach is highly competitive. Furthermore, constructing random expression libraries by fractionation of whole bacterial DNA would likely identify less immunogenic proteins since any plasmid carrying ORFs whose product is toxic will not be successfully expressed.

The 4 months that it takes to develop an ELISA may seem long as compared to the few weeks needed to develop a DNA-based test. However, detecting proteins represents a distinct advantage over detecting unique DNA sequences, and the availability of a serological test may especially prove very useful (i) to confirm positive PCR results (that may be false positive due to PCR contamination) and to better document a given case, (ii) to perform large seroepidemiological studies in order to precise the mode of transmission of a new pathogen and (iii) to investigate the possible role of a new bacterial pathogen in different clinical settings, such as pericarditis and endocarditis, for which valvular/pericardial fluid samples are not easily available. Similarly, when investigating patients with atypical pneumonia, serum samples are easier to obtain than lower respiratory tract specimen, especially when patients present a non-productive cough. Moreover, for several fastidious intracellular bacteria, even highly sensitive PCR tests may fail in detecting the agent at the infection site (i) due to relatively low bacterial load, e.g. sensitivity of only 50% to detect Borrelia in cerebrospinal fluid taken from patient with neurological Lyme disease [38], (ii) due to the presence of inhibiting molecules present in clinical samples, or (iii) due to "sampling bias" of PCRs tests performed on tissue samples, e.g. sensitivity of 60% of PCR on valve samples taken from patients with definite endocarditis [39].

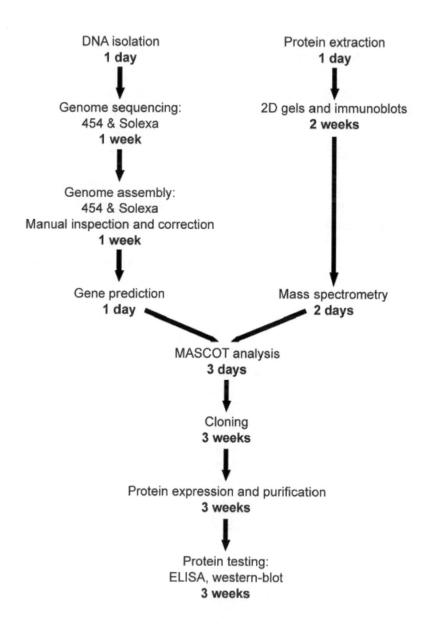

FIGURE 4: Time scale of a dirty genome approach combined with proteomics to develop serological diagnostic tools. Schematic representation of the main steps of genome sequencing, immunogenic proteins identification and testing of candidate proteins in an ELISA. In bold, approximate time necessary to complete each step.

Furthermore, the identification of immunogenic proteins also allows the development of species specific immunohistochemistry, that is useful (i) to confirm the presence of the pathogen in the tissue lesions, (ii) to analyze retrospectively various biopsy samples taken from patients with infection of unknown etiology and (iii) to shed some light on the underlying pathogenesis in vivo, by precising which cells are infected using double staining.

In summary, this work constitutes the proof of principle for a dirty genome approach, i.e. the use of unfinished genome sequences of pathogenic bacteria, coupled with proteomics to rapidly identify immunogenic proteins useful to develop a specific diagnostic test. Indeed, genomic information concerning new emerging pathogens must be placed at the scientific community's disposal as soon as possible, since their retention in order to close all gaps before genome publication is clearly medically counterproductive. This work demonstrated that 454/Solexa combined dirty genomes are sufficient and useful for medical downstream applications.

11.4 METHODS

11.4.1 ETHICS STATEMENT

Human sera used in this work (see below) have been obtained from patients and control subjects, as part of prospective studies. These clinical studies have been accepted by the local ethical committee of the University of Lausanne. Both patients and controls gave their informed consent for various serological investigations including study of their serum reactivity against *Parachlamydia acanthamoebae*.

11.4.2 CULTIVATION AND PURIFICATION OF P. ACANTHAMOEBAE

Parachlamydia acanthamoebae strain Hall's coccus was grown in *Acanthamoeba castellanii* strain ATCC 30010 in peptone yeast-extract glucose broth (PYG) and purified using a sucrose barrier and a gastrographin gradient, as previously described [22].

11.4.3 *CHROMOSOME SEQUENCING AND ASSEMBLY*

Purified *P. acanthamoebae* elementary bodies resuspended in PBS were lysed for DNA extraction with QIAamp DNA mini kit (Qiagen) according to the manufacturer protocol. Genomic DNA was pyrosequenced by two runs of Genome Sequencer 20 [1]. Genomic DNA was also sequenced using Solexa technology in Illumina Genome Analyzer [2]. Solexa sequences were assembled using Edena software [40] with parameter m = 16. Both GS20 runs only or GS20 runs and Solexa contigs were compiled in one assembly using Newbler software V1.1.02 with default parameters except for overlap size (45 nt) and identity score (95%). Differences between 454 and solexa contigs were manually inspected and corrected when necessary. In case of homopolymer discrepancy preference was generally given to Solexa when correcting frameshifts in protein coding region and in potentially non coding regions.

11.4.4 *GENE PREDICTION AND ANNOTATION*

To improve the prediction of incomplete genes at contig ends, a stop encoding tag "CTAGCTAGCTAG" was added at both extremities of each contig. A reference proteome was created with all open reading frames (stop to stop ORF) for peptide identification. Besides, locally installed Glimmer v3.02 [41] trained on long ORFs of the concatenated contigs was then applied to predict gene position on contigs longer than 500 bp. All reported ORFs larger than 100 nt were submitted to BLASTP versus nr database and local InterProScan search. Finally, tRNAscan-SE [42] and RNAmmer [43] were used to find structural RNAs. Genes of special importance for this study were manually annotated. This Whole Genome Shotgun project has been deposited at DDBJ/EMBL/GenBank under the accession ACZE00000000. The version described in this paper is the first version, ACZE01000000.

11.4.5 CRUDE EXTRACT SAMPLE PREPARATION AND 2D GEL ELECTROPHORESIS

Bacterial cells resuspended in PBS were washed in 10 mM Tris, 5 mM MgAc, pH 8.0 and then lysed by 5 cycles of short-pulse sonication in lysis buffer (30 mM Tris, 7 M urea, 2 M Thiourea, 4% CHAPS, pH 8.5). Proteins were recovered by centrifugation at 6'000 g and their concentration determined using a Bradford assay (Quick Start™ Bradford Protein Assay, Biorad, Hercules, USA).

Two dimensional gel electrophoresis was performed as described by Centeno et al. [44] using approximately 150 μg (mini gels) or 600 μg (midi-gels) of total elementary bodies proteins. Proteins were visualized by Coomassie Blue staining or transferred to nitrocellulose for subsequent immunoblot analysis (see Methods S1).

11.4.6 HUMAN SERA

P. acanthamoebae positive human sera were described in previous studies where their positivity was assessed by immunofluorescence and western-blot [17], [45]. Control sera were taken from women with at term uneventful pregnancy [45]. *Chlamydiales* negative sera were tested negative by immunofluorescence for reactivity against various members of the *Chlamydiales* order (*P. acanthamoebae*, *W. chondrophila*, *S. negevensis*, *N. hartmannellae*, *C. trachomatis*, *C. pneumoniae* and *C. psittaci*), *C. pneumoniae* positive sera were positive for *C. pneumoniae* but negative for all other *Chlamydiales* tested. *C. psittaci* positive human serum was taken from a patient who suffered from well-documented psittacosis [46].

11.4.7 ELISA

Proteins E and N cloned and expressed in *E. coli* were purified thanks to a 6His tail (see Methods S1). Then, 96-well ELISA microplates were coated with 100 ng of purified E or N proteins in carbonate buffer pH 9.6 and incubated overnight at 4°C. After blocking with 3% non-fat dry-milk in PBST (PBS + 0.1% Tween 20) during 1 hour at 37°C, plates were washed with PBST and incubated 2 hours at 37°C with serial two-fold dilutions, in PBST+1% non-fat dry-milk, of sera from 2 rabbits immunized with *P. acanthamoebae* and of corresponding pre-immune sera. After 3 subsequent washes with PBST, plates were incubated 1 hour at 37°C with horse-radish peroxidase-conjugated anti-rabbit IgG (Cell Signaling, Allschwill, Switzerland) diluted 1:1000 in PBS + 1% non-fat dry-milk. Plates were washed 5 more times with PBST. O-phenylenediamine dihydrochloride (OPD) in citrate buffer was used as substrate for the peroxydase. After 15 minutes incubation, the optical density was read at 492 nm using an ELISA reader (Multiskan Ascent, Thermo Scientific, Waltham, USA).

11.4.8 ADDITIONAL METHODS

Descriptions of sera from immunized rabbits, immunoblot analysis, mass spectrometry and cloning, expression and purification of immunogenic proteins E and N are available in Methods S1.

REFERENCES

1. Margulies M, Egholm M, Altman WE, Attiya S, Bader JS, et al. (2005) Genome sequencing in microfabricated high-density picolitre reactors. Nature 437: 376–380.
2. Bennett S (2004) Solexa Ltd. Pharmacogenomics 5: 433–438.
3. Pol A, Heijmans K, Harhangi HR, Tedesco D, Jetten MS, et al. (2007) Methanotrophy below pH 1 by a new Verrucomicrobia species. Nature 450: 874–878.
4. Korbel JO, Urban AE, Affourtit JP, Godwin B, Grubert F, et al. (2007) Paired-end mapping reveals extensive structural variation in the human genome. Science 318: 420–426.
5. Albert TJ, Molla MN, Muzny DM, Nazareth L, Wheeler D, et al. (2007) Direct selection of human genomic loci by microarray hybridization. Nat Methods 4: 903–905.
6. Berezikov E, Cuppen E, Plasterk RH (2006) Approaches to microRNA discovery. Nat Genet 38: SupplS2–7.
7. Taylor KH, Kramer RS, Davis JW, Guo J, Duff DJ, et al. (2007) Ultradeep bisulfite sequencing analysis of DNA methylation patterns in multiple gene promoters by 454 sequencing. Cancer Res 67: 8511–8518.
8. Cokus SJ, Feng S, Zhang X, Chen Z, Merriman B, et al. (2008) Shotgun bisulphite sequencing of the Arabidopsis genome reveals DNA methylation patterning. Nature 452: 215–219.

9. La Scola B, Elkarkouri K, Li W, Wahab T, Fournous G, et al. (2008) Rapid comparative genomic analysis for clinical microbiology: the *Francisella tularensis* paradigm. Genome Res 18: 742–750.

10. Rolain JM, Francois P, Hernandez D, Bittar F, Richet H, et al. (2009) Genomic analysis of an emerging multiresistant Staphylococcus aureus strain rapidly spreading in cystic fibrosis patients revealed the presence of an antibiotic inducible bacteriophage. Biol Direct 4: 1.

11. Francois P, Hochmann A, Huyghe A, Bonetti EJ, Renzi G, et al. (2008) Rapid and high-throughput genotyping of Staphylococcus epidermidis isolates by automated multilocus variable-number of tandem repeats: a tool for real-time epidemiology. J Microbiol Methods 72: 296–305.

12. Jarvie T, Harkins T (2008) De novo assembly and genomic structural variation analysis with genome sequencer FLX 3K long-tag paired end reads. Biotechniques 44: 829–831.

13. Corsaro D, Greub G (2006) Pathogenic potential of novel Chlamydiae and diagnostic approaches to infections due to these obligate intracellular bacteria. Clin Microbiol Rev 19: 283–297.

14. Birtles RJ, Rowbotham TJ, Storey C, Marrie TJ, Raoult D (1997) Chlamydia-like obligate parasite of free-living amoebae. Lancet 349: 925–926.

15. Greub G (2009) *Parachlamydia acanthamoebae*, an emerging agent of pneumonia. Clin Microbiol Infect 15: 18–28.

16. Greub G, Berger P, Papazian L, Raoult D (2003) *ParaChlamydiaceae* as rare agents of pneumonia. Emerg Infect Dis 9: 755–756.

17. Greub G, Boyadjiev I, La Scola B, Raoult D, Martin C (2003) Serological hint suggesting that *ParaChlamydiaceae* are agents of pneumonia in polytraumatized intensive care patients. Ann N Y Acad Sci 990: 311–319.

18. Marrie TJ, Raoult D, La Scola B, Birtles RJ, de Carolis E (2001) Legionella-like and other amoebal pathogens as agents of community-acquired pneumonia. Emerg Infect Dis 7: 1026–1029.

19. Casson N, Posfay-Barbe KM, Gervaix A, Greub G (2008) New diagnostic real-time PCR for specific detection of *Parachlamydia acanthamoebae* DNA in clinical samples. J Clin Microbiol 46: 1491–1493.

20. Casson NE, JM, Borel N, Pospischil A, Greub G (2008) A mice model of lung infection by *Parachlamydia acanthamoebae*. Microbial pathogenesis 2008 45: 92–97.

21. Casson NS, K, Klos A, Stehle J-C, Pusztaszeri M, Greub G (2008) Intranasal murine model of infections by *Parachlamydia acanthamoebae*. Proceedings of the Annual Meeting of the Swiss Society for Infectious Diseases 17.

22. Greub G, Mege JL, Raoult D (2003) *Parachlamydia acanthamoebae* enters and multiplies within human macrophages and induces their apoptosis. Infect Immun 71: 5979–5985.

23. Greub G, Mege JL, Gorvel JP, Raoult D, Meresse S (2005) Intracellular trafficking of *Parachlamydia acanthamoebae*. Cell Microbiol 7: 581–589.

24. Borel N, Ruhl S, Casson N, Kaiser C, Pospischil A, et al. (2007) *Parachlamydia* spp. and related Chlamydia-like organisms and bovine abortion. Emerg Infect Dis 13: 1904–1907.

25. Sanchez-Campillo M, Bini L, Comanducci M, Raggiaschi R, Marzocchi B, et al. (1999) Identification of immunoreactive proteins of Chlamydia trachomatis by Western blot analysis of a two-dimensional electrophoresis map with patient sera. Electrophoresis 20: 2269–2279.

26. Sharma J, Bosnic AM, Piper JM, Zhong G (2004) Human antibody responses to a Chlamydia-secreted protease factor. Infect Immun 72: 7164–7171.

27. Sharma J, Zhong Y, Dong F, Piper JM, Wang G, et al. (2006) Profiling of human antibody responses to Chlamydia trachomatis urogenital tract infection using microplates arrayed with 156 chlamydial fusion proteins. Infect Immun 74: 1490–1499.

28. Peters J, Wilson DP, Myers G, Timms P, Bavoil PM (2007) Type III secretion a la Chlamydia. Trends Microbiol 15: 241–251.

29. Haferkamp I, Schmitz-Esser S, Wagner M, Neigel N, Horn M, et al. (2006) Tapping the nucleotide pool of the host: novel nucleotide carrier proteins of *Protochlamydia amoebophila*. Mol Microbiol 60: 1534–1545.

30. Haferkamp I, Schmitz-Esser S, Linka N, Urbany C, Collingro A, et al. (2004) A candidate NAD+ transporter in an intracellular bacterial symbiont related to Chlamydiae. Nature 432: 622–625.

31. Greub G, Collyn F, Guy L, Roten CA (2004) A genomic island present along the bacterial chromosome of the *ParaChlamydiaceae* UWE25, an obligate amoebal endosymbiont, encodes a potentially functional F-like conjugative DNA transfer system. BMC Microbiol 4: 48.

32. Oh JD, Kling-Backhed H, Giannakis M, Xu J, Fulton RS, et al. (2006) The complete genome sequence of a chronic atrophic gastritis *Helicobacter pylori* strain: evolution during disease progression. Proc Natl Acad Sci U S A 103: 9999–10004.

33. Horn M, Collingro A, Schmitz-Esser S, Beier CL, Purkhold U, et al. (2004) Illuminating the evolutionary history of chlamydiae. Science 304: 728–730.

34. Srivastava N, Zeiler JL, Smithson SL, Carlone GM, Ades EW, et al. (2000) Selection of an immunogenic and protective epitope of the PsaA protein of Streptococcus pneumoniae using a phage display library. Hybridoma 19: 23–31.

35. Naidu BR, Ngeow YF, Wang LF, Chan L, Yao ZJ, et al. (1998) An immunogenic epitope of Chlamydia pneumoniae from a random phage display peptide library is reactive with both monoclonal antibody and patients sera. Immunol Lett 62: 111–115.

36. Araoz R, Honore N, Cho S, Kim JP, Cho SN, et al. (2006) Antigen discovery: a postgenomic approach to leprosy diagnosis. Infect Immun 74: 175–182.

37. McKevitt M, Patel K, Smajs D, Marsh M, McLoughlin M, et al. (2003) Systematic cloning of Treponema pallidum open reading frames for protein expression and antigen discovery. Genome Res 13: 1665–1674.

38. Gooskens J, Templeton KE, Claas EC, van Dam AP (2006) Evaluation of an internally controlled real-time PCR targeting the ospA gene for detection of Borrelia burgdorferi sensu lato DNA in cerebrospinal fluid. Clin Microbiol Infect 12: 894–900.

39. Greub G, Lepidi H, Rovery C, Casalta JP, Habib G, et al. (2005) Diagnosis of infectious endocarditis in patients undergoing valve surgery. Am J Med 118: 230–238.

40. Hernandez D, Francois P, Farinelli L, Osteras M, Schrenzel J (2008) De novo bacterial genome sequencing: Millions of very short reads assembled on a desktop computer. Genome Res 18: 802–809.

41. Delcher AL, Bratke KA, Powers EC, Salzberg SL (2007) Identifying bacterial genes and endosymbiont DNA with Glimmer. Bioinformatics 23: 673–679.
42. Lowe TM, Eddy SR (1997) tRNAscan-SE: a program for improved detection of transfer RNA genes in genomic sequence. Nucleic Acids Res 25: 955–964.
43. Lagesen K, Hallin P, Rodland EA, Staerfeldt HH, Rognes T, et al. (2007) RNAmmer: consistent and rapid annotation of ribosomal RNA genes. Nucleic Acids Res 35: 3100–3108.
44. Centeno C, Repici M, Chatton JY, Riederer BM, Bonny C, et al. (2007) Role of the JNK pathway in NMDA-mediated excitotoxicity of cortical neurons. Cell Death Differ 14: 240–253.
45. Baud D, Thomas V, Arafa A, Regan L, Greub G (2007) Waddlia chondrophila, a potential agent of human fetal death. Emerg Infect Dis 13: 1239–1243.
46. Senn L, Greub G (2008) Local newspaper as a diagnostic aid for psittacosis: a case report. Clin Infect Dis 46: 1931–1932.

There are several supplemental files that are not available in this version of the article. To view this additional information, please use the citation information cited on the first page of this chapter.

CHAPTER 12

CORONAVIRUS GENOMICS AND BIOINFORMATICS ANALYSIS

PATRICK C. Y. WOO, YI HUANG, SUSANNA K. P. LAU, AND KWOK-YUNG YUEN

12.1 INTRODUCTION

Traditionally, viruses were characterized and classified by culture, electron microscopy and serological studies. Using these phenotypic methods, coronaviruses were defined as enveloped viruses of 120-160 nm in diameter with a crown-like appearance. The name "coronavirus" is derived from the Greek κορώνα, meaning crown. Based on their antigenic relationships, coronaviruses were classified into three groups. Group 1 and 2 are composed of mammalian coronaviruses and group 3 avian coronaviruses. The invention of and advances in nucleic acid amplification technologies, automated DNA sequencing and bioinformatics tools in the recent two decades have revolutionized the characterization and classification of all kinds of infectious disease agents. Using molecular methods, coronaviruses are classified as positive-sense, single-stranded RNA viruses. Furthermore, the results of using phylogenetic methods for classification also supported the group boundaries of the traditional antigenic classification. Phylogenetic

This chapter was originally published under the Creative Commons Attribution License. Woo PCY, Huang Y, Lau SKP and Yuen K-Y. Coronavirus Genomics and Bioinformatics Analysis. Viruses 2 (8) (2010). doi:10.3390/v2081803.

methods have also enabled the classification of SARS-related coronavirus (SARSr-CoV) as a subgroup of group 2, group 2b, coronavirus; as well as the discovery of group 2c, 2d, 3b and 3c coronaviruses [1-3]. Recently, the Coronavirus Study Group of the International Committee for Taxonomy of Viruses has proposed three genera, *Alphacoronavirus*, *Betacoronavirus* and *Gammacoronavirus*, to replace these three traditional groups of coronaviruses [4].

The first complete genome of coronavirus, mouse hepatitis virus (MHV), was sequenced more than 50 years after it was isolated. Before the SARS epidemic in 2003, there were less than 10 coronaviruses with complete genome sequences available. These include two human coronaviruses (HCoV-229E and HCoV-OC43), four other mammalian coronaviruses [MHV, bovine coronavirus (BCoV), transmissible gastroenteritis virus (TGEV), porcine epidemic diarrhea virus (PEDV)], and one avian coronavirus (IBV). The SARS epidemic that originated from southern China in 2003 has boosted interest in all areas of coronavirus research, most notably, coronavirus biodiversity and genomics [5-7]. After the SARS epidemic, up to April 2010, 15 novel coronaviruses were discovered with their complete genomes sequenced. Among these 15 previously unrecognized coronaviruses were two globally distributed human coronaviruses, human coronavirus NL63 (HCoV-NL63) and human coronavirus HKU1 (HCoV-HKU1) [8-10]; 10 other mammalian coronaviruses, SARS-related *Rhinolophus* bat coronavirus (SARSr-Rh-BatCoV), *Rhinolophus* bat coronavirus HKU2 (Rh-BatCoV HKU2), *Tylonycteris* bat coronavirus HKU4 (Ty-BatCoV HKU4), *Pipistrellus* bat coronavirus HKU5 (Pi-BatCoV HKU5), *Miniopterus* bat coronavirus HKU8 (Mi-BatCoV HKU8), *Rousettus* bat coronavirus HKU9 (Ro-BatCoV HKU9), *Scotophilus* bat coronavirus 512 (Sc-BatCoV 512), *Miniopterus* bat coronavirus 1A/B (Mi-BatCoV 1A/B), equine coronavirus (ECoV) and beluga whale coronavirus SW1 [3,6,11-15]; and three avian coronaviruses, bulbul coronavirus HKU11 (BuCoV HKU11), thrush coronavirus HKU12 (ThCoV HKU12) and munia coronavirus HKU13 (MunCoV HKU13) [2]. Most of these genomes were sequenced using the RNA extracted directly from the clinical specimens, such as nasopharyngeal aspirate or stool, as the template, while the viruses themselves were still non-cultivable [2,3,6,11-15]. This provided more accurate analysis of the in situ viral genomes avoiding mutational bias during in vitro viral replication. These sequence efforts have resulted in a marked increase in the number of coronavirus genomes and have given us an unprecedented opportunity to understand this family of virus at the genomic and in silico levels. These understandings have also led to generation of further hypotheses and experiments in the laboratory. In this article, we reviewed our current understanding on the genomics and bioinformatics analysis of coronaviruses. Details of the bioinformatics tools will not be discussed.

12.2 GENOMICS

Coronaviruses possess the largest genomes [26.4 kb (ThCoV HKU12) to 31.7 kb (SW1)] among all known RNA viruses (Figure 1) [2,13,16]. The large genome has given this family of virus extra plasticity in accommodating and modifying genes. The G + C contents of coronavirus genomes vary from 32% (HCoV-HKU1) to 43% (Pi-BatCoV HKU5 and MunCoV HKU13) (Table 1) [2,3,10]. Both the 5' and 3' ends of coronavirus genomes contain short untranslated regions. For the coding regions, the genome organizations of all coronaviruses are similar, with the characteristic gene order 5'-replicase ORF1ab, spike (S), envelope (E), membrane (M), nucleocapsid (N)-3', although variable numbers of additional ORFs are present in each subgroup of coronavirus (Table 1, Figure 1). A transcription regulatory sequence (TRS) motif is present at the 3' end of the leader sequence preceding most ORFs (Table 1). The TRS motifs are thought to be important for a "copy-choice" mechanism that mediates the unique random template switching during RNA replication, resulting in a high frequency of homologous RNA recombination in coronaviruses [17].

12.2.1 ORF1AB

ORF1ab of coronaviruses occupy about two thirds of their genomes. It encodes the replicase polyprotein and is translated from ORF1a (11826 to 13425 nt) and ORF1b (7983 to 8157 nt). In all coronaviruses, a slippery sequence (UUUAAAC), followed by sequences that form a putative pseudoknot structure, are present at the junction between ORF1a and ORF1b. Translation occurs by a -1 RNA-mediated ribosomal frame-shift at the end of the slippery sequence. Instead of reading the transcript as UUUAAACGGG, it will be read as UUUAAACCGGG. The replicase polyprotein is cleaved by papain-like protease(s) (PL^{pro}) and 3C-like pro-tease (3CL^{pro}), proteins encoded by ORF1ab of the coronavirus genome, at consensus cleavage sites, into 15 to 16 non-structural proteins (nsps) named nsp1, nsp2, nsp3, etc (Table 1). As the number of coronavirus ge-nomes is expanding, novel cleavage sites have been discovered [3,18]. Some of these non-structural proteins encode proteins of essential func-tions, such as PL^{pro} (nsp3), 3CL^{pro} (nsp5), RNA-dependent RNA poly-merase (Pol) (nsp12) and helicase (nsp13) (Figure 1). The genomes of all known members of *Alphacoronavirus* and *Betacoronavirus* subgroup A possess two PL^{pro} (PL1^{pro} and PL2^{pro}), while those of all known members of *Betacoronavirus* subgroup B, C and D and *Gammacoronavirus* possess only one PL^{pro} (Table 1, Figure 1). The gene sequences that encode these conserved proteins are frequently used for phylogenetic analysis.

   In addition to the nsps with essential functions, bioinformatics analysis of some other nsps revealed their putative functions. Downstream to PL^{pro} or PL1^{pro} in nsp3 is the X domain which contains putative ADP-ribose 1"-phosphatase (ADRP) activity [1]. In other microorganisms, such as *Saccharomyces cerevisiae* and other eukaryotes, ADRP and its function-ally related enzyme cyclic nucleotide phosphodiesterase (CPDase), were important for tRNA processing [19]. ADP-ribose 1",2"-cyclic phosphate (Appr>p) is produced as a result of tRNA splicing. Appr>p is in turn con-verted to ADP-ribose 1"-phosphate (Appr-1"p) by CPDase and Appr-1"p is then further processed by ADRP. As for nsp13, nsp14 and nsp15, they possess a putative 3'-to-5' exonuclease (ExoN) domain of the DEDD su-perfamily [1], a putative poly(U)-specific endoribonuclease (XendoU) domain, and a putative S-adenosylmethionine-dependent ribose 2'-O-methyltransferase (2'-O-MT) domain of the RrmJ family respectively [1]. ADRP, CPDase, ExoN, XendoU and 2'-O-MT are enzymes in RNA pro-cessing pathways. Contrary to the pre-tRNA splicing pathway that ADRP and CPDase belong to, ExoN, XendoU and 2'-O-MT are enzymes in a small nucleolar RNA processing and utilization pathway.

TABLE 1: Genome comparison of coronaviruses.

Viruses	Hosts	G+C contents	Transcription regulatory sequences	No. of nsp in ORF1ab	No. of papain-like proteases in ORF1ab	No. of small ORFs between ORF1ab and N	Presence of conserved S cleavage site	No. of small ORFs downstream to N
Alphacoronavirus								
Transmissible gastroenteritis virus	Pigs	0.38	CUAAAC	16	2	2	N	1
Porcine respiratory coronavirus	Pigs	0.37	CUAAAC	16	2	1	N	1
Feline coronavirus	Cats	0.39	CUAAAC	16	2	4	N	2
Human coronavirus 229E	Humans	0.38	CUAAAC	16	2	2	N	-
Human coronavirus NL63	Humans	0.34	CUAAAC	16	2	1	N	-
Porcine epidemic diarrhea virus	Pigs	0.42	CUAAAC	16	2	1	N	-
Scotophilus bat coronavirus 512	Lesser Asiatic yellow house bats	0.40	CUAAAC	16	2	1	N	1
Rhinolophus bat coronavirus HKU2	Chinese horseshoe bats	0.39	CUAAAC	16	2	1	N	1
Miniopterus bat coronavirus HKU8	Bent-winged bats	0.42	CUAAAC	16	2	1	N	1
Miniopterus bat coronavirus 1A	Bent-winged bats	0.38	CUAAAC	16	2	1	N	-

TABLE 1: *Cont.*

Viruses	Hosts	G+C contents	Transcription regulatory sequences	No. of nsp in ORF1ab	No. of papain-like proteases in ORF1ab	No. of small ORFs between ORF1ab and N	Presence of conserved S cleavage site	No. of small ORFs downstream to N
Miniopterus bat coronavirus 1B	Bent-winged bats	0.39	CUAAAC	16	2	1	n	-
Betacoronavirus								
Subgroup A								
Human coronavirus OC43	Humans	0.37	CUAAAC	16	2	1	Y	-
Bovine coronavirus	Cows	0.37	CUAAAC	16	2	3	Y	-
Porcine hemagglutinating encephalomyelitis virus	Pigs	0.37	CUAAAC	16	2	2	Y	-
Equine coronavirus	Horses	0.37	CUAAAC	16	2	2	Y	-
Human coronavirus HKU1	Humans	0.32	CUAAAC	16	2	1	Y	-
Mouse hepatitis virus	Mice	0.42	CUAAAC	16	2	2	Y	-
Subgroup B								
Human SARS related coronavirus	Humans	0.41	ACGAAC	16	1	7	N	-
SARS-related Rhinolophus bat coronavirus HKU3	Chinese horseshoe bats	0.41	ACGAAC	16	1	5	N	-
Subgroup C								
Tylonycteris bat coronavirus HKU4	Lesser bamboo bats	0.38	ACGAAC	16	1	4	N	-

TABLE 1: *Cont.*

Viruses	Hosts	G+C contents	Transcription regulatory sequences	No. of nsp in ORF1ab	No. of papain-like proteases in ORF1ab	No. of small ORFs between ORF1ab and N	Presence of conserved S cleavage site	No. of small ORFs downstream to N
Pipistrellus bat coronavirus HKU5	Japanese pipistrelle bats	0.43	ACGAAC	16	1	4	N	-
Subgroup D								
Rousettus bat coronavirus HKU9	Leschenault's rousette bats	0.41	ACGAAC	16	1	1	N	2
Gammacoronavirus								
Infectious bronchitis virus	Chickens	0.38	CUUAACAA	15	1	4	Y	-
Turkey coronavirus	Turkeys	0.38	CUUAACAA	15	1	5	Y	-
Beluga whale coronavirus	Beluga whales	0.39	AAACA	15	1	8	N	-
Deltacoronavirus								
Bulbul coronavirus HKU11	Chinese bulbuls	0.39	ACACCA	15	1	1	N	3
Thrush coronavirus HKU12	Gray-backed thrushes	0.38	ACACCA	15	1	1	N	3
Munia coronavirus HKU13	White-rumped munias	0.43	ACACCA	15	1	1	N	3

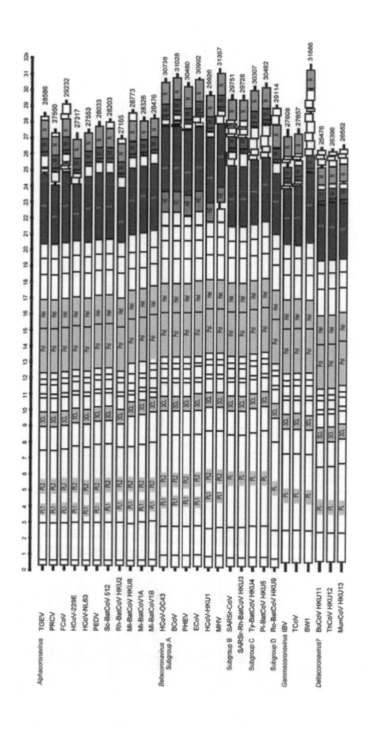

FIGURE 1: Genome organizations of members in different genera of the *Coronaviridae* family. PL1, papain-like protease 1; PL2, papain-like protease 2; PL, papain-like protease; 3CL, chymotrypsin-like protease; Pol, RNA-dependent RNA polymerase; Hel, helicase; HE, haemagglutinin esterase; S, spike; E, envelope; M, membrane; N, nucleocapsid. TGEV, porcine transmissible gastroenteritis virus (NC_002306); PRCV, porcine respiratory coronavirus (DQ811787); FCoV, feline coronavirus (NC_012937); HCoV-229E, human coronavirus 229E (NC_002645); HCoV-NL63, human coronavirus NL63 (NC_005831); PEDV, porcine epidemic diarrhea virus (NC_003436); Sc-BatCoV 512, *Scotophilus* bat coronavirus 512 (NC_009657); Rh-BatCoV-HKU2, *Rhinolophus* bat coronavirus HKU2 (NC_009988); Mi-BatCoV-HKU8, *Miniopterus* bat coronavirus HKU8 (NC_010438); Mi-BatCoV 1A, *Miniopterus* bat coronavirus 1A (NC_010437); Mi-BatCoV 1B, *Miniopterus* bat coronavirus 1B (NC_010436); HCoV-OC43, human coronavirus OC43 (NC_005147); BCoV, bovine coronavirus (NC_003045); PHEV, porcine hemagglutinating encephalomyelitis virus (NC_007732); HCoV-HKU1, human coronavirus HKU1 (NC_006577); MHV, mouse hepatitis virus (NC_006852); ECoV, equine coronavirus (NC_010327); SARSr-CoV, human SARS related coronavirus (NC_004718); SARSr-Rh-BatCoV HKU3, SARS-related *Rhinolophus* bat coronavirus HKU3 (NC_009694); Ty-BatCoV-HKU4, *Tylonycteris* bat coronavirus HKU4 (NC_009019); Pi-BatCoV-HKU5, *Pipistrellus* bat coronavirus HKU5 (NC_009020); Ro-BatCoV-HKU9, *Rousettus* bat coronavirus HKU9 (NC_009021); IBV, infectious bronchitis virus (NC_001451); TCoV, turkey coronavirus (NC_010800); SW1, beluga whale coronavirus (NC_010646); BuCoV HKU11, bulbul coronavirus HKU11 (FJ376620); ThCoV HKU12, thrush coronavirus HKU12 (NC_011549); MunCoV HKU13, munia coronavirus HKU13 (NC_011550).

12.2.2 HAEMAGGLUTININ ESTERASE

In all members of *Betacoronavirus* subgroup A, a haemagglutinin esterase (HE) gene, which encodes a glycoprotein with neuraminate O-acetyl-esterase activity and the active site FGDS, is present downstream to ORF1ab and upstream to S gene (Figure 1). The HE gene of coronavirus is believed to be acquired from influenza C virus, and is the most notable example of acquisition of new genes from non-coronavirus RNA donors by heterologous recombination [20]. The presence of HE genes exclusively in members of *Betacoronavirus* subgroup A, but not members of *Betacoronavirus* subgroup B, C and D suggested that the recombination had probably occurred in the ancestor of members of *Betacoronavirus* subgroup A, after diverging from the ancestor of other subgroups of *Betacoronavirus*.

12.2.3 SPIKE

The S proteins are responsible for the "spikes" present on the surface of coronaviruses and give this family of virus the characteristic crown-like appearance under electron microscopy. The S proteins are type I membrane glycoproteins with signal peptides. The S proteins are used for receptor binding and viral entry, and are the proteins with the most variable sequences in the coronavirus genomes. In some coronaviruses, the S proteins are cleaved into the S1 and S2 domains at consensus cleavage site (RRSRR of BCoV, RRSR of HCoV-OC43, RRKRR of HCoV-HKU1, RSRR of PHEV, RRADR of MHV, RRFRR of SDAV and RRFRR of IBV) (Table 1), with the sequences of the S1 domains much more variable than the S2 domains. In all coronaviruses, most of the S protein is exposed on the outside of the virus, with a short transmembrane domain at the C terminus, followed by a short cytoplasmic tail rich in cysteine residues. Two heptad repeats are present at the C termini of the extracellular parts of the S proteins. At the moment, no bioinformatics tool is available for accurate prediction of the receptor by analyzing the amino acid sequences of the S proteins of the corresponding coronaviruses.

12.2.4 ENVELOPE AND MEMBRANE

The E and M proteins are small transmembrane proteins associated with the envelope of all coronaviruses. In some coronaviruses, such as MHV and SDAV and possibly HCoV-HKU1, the translation of the E protein is cap-independent, via an internal ribosomal entry site. Although these two genes are conserved among all coronaviruses, they are not good targets for phylogenetic studies because of their short sequences.

12.2.5 NUCLEOCAPSID

Similar to the conserved proteins encoded by ORF1ab, the N gene is also another common target for phylogenetic analysis. Due to its immunogenicity, it is also a common target for cloning and generation of recombinant proteins for serological assays.

12.2.6 OTHER SMALL ORFS

Variable numbers of small ORFs are present between the various conserved genes in different lineages in the *Coronaviridae* family (Table 1, Figure 1). In some coronaviruses, small ORFs are present downstream to the N gene (Table 1, Figure 1). Most of these small ORFs are of unknown function. One exception is the small ORFs downstream to N in feline infectious peritonitis virus (FIPV) and TGEV, which are important for virulence and viral replication/assembly respectively [21-23]. Another notable exception is the 3a protein of SARSr-CoV, which forms a transmembrane homotetramer complex with ion channel function and modulates virus release [24]. For some of these small ORFs, such as ORF3a and ORF8 of SARSr-CoV genomes, their sequences are as highly variable as those of the S proteins. In particular, the most significant difference between human SARSr-CoV and civet SARSr-CoV genomes was a 29-bp deletion in the ORF8 of human SARSr-CoV [25].

12.3 PHYLOGENY

The first impression of the phylogenetic position of a strain or species of coronavirus is usually acquired by constructing a phylogenetic tree using a short fragment of a conserved gene, such as Pol or N. However, this can sometimes be misleading because the results of phylogenetic analysis using different genes or characters can be different. When SARSr-CoV was first discovered, it was proposed that it constituted a fourth group of coronavirus [26,27]. However, analyses of the amino-terminal domain of S of SARSr-CoV revealed that 19 out of the 20 cysteine residues were spatially conserved with those of the consensus sequence for *Betacoronavirus* [28]. On the other hand, only five of the cysteine residues were spatially conserved with those of the consensus sequences in *Alphacoronavirus* and

Gammacoronavirus [28]. Furthermore, subsequent phylogenetic analysis using both complete genome sequence and proteomic approaches, it was concluded that SARSr-CoV is probably an early split-off from the *Betacoronavirus* lineage [1], and SARSr-CoV was subsequently classified as *Betacoronavirus* subgroup B and the historical *Betacoronavirus* as *Betacoronavirus* subgroup A. Therefore, the phylogenetic position of a coronavirus is best appreciated and confirmed by constructing phylogenetic trees using different genes in the coronavirus genome. The most commonly used genes along the coronavirus genome for phylogenetic studies include chymotrypsin-like protease, Pol (Figure 2), helicase, S and N, because these genes are present in all coronavirus genomes and are of significant length. The envelope and membrane genes, although present in all coronavirus genomes, are too short for phylogenetic studies. It is noteworthy that the cluster formed by the three novel avian coronaviruses BuCoV HKU11, ThCoV HKU12 and MunCoV HKU13, which was originated proposed as group 3c [2], might represent a new coronavirus genus provisionally designated *Deltacoronavirus* (Figure 2).

Using this approach of multiple gene phylogenetic studies, unique phylogeny of individual gene that may have biological significance may be discovered. During our phylogenetic study on Rh-BatCoV HKU2, another coronavirus that has was also found in the stool samples of Chinese horseshoe bats, its unique S protein phylogenetically distinct from the rest of the genome was discovered [15]. The S protein of Rh-BatCoV HKU2 is the shortest among S proteins of all coronaviruses and had less than 30% amino acid identities to those of all known coronaviruses, in contrast to other genes that showed higher amino acid identities to the corresponding genes in other members of *Alphacoronavirus*. When the S protein of Rh-BatCoV HKU2 is aligned with those of other members of *Alphacoronavirus,* many of the amino acid residues conserved among and specific to *Alphacoronavirus* were not found. Rather, the S protein of Rh-BatCoV HKU2 shares the two conserved regions of deletions both of 14 amino acids among members of *Betacoronavirus* in its C-terminus, suggesting that this segment of the S protein of Rh-BatCoV HKU2 may have co-evolved with the corresponding regions in *Betacoronavirus*. Most interestingly, a short peptide of 15 amino acids in the S protein of Rh-BatCoV HKU2 was found to be homologous to a corresponding peptide within the RBM in the

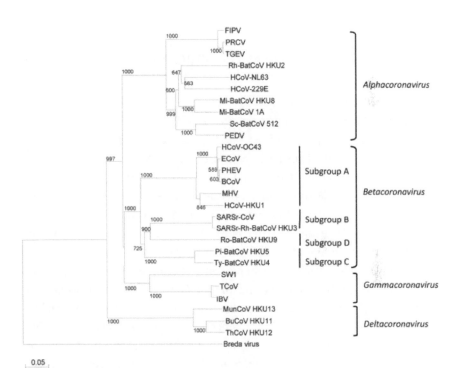

FIGURE 2: Phylogenetic analysis of RNA-dependent RNA polymerases (Pol) of coronaviruses with complete genome sequences available. The tree was constructed by the neighbor-joining method and rooted using *Breda* virus polyprotein (YP_337905). Bootstrap values were calculated from 1000 trees. 1118 amino acid positions in Pol were included. The scale bar indicates the estimated number of substitutions per 20 amino acids. All abbreviations for the coronaviruses were the same as those in Figure 1.

S1 domain of SARSr-CoV. A similar peptide was also observed in SARSr-Rh-BatCoV, but not in any other known coronaviruses. These suggested that there is a common evolutionary origin in the S protein of SARSr-CoV, SARSr-Rh-BatCoV and Rh-BatCoV HKU2, and Rh-BatCoV HKU2 might have acquired its unique S protein from a yet unidentified coronavirus through recombination.

12.4 EVOLUTIONARY RATE AND DIVERGENCE

In 1992, Sanchez et al. analyzed 13 enteric and respiratory TGEV related isolates and estimated the mutation rate of TGEV to be 7×10^{-4} nucleotide substitutions per site per year [29]. In 2005, using linear regression, maximum likelihood and Bayesian inference methods, Vijgen et al. estimated the rate of evolution in BCoV to be 4.3 (95% confidence internal 2.7 to 6.0) $\times 10^{-4}$ nucleotide substitutions per site per year [30]. The estimation of time of divergence was first extensively used in coronaviruses after the SARS epidemic for estimating the date of interspecies jumping of SARSr-CoV from civets to humans and that from BCoV to HCoV-OC43 [31,32]. Subsequently, when various novel human and animal coronaviruses were discovered, evolutionary rates and divergence time in the *Coronaviridae* family were estimated by various groups using different approaches [31,33-35]. Although Bayesian inference in BEAST is probably the most widely accepted approach and was used by most researchers, the use of different genes (ORF1ab, helicase, S and N genes) and datasets by different groups have resulted in considerable difference in the estimated history of coronaviruses. It was found that the S and N genes of PHEV, BCoV and HCoV-OC43 evolved at different rates, and the divergence time of the PHEV lineage and the HCoV-OC43 and BCoV lineage based on these two rates were 100 years different [31]. One group, using the helicase gene for analysis, has estimated the life history of coronaviruses to be as short as about 420 years [35].

Recently, we used the uncorrelated exponentially distributed relaxed clock model (UCED) in BEAST version 1.4 [36] to estimate the time of divergence of SARSr-CoV based on an alignment of a large set of SARSr-Rh-BatCoV ORF1 sequences collected over a period of five years. Under

TABLE 2: Studies on estimation of dates of divergence of SARSr-CoV.

References	Gene	No. of SARSr-CoV strains Human Civet Bat Rp3SARSr-Rh-BatCoV	Estimated mean substitution rate (no. of substitutions per site per year)	Methods for estimating TMRCA	TMRCA of human/civet SARSr-CoV (95% HPD)	TMRCA of (human/civet)/ Bat Rp3 SARSr-CoV (95% HPD)	TMRCA of (human/civet/Bat Rp3 SARSr-CoV)/ SARSr-Rh-BatCoV (95% HPD)
Zeng et al. 2003 [38]	Spike	139	-	Linear regression	Dec 2002 (Sep 2002, Jan 2003)	-	-
Salemi et al. 2004 [39]	ORF1ab	10	$4/35 \times 10^{-4b}$	Molecular clock model	-	-	-
Zhao et al. 2004 [40]	Genome	16	$8\text{-}23.8 \times 10^{-4}$	Three strategies described by the author	Spring 2002	-	-
Song et al. 2005 [32]	CDSsa	3 5	2.92×10^{-3}	Linear regression	Nov 2002	-	-
Vijaykrishna et al. 2007 [35]	Helicase	3 3 1 5	2.0×10^{-2}, 1.7×10^{-2c}	Relaxed clock model	1999 (1990-2003)	1986 (1964-2002)	1961 (1918-1995)
Hon et al. 2008 [33]	ORF1ab	13 6 1 4	2.79×10^{-3}	Various clock models	2002.63 (2002.14-2002.96)	1998.51 (1993.55-2001.32)	~1985d
Lau et al. 2010 [37]	ORF1ab	8 8 1 15	2.82×10^{-3}	Relaxed clock model	2001 (1999.16-2002.14)	1995.10 (1986.53-2000.13)	1972.39 (1935.28-1990.63)

[a] Concatenated CDS of ORF1ab, S, E, M and N. [b] The rate for all sites is 4×10^{-4}. The rate for variable sites is 35×10^{-4}. [c] Two numbers present the estimated rate of SARSr-Rh-BatCoV lineage and the estimated rate of human/civet/bat SARSr-CoV lineage respectively. [d] The date obtained from the figure of the reference but was not mentioned in the reference's text.

this model, the rates were allowed to vary at each branch drawn independently from an exponential distribution. Using this model and large dataset, the time of emergence of SARSr-CoV was at 1972, about 31 years before the SARS epidemic; that of SARSr-CoV in civet was at 1995, about eight years before the SARS epidemic; and the most recent common ancestor date of human and civet SARSr-CoV was estimated to be 2001.36, which was comparable to the dates estimated by other groups (Table 2) [37].

12.5 RECOMBINATION ANALYSIS

As a result of their unique random template switching during RNA replication, thought to be mediated by a "copy-choice" mechanism, coronaviruses have a high frequency of homologous RNA recombination [41,42]. Recombination in coronaviruses was first recognized between different strains of MHV and subsequently in other coronaviruses such as IBV, between MHV and BCoV, and between feline coronavirus (FCoV) type I and canine coronavirus (CCoV) [43-46]. As shown below, such recombination can result in the generation of coronavirus species or different genotypes within a coronavirus species. In our experience, the possibility of homologous RNA recombination and the possible part of the genome that recombination has taken place are usually first appreciated using bootscan analysis or phylogenetic analysis using different parts of the coronavirus genome. Other methods for recombination analysis, such as those in the RDP3 package, are also available. Then, the exact site of homologous RNA recombination would be best revealed by multiple sequence alignment.

The best documented example of generation of coronavirus species through homologous RNA recombination is the generation of FCoV type II by double recombination between FCoV type I and CCoV. It was first observed that the sequence of the S protein in FCoV type II was closely related to that of CCoV [47,48] but the sequence downstream of the E gene in FCoV type II was more closely related to that of FCoV type I strain than to CCoV [49,50]. This observation suggested that there might have been a homologous RNA recombination event between the genomes of CCoV and FCoV type I, resulting in the generation of FCoV type II. Further analysis by multiple sequence alignments pinpointed the site of recombination to a region in the E gene. A few years later, an additional recombination region in the Pol gene was also discovered, and it was concluded that FCoV type II originated from two recombination events between the genomes of CCoV and FCoV type I [43].

As for the generation of different genotypes in a coronavirus species through homologous RNA recombination, the best documented example is HCoV-HKU1. The possibility of homologous RNA recombination was first suspected when a few strains of HCoV-HKU1 showed differential clustering when the Pol, S and N genes were used for phylogenetic tree construction [51]. This observation has led to our subsequent study on complete genome sequencing of 22 strains of HCoV-HKU1. Recombination analysis by bootscan analysis and phylogenetic analysis using different parts of the 22 complete genomes revealed extensive recombination in different parts of the genomes, resulting in the generation of three genotypes, A, B and C, of HCoV-HKU1 [52]. Using multiple sequence alignment, two sites of recombination were pinpointed. The first one was observed in a stretch of 143 nucleotides near the 3' end of nsp6, where recombination between HCoV-HKU1 genotype B and genotype C has generated genotype A; and the second one in another stretch of 29 nucleotides near the 3' end of nsp16, where recombination between HCoV-HKU1 genotype A and genotype B has generated genotype C [52].

12.6 CODON USAGE BIAS

Recently, using the complete genome sequences of the 19 coronavirus genomes, we analyzed the codon usage bias in coronaviruses as well as selection of CpG suppressed clones by the immune system and cytosine deamination being the two major independent biochemical and biological selective forces that has shaped such codon usage bias [53]. In the study, we showed that the mean CpG relative abundance in the coronavirus genomes is markedly suppressed [53]. However, we observed that only CpG containing codons in the context of purine-CpG (ACG and GCG), pyrimidine-CpG (UCG and CCG) and CpG-purine (CGA and CGG); but not CpG-pyrimidine (CGU and CGC); are suppressed. However, when trinucleotide frequencies were analyzed in the 19 coronavirus genomes, all the eight trinucleotides with CpG were suppressed [53]. These indicate that another force that has led to an increase use of CGU and CGC as codons for arginine, but does not act on trinucleotides over the whole genome in general, is probably present. Furthermore, this force is probably unrelated to the relative abundance of the corresponding tRNA molecules in the hosts of the coronaviruses, as the pattern of bias in the hosts is not the same as that in the coronaviruses.

In addition to CpG suppression, marked cytosine deamination is also observed in all the 19 coronavirus genomes [53]. Using the six amino acids that are only encoded by NNC or NNU (asparagine, histidine, aspartic acid, tyrosine, cysteine and phenylalanine), hence excluding most other pressures that may affect the relative abundance of cytosine and uracil, it was observed that all NNU are markedly over represented with usage fractions of more than 0.700, whereas the usage fractions of all NNC are less than 0.300 [53]. For all codons that encode the same amino acid and with either U or C in any position, the usage fraction of the codon that uses U is always higher than the one that uses C in all coronaviruses. Furthermore, the percentage of U showed strong inverse relationships with the percentage of C in the coronavirus genomes. These suggest that cytosine deamination is another important biochemical force that shaped coronavirus evolution.

Interestingly, among all the 19 coronaviruses, HCoV-HKU1 showed the most extreme codon usage bias. HCoV-HKU1 is the only coronavirus that had effective number of codons outside the mean ± 2 standard deviations range. In addition, HCoV-HKU1 also possessed the lowest G + C content, highest GC skew, lowest percentages of G and C and highest percentage of U among all coronavirus genomes. Furthermore, HCoV-HKU1 showed extremely high NNU/NNC ratio of 8.835. The underlying mechanism for the extreme codon usage bias, cytosine deamination and G + C content in HCoV-HKU1 is intriguing.

12.7 DATABASE

Rapid and accurate batch sequence retrieval is always the cornerstone and bottleneck for all kinds of comparative genomics and bioinformatics analysis. During the process of batch sequence retrieval for comparative genomics and other bioinformatics analysis of the coronavirus genomes that we have sequenced, we encountered a number of major problems about the coronavirus sequences in GenBank and other coronavirus databases. First, in GenBank, the non-structural proteins encoded by ORF1ab are not annotated. Second, in all databases, the annotations for the non-structural proteins encoded by ORFs downstream to ORF1ab are often confusing because they are not annotated using a standardized system. Third, multiple accession numbers are often present for reference sequences. These problems will often lead to confusion during sequence retrieval. Fourth, coronaviruses, especially SARSr-CoV, amplified from different specimens may contain the same gene or genome sequences, which will lead to redundant work when they are analyzed. In view of these problems, we have developed a comprehensive database, CoVDB, of annotated coronavirus genes and genomes, which offers rapid, efficient and user-friendly batch sequence retrieval and analysis [54]. In CoVDB, first, annotations on all non-structural proteins in the polyprotein encoded by ORF1ab of every single sequence were performed. Second, annotation was performed for the non-structural proteins encoded by ORFs downstream to ORF1ab using a standardized system. Third, all sequences with identical nucleotide sequences were labeled and one can choose to show or not to show strains with identical sequences. Fourth, this database contains not only complete coronavirus genome sequences, but also incomplete genomes and their genes. This is useful because some genes of coronaviruses, such as Pol, S and N, are sequenced much more frequently than others because they are either most conserved or least conserved, and therefore are particularly important for primers design for RT-PCR assays and evolutionary studies.

12.8 CONCLUDING REMARKS

After the SARS epidemic, there has been a marked increase in the number of coronaviruses discovered and coronavirus genomes being sequenced. This increase in the number of coronavirus species and genomes, a comprehensive and user-friendly database for efficient sequence retrieval, and the ever improving bioinformatics tools have enabled us to perform meaningful genomic, phylogenetic, evolutionary rate and divergence, recombination, and other bioinformatics analyses on the *Coronaviridae* family. Three genera, *Alphacoronavirus, Betacoronavirus* and *Gammacoronavirus*, have been used to replace the traditional group 1, 2 and 3 coronaviruses. A fourth genus, *Deltacoronavirus*, which includes BuCoV HKU11, ThCoV HKU12 and MunCoV HKU13, is likely to emerge. Under this new classification system, bat coronaviruses dominate the *Alphacoronavirus* and *Betacoronavirus* genera and bird coronaviruses dominate the *Gammacoronavirus* and *Deltacoronavirus* genera. This huge diversity of coronaviruses in bats and birds has made them excellent gene pools for coronaviruses in these four genera [55].

REFERENCES

1. Snijder, E.J.; Bredenbeek, P.J.; Dobbe, J.C.; Thiel, V.; Ziebuhr, J.; Poon, L.L.; Guan, Y.; Rozanov, M.; Spaan, W.J.; Gorbalenya, A.E. Unique and conserved features of genome and proteome of SARS-coronavirus, an early split-off from the coronavirus group 2 lineage. J. Mol. Biol. 2003, 331, 991-1004.
2. Woo, P.C.; Lau, S.K.; Lam, C.S.; Lai, K.K.; Huang, Y.; Lee, P.; Luk, G.S.; Dyrting, K.C.; Chan, K.H.; Yuen, K.Y. Comparative analysis of complete genome sequences of three avian coronaviruses reveals a novel group 3c coronavirus. J. Virol. 2009, 83, 908-917.
3. Woo, P.C.; Wang, M.; Lau, S.K.; Xu, H.; Poon, R.W.; Guo, R.; Wong, B.H.; Gao, K.; Tsoi, H.W.; Huang, Y.; Li, K.S.; Lam, C.S.; Chan, K.H.; Zheng, B.J.; Yuen, K.Y. Comparative analysis of twelve genomes of three novel group 2c and group 2d coronaviruses reveals unique group and subgroup features. J. Virol. 2007, 81, 1574-1585.
4. ICTV Virus Taxonomy: 2009 Release. Available online: http://ictvonline.org/ virusTaxonomy.asp?version=2009 (accessed on 1 August 2010).
5. Liu, S.; Chen, J.; Chen, J.; Kong, X.; Shao, Y.; Han, Z.; Feng, L.; Cai, X.; Gu, S.; Liu, M. Isolation of avian infectious bronchitis coronavirus from domestic peafowl (Pavo cristatus) and teal (Anas). J. Gen. Virol. 2005, 86, 719-725.
6. Tang, X.C.; Zhang, J.X.; Zhang, S.Y.; Wang, P.; Fan, X.H.; Li, L.F.; Li, G.; Dong, B.Q.; Liu, W.; Cheung, C.L.; Xu, K.M.; Song, W.J.; Vijaykrishna, D.; Poon, L.L.; Peiris, J.S.; Smith, G.J.; Chen, H.; Guan, Y. Prevalence and genetic diversity of coronaviruses in bats from China. J. Virol. 2006, 80, 7481-7490.
7. Woo, P.C.; Lau, S.K.; Li, K.S.; Poon, R.W.; Wong, B.H.; Tsoi, H.W.; Yip, B.C.; Huang, Y.; Chan, K.H.; Yuen, K.Y. Molecular diversity of coronaviruses in bats. Virology 2006, 351, 180-187.
8. Fouchier, R.A.; Hartwig, N.G.; Bestebroer, T.M.; Niemeyer, B.; de Jong, J.C.; Simon, J.H.; Osterhaus, A.D. A previously undescribed coronavirus associated with respiratory disease in humans. Proc. Natl. Acad. Sci. U. S. A. 2004, 101, 6212-6216.

9. van der Hoek, L.; Pyrc, K.; Jebbink, M.F.; Vermeulen-Oost, W.; Berkhout, R.J.; Wol-
    thers, K.C.; Wertheim-van Dillen, P.M.; Kaandorp, J.; Spaargaren, J.; Berkhout, B.
    Identification of a new human coronavirus. Nat. Med. 2004, 10, 368-373.

10. Woo, P.C.; Lau, S.K.; Chu, C.M.; Chan, K.H.; Tsoi, H.W.; Huang, Y.; Wong, B.H.;
    Poon, R.W.; Cai, J.J.; Luk, W.K.; Poon, L.L.; Wong, S.S.; Guan, Y.; Peiris, J.S.;
    Yuen, K.Y. Characterization and complete genome sequence of a novel coronavirus,
    coronavirus HKU1, from patients with pneumonia. J. Virol. 2005, 79, 884-895.

11. Lau, S.K.; Woo, P.C.; Li, K.S.; Huang, Y.; Tsoi, H.W.; Wong, B.H.; Wong, S.S.;
    Leung, S.Y.; Chan, K.H.; Yuen, K.Y. Severe acute respiratory syndrome coronavi-
    rus-like virus in Chinese horseshoe bats. Proc. Natl. Acad. Sci. U. S. A. 2005, 102,
    14040-14045.

12. Chu, D.K.; Peiris, J.S.; Chen, H.; Guan, Y.; Poon, L.L. Genomic characterizations of
    bat coronaviruses (1A, 1B and HKU8) and evidence for co-infections in Miniopterus
    bats. J. Gen. Virol. 2008, 89, 1282-1287.

13. Mihindukulasuriya, K.A.; Wu, G.; St Leger, J.; Nordhausen, R.W.; Wang, D. Identi-
    fication of a novel coronavirus from a beluga whale by using a panviral microarray.
    J. Virol. 2008, 82, 5084-5088.

14. Zhang, J.; Guy, J.S.; Snijder, E.J.; Denniston, D.A.; Timoney, P.J.; Balasuriya, U.B.
    Genomic characterization of equine coronavirus. Virology 2007, 369, 92-104.

15. Lau, S.K.; Woo, P.C.; Li, K.S.; Huang, Y.; Wang, M.; Lam, C.S.; Xu, H.; Guo, R.;
    Chan, K.H.; Zheng, B.J.; Yuen, K.Y. Complete genome sequence of bat coronavi-
    rus HKU2 from Chinese horseshoe bats revealed a much smaller spike gene with
    a different evolutionary lineage from the rest of the genome. Virology 2007, 367,
    428-439.

16. Lai, M.M.; Perlman, S.; L., A. Coronaviridae. In Fields virology, 5th ed.; Knipe,
    D.M., Howley, P.M., Eds.; Lippincott Williams and Wilkins: Philadelphia, PA, USA,
    2007; pp. 1305-1335.

17. Lai, M.M.; Baric, R.S.; Makino, S.; Keck, J.G.; Egbert, J.; Leibowitz, J.L.; Stohl-
    man, S.A. Recombination between nonsegmented RNA genomes of murine corona-
    viruses. J. Virol. 1985, 56, 449-456.

18. Woo, P.C.; Huang, Y.; Lau, S.K.; Tsoi, H.W.; Yuen, K.Y. In silico analysis of OR-
    F1ab in coronavirus HKU1 genome reveals a unique putative cleavage site of coro-
    navirus HKU1 3C-like protease. Microbiol. Immunol. 2005, 49, 899-908.

19. Nasr, F.; Filipowicz, W. Characterization of the Saccharomyces cerevisiae cyclic
    nucleotide phosphodiesterase involved in the metabolism of ADP-ribose 1",2"-cy-
    clic phosphate. Nucleic Acids Res. 2000, 28, 1676-1683.

20. Luytjes, W.; Bredenbeek, P.J.; Noten, A.F.; Horzinek, M.C.; Spaan, W.J. Sequence
    of mouse hepatitis virus A59 mRNA 2: indications for RNA recombination between
    coronaviruses and influenza C virus. Virology 1988, 166, 415-422.

21. Haijema, B.J.; Volders, H.; Rottier, P.J. Live, attenuated coronavirus vaccines
    through the directed deletion of group-specific genes provide protection against fe-
    line infectious peritonitis. J. Virol. 2004, 78, 3863-3871.

22. Olsen, C.W. A review of feline infectious peritonitis virus: molecular biology, im-
    munopathogenesis, clinical aspects, and vaccination. Vet. Microbiol. 1993, 36, 1-37.

23. Tung, F.Y.; Abraham, S.; Sethna, M.; Hung, S.L.; Sethna, P.; Hogue, B.G.; Brian,
    D.A. The 9-kDa hydrophobic protein encoded at the 3' end of the porcine transmis-

sible gastroenteritis coronavirus genome is membrane-associated. Virology 1992, 186, 676-683.

24. Lu, W.; Zheng, B.J.; Xu, K.; Schwarz, W.; Du, L.; Wong, C.K.; Chen, J.; Duan, S.; Deubel, V.; Sun, B. Severe acute respiratory syndrome-associated coronavirus 3a protein forms an ion channel and modulates virus release. Proc. Natl. Acad. Sci. U. S. A. 2006, 103, 12540-12545.

25. Guan, Y.; Zheng, B.J.; He, Y.Q.; Liu, X.L.; Zhuang, Z.X.; Cheung, C.L.; Luo, S.W.; Li, P.H.; Zhang, L.J.; Guan, Y.J.; Butt, K.M.; Wong, K.L.; Chan, K.W.; Lim, W.; Shortridge, K.F.; Yuen, K.Y.; Peiris, J.S.; Poon, L.L. Isolation and characterization of viruses related to the SARS coronavirus from animals in southern China. Science 2003, 302, 276-278.

26. Marra, M.A.; Jones, S.J.; Astell, C.R.; Holt, R.A.; Brooks-Wilson, A.; Butterfield, Y.S.; Khattra, J.; Asano, J.K.; Barber, S.A.; Chan, S.Y.; et al. The Genome sequence of the SARS-associated coronavirus. Science 2003, 300, 1399-1404.

27. Rota, P.A.; Oberste, M.S.; Monroe, S.S.; Nix, W.A.; Campagnoli, R.; Icenogle, J.P.; Penaranda, S.; Bankamp, B.; Maher, K.; Chen, M.H.; et al. Characterization of a novel coronavirus associated with severe acute respiratory syndrome. Science 2003, 300, 1394-1399.

28. Eickmann, M.; Becker, S.; Klenk, H.D.; Doerr, H.W.; Stadler, K.; Censini, S.; Guidotti, S.; Masignani, V.; Scarselli, M.; Mora, M.; Donati, C.; Han, J.H.; Song, H.C.; Abrignani, S.; Covacci, A.; Rappuoli, R. Phylogeny of the SARS coronavirus. Science 2003, 302, 1504-1505.

29. Sanchez, C.M.; Gebauer, F.; Sune, C.; Mendez, A.; Dopazo, J.; Enjuanes, L. Genetic evolution and tropism of transmissible gastroenteritis coronaviruses. Virology 1992, 190, 92-105.

30. Vijgen, L.; Keyaerts, E.; Moes, E.; Thoelen, I.; Wollants, E.; Lemey, P.; Vandamme, A.M.; Van Ranst, M. Complete genomic sequence of human coronavirus OC43: molecular clock analysis suggests a relatively recent zoonotic coronavirus transmission event. J. Virol. 2005, 79, 1595-1604.

31. Vijgen, L.; Keyaerts, E.; Lemey, P.; Maes, P.; Van Reeth, K.; Nauwynck, H.; Pensaert, M.; Van Ranst, M. Evolutionary history of the closely related group 2 coronaviruses: porcine hemagglutinating encephalomyelitis virus, bovine coronavirus, and human coronavirus OC43. J. Virol. 2006, 80, 7270-7274.

32. Song, H.D.; Tu, C.C.; Zhang, G.W.; Wang, S.Y.; Zheng, K.; Lei, L.C.; Chen, Q.X.; Gao, Y.W.; Zhou, H.Q.; Xiang, H.; et al. Cross-host evolution of severe acute respiratory syndrome coronavirus in palm civet and human. Proc. Natl. Acad. Sci. U. S. A. 2005, 102, 2430-2435.

33. Hon, C.C.; Lam, T.Y.; Shi, Z.L.; Drummond, A.J.; Yip, C.W.; Zeng, F.; Lam, P.Y.; Leung, F.C. Evidence of the recombinant origin of a bat severe acute respiratory syndrome (SARS)-like coronavirus and its implications on the direct ancestor of SARS coronavirus. J. Virol. 2008, 82, 1819-1826.

34. Pyrc, K.; Dijkman, R.; Deng, L.; Jebbink, M.F.; Ross, H.A.; Berkhout, B.; van der Hoek, L. Mosaic structure of human coronavirus NL63, one thousand years of evolution. J. Mol. Biol. 2006, 364, 964-973.

35. Vijaykrishna, D.; Smith, G.J.; Zhang, J.X.; Peiris, J.S.; Chen, H.; Guan, Y. Evolutionary insights into the ecology of coronaviruses. J. Virol. 2007, 81, 4012-4020.

36. BEAST Home Page. http://beast.bio.ed.ac.uk/Main_Page (accessed on 1 August 2010).

37. Lau, S.K.; Li, K.S.; Huang, Y.; Shek, C.T.; Tse, H.; Wang, M.; Choi, G.K.; Xu, H.; Lam, C.S.; Guo, R.; Chan, K.H.; Zheng, B.J.; Woo, P.C.; Yuen, K.Y. Ecoepidemiology and complete genome comparison of different strains of severe acute respiratory syndrome-related Rhinolophus bat coronavirus in China reveal bats as a reservoir for acute, self-limiting infection that allows recombination events. J. Virol. 2010, 84, 2808-2819.

38. Zeng, F.; Chow, K.Y.; Leung, F.C. Estimated timing of the last common ancestor of the SARS coronavirus. N. Engl. J. Med. 2003, 349, 2469-2470.

39. Salemi, M.; Fitch, W.M.; Ciccozzi, M.; Ruiz-Alvarez, M.J.; Rezza, G.; Lewis, M.J. Severe acute respiratory syndrome coronavirus sequence characteristics and evolutionary rate estimate from maximum likelihood analysis. J. Virol. 2004, 78, 1602-1603.

40. Zhao, Z.; Li, H.; Wu, X.; Zhong, Y.; Zhang, K.; Zhang, Y.P.; Boerwinkle, E.; Fu, Y.X. Moderate mutation rate in the SARS coronavirus genome and its implications. BMC Evol. Biol. 2004, 4, 21.

41. Lai, M.M. RNA recombination in animal and plant viruses. Microbiol. Rev. 1992, 56, 61-79.

42. Pasternak, A.O.; Spaan, W.J.; Snijder, E.J. Nidovirus transcription: how to make sense...? J. Gen. Virol. 2006, 87, 1403-1421.

43. Herrewegh, A.A.; Smeenk, I.; Horzinek, M.C.; Rottier, P.J.; de Groot, R.J. Feline coronavirus type II strains 79-1683 and 79-1146 originate from a double recombination between feline coronavirus type I and canine coronavirus. J. Virol. 1998, 72, 4508-4514.

44. Keck, J.G.; Matsushima, G.K.; Makino, S.; Fleming, J.O.; Vannier, D.M.; Stohlman, S.A.; Lai, M.M. In vivo RNA-RNA recombination of coronavirus in mouse brain. J. Virol. 1988, 62, 1810-1813.

45. Kottier, S.A.; Cavanagh, D.; Britton, P. Experimental evidence of recombination in coronavirus infectious bronchitis virus. Virology 1995, 213, 569-580.

46. Lavi, E.; Haluskey, J.A.; Masters, P.S. The pathogenesis of MHV nucleocapsid gene chimeric viruses. Adv. Exp. Med. Biol. 1998, 440, 537-541.

47. Motokawa, K.; Hohdatsu, T.; Aizawa, C.; Koyama, H.; Hashimoto, H. Molecular cloning and sequence determination of the peplomer protein gene of feline infectious peritonitis virus type I. Arch. Virol. 1995, 140, 469-480.

48. Wesseling, J.G.; Vennema, H.; Godeke, G.J.; Horzinek, M.C.; Rottier, P.J. Nucleotide sequence and expression of the spike (S) gene of canine coronavirus and comparison with the S proteins of feline and porcine coronaviruses. J. Gen. Virol. 1994, 75 (Pt 7), 1789-1794.

49. Herrewegh, A.A.; Vennema, H.; Horzinek, M.C.; Rottier, P.J.; de Groot, R.J. The molecular genetics of feline coronaviruses: comparative sequence analysis of the ORF7a/7b transcription unit of different biotypes. Virology 1995, 212, 622-631.

50. Motokawa, K.; Hohdatsu, T.; Hashimoto, H.; Koyama, H. Comparison of the amino acid sequence and phylogenetic analysis of the peplomer, integral membrane and nucleocapsid proteins of feline, canine and porcine coronaviruses. Microbiol. Immunol. 1996, 40, 425-433.

51. Woo, P.C.; Lau, S.K.; Tsoi, H.W.; Huang, Y.; Poon, R.W.; Chu, C.M.; Lee, R.A.; Luk, W.K.; Wong, G.K.; Wong, B.H.; Cheng, V.C.; Tang, B.S.; Wu, A.K.; Yung, R.W.; Chen, H.; Guan, Y.; Chan, K.H.; Yuen, K.Y. Clinical and molecular epidemiological features of coronavirus HKU1-associated community-acquired pneumonia. J. Infect. Dis. 2005, 192, 1898-1907.

52. Woo, P.C.; Lau, S.K.; Yip, C.C.; Huang, Y.; Tsoi, H.W.; Chan, K.H.; Yuen, K.Y. Comparative analysis of 22 coronavirus HKU1 genomes reveals a novel genotype and evidence of natural recombination in coronavirus HKU1. J. Virol. 2006, 80, 7136-7145.

53. Woo, P.C.; Wong, B.H.; Huang, Y.; Lau, S.K.; Yuen, K.Y. Cytosine deamination and selection of CpG suppressed clones are the two major independent biological forces that shape codon usage bias in coronaviruses. Virology 2007, 369, 431-442.

54. Huang, Y.; Lau, S.K.; Woo, P.C.; Yuen, K.Y. CoVDB: a comprehensive database for comparative analysis of coronavirus genes and genomes. Nucleic Acids Res. 2008, 36, D504-511.

55. Woo, P.C.; Lau, S.K.; Huang, Y.; Yuen, K.Y. Coronavirus diversity, phylogeny and interspecies jumping. Exp. Biol. Med. (Maywood) 2009, 234, 1117-1127.

PART V

COMPANION DIAGNOSTICS

CHAPTER 13

APPLICATIONS OF NEXT-GENERATION SEQUENCING TECHNOLOGIES TO DIAGNOSTIC VIROLOGY

LUISA BARZON, ENRICO LAVEZZO, VALENTINA MILITELLO, STEFANO TOPPO, AND GIORGIO PALÙ

13.1 INTRODUCTION

Novel DNA sequencing techniques, referred to as "next-generation" sequencing (NGS), provide high speed and throughput that can produce an enormous volume of sequences. The most important advantage provided by these platforms is the determination of the sequence data from single DNA fragments of a library that are segregated in chips, avoiding the need for cloning in vectors prior to sequence acquisition.

The first next-generation high-throughput sequencing technology, the 454 FLX pyrosequencing platform (http://www.454.com/), which was developed by 454 Life Sciences and later bought by Roche, became available in 2005. In early 2007, Illumina released the Genome Analyzer (http://www.illumina.com), developed by Solexa GA, and more recently, SOLiD was released by Applied Biosystems (http://www.appliedbiosystems.

This chapter was originally published under the Creative Commons Attribution License. Barzon L, Lavezzo E, Militello V, Toppo S, and Palù G. Applications of Next-Generation Sequencing Technologies to Diagnostic Virology. International Journal of Molecular Sciences 12 (2011). doi:10.3390/ijms12117861.

com). This field is in rapid expansion and novel and improved platforms are continuously being developed and released, like Heliscope by Helicos (http://www.helicosbio.com/), Ion Torrent PGM by Life Technologies (http://www.iontorrent.com/) and a real-time sequencing platform by Pacific Biosciences (http://www.pacificbiosciences.com/).

While the platform developed by Pacific Biosciences, as well as other novel sequencing platforms, are referred as "third-generation" because they sequence processively single large DNA molecules without the need to halt between read steps, 454 pyrosequencing, Illumina GA and SOLiD methods represent the "second generation" systems, able to sequence populations of amplified template-DNA molecules with a typical "wash-and-scan" technique [1]. Given these criteria, Ion Torrent PGM and Heliscope sit between "second-" and "third-generation" technologies, since they do not completely fulfill the features assigned to each category.

These NGS methods have different underlying biochemistries and differ in sequencing protocol (sequencing by synthesis for 454 pyrosequencing, Illumina GA, Ion Torrent PGM and Heliscope, sequencing by ligation for SOLiD), throughput, and for sequence length (Table 1). Thus, the SOLiD system may be more suitable for applications that require a very high throughput of sequences, but not long reads, such as whole genome re-sequencing or RNA-sequencing projects, while both 454 and Illumina provide data suitable for de novo assembly and the relative long length of 454 FLX (and its smaller version GS Junior) reads allows deep sequencing of amplicons, with applications in microbial and viral metagenomics and analysis of viral quasispecies, as described in this review. The technical features of NGS methods (reviewed in refs. [2,3]) will not be described in this review, which is focused on the diagnostic applications of NGS in clinical virology.

13.2 APPLICATIONS OF NGS TECHNOLOGIES TO DIAGNOSTIC VIROLOGY

NGS technologies are currently used for whole genome sequencing, investigation of genome diversity, metagenomics, epigenetics, discovery of non-coding RNAs and protein-binding sites, and gene-expression profiling by

TABLE 1: Features of "next-generation" sequencing (NGS) platforms.

	Maximum Throughput Mb/run	Mean Length (nucleotide)	Error rate*	Applications	Main source of errors
454 FLX	700	~800 (for shotgun experiments) ~400 (for amplicon experiments)	10^{-3}–10^{-4}	De novo genome sequencing and resequencing, target resequencing, genotyping, metagenomics	Intensity cutoff, homopolymers, signal cross-talk interference among neighbors, amplification, mixed beads
Illumina	6,000	~100	10^{-2}–10^{-3}	Genome resequencing, quantitative transcriptomics, genotyping, metagenomics	Signal interference among neighboring clusters, homopolymers, phasing, nucleotide labeling, amplification, low coverage of AT rich regions
SOLiD	20,000	~50	10^{-2}–10^{-3}	Genome resequencing, quantitative transcriptomics, genotyping	Signal interference among neighbours, phasing, nucleotide labeling, signal degradation, mixed beads, low coverage of AT rich regions
Helicos	21,000–35,000	~35	10^{-2}	Non amplifiable samples, PCR free and unbiased quantitative analyses	Polymerase employed, molecule loss, low intensities
Ion Torrent PGM	1,000	~200	3×10^{-2}	De novo genome sequencing and resequencing, target resequencing, genotyping, RNA-seq on low-complexity transcriptome, metagenomics	Homopolymers, amplification
GS Junior	~35	~400	10^{-3}–10^{-4}	Target resequencing (amplicons), genotyping	Intensity cutoff, homopolymers, signal cross-talk interference among neighbors, amplification, mixed beads

Error rate considering only substitutions and not insertions/deletions.

RNA sequencing (reviewed in refs. [2–6]). Typical applications of NGS methods in microbiology and virology, besides high-throughput whole genome sequencing, are discovery of new microorganisms and viruses by using metagenomic approaches, investigation of microbial communities in the environment and in human body niches in healthy and disease conditions, analysis of viral genome variability within the host (i.e., quasispecies), detection of low-abundance antiviral drug-resistance mutations in patients with human immunodeficiency virus (HIV) infection or viral hepatitis, as outlined in this review article.

13.2.1 DETECTION OF UNKNOWN VIRAL PATHOGENS AND DISCOVERY OF NOVEL VIRUSES

The human population is exposed to an increasing burden of infectious diseases caused by the emergence of new previously unrecognized viruses. Climate changes, globalization, settlements near animal and livestock habitats, and the increased number of immunocompromised people probably contribute to the emergence and spread of new infections [7]. In addition, several clinical syndromes are suspected to be of viral etiology, but the causing agent cannot be isolated and recognized by traditional culture and molecular methods. Thus, there is the need to improve methods for the identification of unsuspected viral pathogens or new viruses. Subtractive techniques, such as representational difference analysis or random sequencing of plasmid libraries of nuclease resistant fragments of viral genomes, have led in the past to the discovery of several viruses, including human herpesvirus type 8 [8], human GB virus [9], Torque Teno Virus [10], bocavirus [11], human parvovirus 4 [12], WU polyomavirus [13] and KI polyomavirus [14]. These techniques are poorly sensitive and time-consuming, and thus are unsuitable for large scale analysis. For these purposes, NGS-based methods have been developed. However, traditional cloning and sequencing methods can be relatively simple and sensitive for the discovery of new viruses when used for the analysis of otherwise sterile samples, and may represent an alternative to NGS. One of these methods is termed VIDISCA (Virus Discovery cDNA Amplified Fragment Length Polymorphism Analysis) and may be applied to sterile specimens, such as cell culture supernatants [15]. In this method, samples are ultracentrifuged for viral particle enrichment and treated by DNase and RNase to digest away cellular nucleic acids. Capsid-protected viral nucleic acids are then purified, converted to double stranded DNA, digested with restriction enzymes and ligated to oligonucleotide adaptors, which are used as primer binding sites for comparative PCR [15]. This method was described originally in the context of the discovery of severe acute respiratory syndrome coronavirus (SARS-CoV) in 2004 [16]. Microarray-based diagnostic assays have also been used to characterize previously unknown viruses, such as SARS-CoVs [17], but require information on the genome of the virus or closely related viruses that are under investigation [18].

High throughput NGS techniques represent a powerful tool which can be applied to metagenomics-based strategies for the detection of unknown disease-associated viruses and for the discovery of novel human viruses [19,20]. Compared with microarray-based assays, NGS methods offer the advantage of higher sensitivity and the potential to detect the full spectrum of viruses, including unknown and unexpected viruses.

One of the first applications of NGS for pathogen discovery was the investigation of three patients who died of a febrile illness a few weeks after transplantation of solid organs from a single donor and for whom conventional microbiological and molecular tests, as well as microarray analysis for a wide range of infectious agents, had not been informative [21]. In this study, RNA was purified from blood, cerebrospinal fluid and tissue specimens from transplant recipients and, after digestion with DNase to eliminate human DNA, RNA was reverse-transcribed and amplified with random primers. Amplification products were pooled and sequenced with the use of the 454 pyrosequencing platform. After subtraction of sequences of vertebrates and highly repetitive sequences, contiguous sequences were assembled and compared with motifs represented in databases of microbes, leading to the identification of putative protein sequences which were consistent with an Old World arenavirus. Additional sequence analysis showed that it was a new arenavirus related to lymphocytic choriomeningitis viruses. Further serological and immunohistochemical analyses documented that the virus was transmitted through organ transplantation [21].

A similar strategy, based on unbiased high-throughput sequencing using 454 pyrosequencing for the direct diagnosis of viral infections in clinical specimens, has been used in different diagnostic settings, such as the investigation of patients during seasonal influenza and norovirus outbreaks [22], the identification of an astrovirus as a causative agent for encephalitis in a boy with agammaglobulinemia, after conventional methods had failed to identify an infectious agent [23], and the identification of a hemorrhagic fever-associated arenavirus from South Africa (Lujo virus) [24].

When implemented into virus-discovery methods based on shotgun sequencing, next-generation technologies greatly enhance turnaround time and sensitivity. For example, the 454 system was implemented into a virus discovery assay based on an improved version of the VIDISCA protocol to minimize rRNA contamination [25]. Likewise, the association of NGS techniques with rolling circle amplification (RCA), another method for virus discovery, could greatly increase its performance. RCA employs the PhiX29 polymerase to selectively amplify small double stranded DNA (dsDNA) molecules and is used to amplify circular genomes of DNA viruses and bacteria plasmids [26]. Recently, RCA led to the identification and whole genome sequencing of novel human papillomaviruses and polyomaviruses [27], including human polyomaviruses 6 and 7 (HPyV6 and HPyV7), detected in cutaneous swab specimens of healthy persons [28], and trichodysplasia spinulosa–associated polyomavirus (TSPyV), detected in skin lesions from immunocompromised patients [29].

Besides 454 pyrosequencing, short-read-based metagenomic methods using the Illumina GA platform have also been used to detect unknown viruses in clinical specimens. The Illumina GA platform allowed to identify influenza A viruses from swab specimens and de novo assembly of its genome [30–32]. It also led to the detection of viral pathogens in nasopharyngeal aspirate samples from patients with acute lower respiratory tract infections [33], such as a new enterovirus, named enterovirus 109 (EV109) detected in a cohort of Nicaraguan children with viral respiratory illness [34].

A comparative study of the analytical sensitivity of the two platforms, 454 pyrosequencing and Illumina GA, for the detection of viruses in biological samples was done on a set of samples which were artificially spiked with eleven different viruses [35]. The Illumina method had a much greater sensitivity than 454, approaching that of optimized quantitative real-time PCR. However, at low viral concentration in the specimen, the number of reads generated by the Illumina platform was too small for de novo assembly of viral genome sequences [35].

Vector-borne viruses and zoonotic viruses represent another important and challenging field for viral discovery. The feasibility of detecting arthropod-borne viruses was explored in *Aedes aegypti* mosquitoes experimentally infected with dengue virus and pooled with noninfected mosquitoes to simulate samples derived from ongoing arbovirus surveillance programs [36]. Total RNA was purified from mosquito pools, reverse-transcribed using random primers and subjected to 454 pyrosequencing, which led to the correct identification of infected mosquito pools [36].

Another interesting strategy to discover arthropod-borne viruses exploits the property of invertebrates to respond to infection by processing viral RNA genomes into siRNAs of discrete sizes. A recent study on small RNA libraries sequenced by NGS platforms [37] showed that viral small silencing RNAs produced by invertebrate animals are overlapping in sequence and can assemble into long contiguous fragments of the invading viral genome. Based on this finding, an approach of virus discovery in invertebrates by deep sequencing and assembly of total small RNAs was developed and applied to the analysis of contigs (i.e., a contiguous length of genomic sequences in which the order of bases is known to a high confidence level) assembled from published small RNA libraries. Five previously undescribed viruses from cultured Drosophila cells and adult mosquitoes were discovered, including three with a positive-strand RNA genome and two with a dsRNA genome [37]. This strategy for virus discovery based on deep sequencing of small RNAs has been also successfully used in plant virology [38].

Bats are reservoirs for emerging zoonotic viruses that cause diseases in humans and livestock, including lyssaviruses, filoviruses, paramyxoviruses, and SARS-CoV. In a surveillance study focused on the discovery of bat-transmitted pathogens, gastrointestinal tissue obtained from bats was analyzed by coronavirus consensus PCR and unbiased high-throughput pyrosequencing that revealed the presence of sequences of a new coronavirus, related to those of SARS-CoV [39].

13.2.2 DETECTION OF TUMOR VIRUSES

Computational subtraction analysis of data obtained using conventional shotgun sequencing methods has been used to identify viral sequences (e.g., HBV, HCMV, human papillomaviruses 18 and 16, HHV8, HCV, EBV and human spumavirus) in EST libraries derived from normal and cancerous tissues [40] and in post-transplant lymphoproliferative disorder tissue [41]. In these studies, computational subtraction analysis relied on sequence data gathered for other purposes as the yield of viral sequences was very low due to the predominance of human sequences. However, exploiting the great amount of sequencing data achievable by NGS methods, computational subtraction analysis could become a method of choice for viral discovery. This approach has been used for the discovery of a new polyomavirus associated with most cases of Merkel cell carcinoma (MCC) [42]. MCC is a rare and aggressive human skin cancer that typically affects elderly and immunosuppressed individuals, a feature which was suggestive of an infectious origin. RNA was purified from MCC samples and analyzed by 454 pyrosequencing. Digital transcriptome subtraction of all human sequences led to the detection of a fusion transcript between a human receptor tyrosine phosphatase and a Large T antigen sequence related to murine polyomaviruses. This sequence was used as starting point for whole genome sequencing and characterization of this previously unknown polyomavirus that was called Merkel cell polyomavirus (MCPyV). The presence of the virus in 80% MCC tissues but only in about 10% of control tissues from various body sites, including the skin, and the demonstration that, in MCPyV-positive MCCs, viral DNA was integrated within the tumor genome in a clonal pattern, strongly suggested the etiological role of the virus in the pathogenesis of MCC [42].

In a NGS study of the skin virome of a patient with MCC in comparison with healthy controls [43], another human polyomavirus strain was detected, which was nearly identical to the recently discovered HPyV9 polyomavirus [44] and closely related to the lymphotropic polyomavirus (LPV). Likewise, unbiased high-throughput sequencing or deep sequencing of amplicons generated with consensus primers targeting regions of the viral genome conserved within viral families, like the tumor-associated *Polyomaviridae* and *Papillomaviridae*, allowed the discovery and characterization of many new polyomavirus and papillomavirus genotypes in several animal species.

The *Papillomaviridae* family includes several viral species and at least 189 completely characterized papillomavirus types and putative new types are continuously found [45]. High throughput 454 pyrosequencing of amplicons generated by consensus PCR of a conserved region of viral genome was used to detect and genotype HPV in cervical cytology specimens [46]. The method allowed the detection of HPV types which were present in low amount in multiple infections and had the potentiality to detect a broad spectrum of HPV types, subtypes, and variants [46]. A similar approach was used to detect and genotype cutaneous HPV types in a large series of squamous cell carcinoma of the skin and other skin lesions [47]. Several different HPV types were detected, including novel putative cutaneous HPVs [47].

Investigation of retrovirus and retroviral vector integration sites in host cell chromosomes is another field of viral oncology which received a great contribution from NGS technologies. The use of viral vectors that integrate in host genome for gene transfer may cause malignant transformation due to activation of host proto-oncogenes or inactivation of tumor-suppressor genes, as a consequence of viral vector integration within these genes [48–50]. Deep sequencing technology has been used to map the integration sites of retroviruses and HIV [51], as well as retroviral and HIV-based vectors for gene therapy and cell reprogramming [52–54]. Deep sequencing methods for detection of retrovirus integration are based on 454 pyrosequencing of products of ligation-mediated PCR (LM-PCR) [55,56] or linear amplification–mediated PCR (LAM-PCR) [57]. Both LM-PCR and LAM-PCR use restriction enzymes to fragment the DNA of interest containing proviruses. Then, digested DNA is ligated with a compatible linker and amplified by PCR using primers that anneal in the LTR and in the linker sequence. Nested primers containing linkers for the 454 protocol are then used for a second PCR, which is processed by 454 high-throughput sequencing. A LAM-PCR method without the use of restriction enzymes was also developed for high throughput sequencing [58]. Recently, a new method was developed for recovering sites of integrated DNA based on the bacterial transposase MuA. The transposase is used to introduce adaptors into genomic DNA to allow PCR amplification and analysis by 454 pyrosequencing. This method could avoid the bias associated with restriction enzymes and recovered integration sites in a near random fashion. It provided a measure of cell clonal abundance, which is crucial for detecting expansion of cell clones that may be a prelude to malignant transformation [59].

13.2.3 CHARACTERIZATION OF THE HUMAN VIROME

The human microbiome is the entire population of microbes (i.e., bacteria, fungi, and viruses) that colonize the human body. Metagenomics refers to culture-independent studies of the collective set of genomes of mixed microbial communities and applies to explorations of all microbial genomes in consortia that reside in environmental niches, in plants, or in animal hosts, including the human body [60–62]. The "metagenome" of microbial communities that occupy human body niches is estimated to have a gene content approximately 100-fold greater than the human genome [63]. These diverse and complex collections of genes encode a wide array of biochemical and physiological functions that may be relevant in healthy and disease conditions.

Metagenomics strategies are generally based on whole genome shotgun sequencing of nucleic acids purified from a specimen. In case of bacteria metagenomics, analysis can be simplified by exploiting universal and conserved targets, such as 16S rRNA genes, which have both conserved regions that can be targeted by PCR primers, and intervening variable sequences that facilitate genus and species identification [60,61]. At variance, no conserved ubiquitous viral sequences are available for broad amplification of viral genomes and methods to enrich samples with viral particles can only be used. In addition, viral metagenomics analyses, which have been applied so far mostly in environmental samples like fresh water, reused wastewater, and ocean water [64–67], have shown that many of the detected viral sequences are unique and represent unknown viral species. Thus, viral sequences may be missed even by shotgun sequencing [68].

A recent study [69] developed a bioinformatic annotation strategy for identification and quantitative description of human pathogenic viruses in virome data sets and applied this strategy to annotate sequences of viral

DNA and RNA (cDNA) extracted from sewage sludge residuals resulting from municipal wastewater treatment (biosolids), which were obtained by 454 pyrosequencing. In this experimental model, within the 51,925 annotated sequences, 94 DNA and 19 RNA sequences were identified as human viruses. Virus diversity included environmentally transmitted agents such as parechovirus, coronavirus, adenovirus and aichi virus, as well as viruses associated with chronic human infections, such as human herpesviruses and hepatitis C virus [69].

In the diagnostic setting, metagenomic approaches could be used for systematic analysis of samples collected from patients with unexplained illness, especially in the context of outbreaks and epidemics [70,71]. As mentioned in the above section, application of high throughput NGS methods in viral metagenomics can greatly enhance the chances to identify viruses in clinical samples, including viruses that are too divergent from known viruses to be detected by PCR or microarray techniques (reviewed in ref. [20]). An attractive application of metagenomic approaches is the study of influenza, given the constant threat of antigenic drift and shift. Deep sequencing strategies can be used to monitor the emergence of mutations that confer virulence or resistance to antiviral drugs, to detect influenza viruses in clinical samples, and to identify viral quasispecies [22,31,32]. In addition, deep sequencing of clinical samples allows to identify and characterize not only novel pathogens but also the microbiota and host response to infection [32].

The study of the human virome includes also the description of viral communities—including bacteriophages—in human body and their relationship with health and disease. Examples are the characterization of fecal viromes (mainly phages) and their relations with bacterial metagenome [72] and the characterization of the virome in the skin of healthy individuals [28].

13.2.4 FULL-LENGTH VIRAL GENOME SEQUENCING

Like viral metagenomics, sequencing of full-length viral genomes is a difficult task due to the presence of contaminating nucleic acids of the host cell and other agents in viral isolates. In fact, preparation of a simple shotgun sequencing DNA library, the most comprehensive approach, or of a library of cDNA synthesized from RNA with random priming, results in a huge amount of host specific instead of a comprehensive representation of the viral sequences, even in the presence of a very high viral load [21,31,73]. Very high throughput sequencing techniques, such as SOLiD platform, could be used to obtain sufficient sequence coverage [74], but the length of reads might be too short to allow de novo assembly of viral genomes and methods that provide longer reads, like 454 and Illumina technology, might be preferable [31,32]. Several techniques have been used to enrich virions or viral nucleic acids from cell culture or from host tissue and fluids before extracting the genomic DNA/RNA, in order to limit the contamination from host nucleic acids. One of these methods is ultracentrifugation, but this procedure may be very time-consuming and laborious with uncertain outcome [75]. Other methods are based on enrichment of viral nucleic acids by using capture probes or PCR amplification targeting conserved genome segments [76,77] or, vice versa, by depletion of host nucleic acids by probing total RNA with labeled host nucleic acid [78]. Other approaches could be enrichment of dsRNA virus genomes [79] or circular dsDNA viral genomes by RCA [28,29].

13.2.5 INVESTIGATION OF VIRAL GENOME VARIABILITY AND CHARACTERIZATION OF VIRAL QUASISPECIES

High mutation rates inherent to replication of RNA viruses create a wide variety of mutants that are present in virus populations, which are often referred to as quasispecies [80]. The diffuse, "cloud-like" nature of viral populations allows them to rapidly adapt to changing replicative environments by selecting preexisting variants with better fitness [81,82]. Thus, many important virus properties cannot be explained by a mere consensus sequence, but require knowledge about the microvariants present in viral populations. These sequence variants may be critically relevant to viral evolution and spread, virulence, evasion of the immune response, antiviral drug resistance, and vaccine development and manufacture.

The use of deep sequencing data for mutation analysis in viral genomes has required the development of computational methods for estimation of the quality of sequences and for error correction, algorithms for sequence alignment and haplotype reconstruction, statistical models to infer the frequencies of the haplotypes in the population, for comparative analysis and for their visualization [83–86].

Among RNA viruses, HIV quasispecies have been extensively investigated because of their relevance for vaccine design and response to antiviral drug therapy [87]. Within infected individuals, HIV is highly heterogeneous owing to rapid turnover rates, high viral load, and a replication mediated by the error-prone reverse transcriptase enzyme that lacks proofreading activity. High variability is also the consequence of recombination, which can shuttle mutations between viral genomes and lead to major antigenic shifts or alterations in virulence [88]. An example of application of NGS for analysis of HIV quasispecies is the use of massive parallel 454 pyrosequencing with the shotgun approach to characterize the full length genome of an HIV-1 BF recombinant and its quasispecies heterogeneity in a patient who died from multiorgan failure during seroconversion [89]. Another fascinating application of deep sequencing in HIV research is the use of the 454 pyrosequencing methods to analyze the variable regions of heavy and light chains of neutralizing antibodies against HIV in the blood obtained from HIV-1-infected individuals, in order to understand how broadly neutralizing antibodies develop [90]. But the most relevant application of NGS in HIV diagnostics is the detection of anti-viral drug resistant minor variants, which will be discussed in the next section.

Analysis of full-length viral genome and quasispecies was also applied to other RNA viruses. Deep sequencing with the Illumina platform on total RNAs extracted from the lung of a patient who died of viral pneumonia due to pandemic 2009 influenza A virus (A/H1N1/2009) revealed nucleotide heterogeneity on hemagglutinin as quasispecies, leading to amino acid changes on antigenic sites which could be relevant for antigenic drift [31].

Mutations of human rhinovirus (HRV) genome were explored in a lung transplant recipient infected with the same HRV strain for more than two years [91]. Analysis of complete HRV genome sequences by both classical and Illumina ultra-deep sequencing of samples collected at different time points in the upper and lower respiratory tracts showed that HRV populations in the upper and lower respiratory tract were phylogenetically indistinguishable over the course of infection, likely because of constant viral population mixing. Nevertheless, signatures of putative adaptation to lower airway conditions appeared after several months of infection, with the occurrence of specific changes in the 5'UTR polypyrimidine tract and the VP2 immunogenic site 2 of HRV genome, which might have been relevant for viral growth at lower airway conditions [91].

Populations of DNA viruses are considered less complex and variable when compared to RNA viruses. However, data from deep sequencing of DNA virus genomes have revealed that complex mixtures of viral genotypes may be present in infected subjects and that positive selection could have contributed to the divergence of different strains. This is the case of human cytomegalovirus (HCMV), which establishes lifelong latent infections in humans and may reactivate and cause severe life-threatening disease in immunocompromised patients. High intra-host variability of HCMV genome was demonstrated in lung transplant recipients by deep sequencing of the amplicons of three variable HCMV genes [92] and in neonates with congenital HCMV infection by deep sequencing of long range, overlapping amplicons covering the entire HCMV genome [93]. Since PCR amplification and sequencing can introduce errors in their own, which could be misinterpreted as mutation or polymorphisms, deep sequencing studies have to develop protocols and algorithms to estimate experimental error and to filter false positive results. In the studies reported here on HCMV genome variability, experimental error rate was estimated by using arbitrary criteria [92] or an algorithm based on experimental data obtained from deep sequencing analysis of a control HCMV genome cloned in a BAC vector [93].

Deep sequencing showed also variability of herpes simplex virus 1 (HSV-1) genome and allowed to demonstrate virulence genes. Using Illumina high-throughput sequencing, genome sequences of both a laboratory strain (F) and a low-passage clinical isolate (H129) were obtained and compared with the available genome sequence of a more virulent isolate of HSV-1 (strain 17) [94]. The HSV-1 H129 strain, isolated from the brain of an encephalitic patient, is the only virus known to transit neural circuits exclusively in an anterograde direction [95]. Whole genome sequencing demonstrated many protein-coding variations between strains F and H129 and the genome reference strain 17 and some genes were proposed to be responsible of the anterograde mutant phenotype of strain H129, including the neurovirulence protein ICP34.5, while a frameshift mutation in the UL13 kinase could account for decreased neurovirulence of strain F [94].

13.2.6 MONITORING ANTIVIRAL DRUG RESISTANCE

Deep sequencing by NGS techniques is being increasingly used in the clinical practice to detect low abundance drug resistant HIV variants and, with the recent availability of new drugs active against hepatitis C virus (HCV), also for the detection of HCV minor variants.

Conventional direct sequencing of RT-PCR products (referred to as "population sequencing") is the gold standard in HIV resistance testing and is used to detect drug-resistance mutations in the molecular targets of HIV-1 therapy, i.e., reverse transcriptase, protease, integrase, and V3 loop of the HIV env gene. A major limitation of direct PCR sequencing, however, is its inability to detect drug-resistant variants present in less than 20–25% of the heterogeneous virus population existing in a patient's plasma sample [96]. Several studies have shown that minor drug-resistant variants that are not detected by population-based sequencing are clinically relevant in that they are often responsible for the virological failure of a new antiretroviral treatment regimen [97–99].

Clonal sequencing of RT-PCR products by 454 pyrosequencing offers the advantage of high sensitivity for minor variants and a relatively long sequence length that facilitates the characterization of the linkage amongst resistance mutations and avoids the risk to miss mutations due to sequence variation around the site under investigation. The application of 454 sequencing-based resistance testing in clinical setting, however, requires careful consideration of potential technical errors that can be introduced in the experimental protocol and in data analysis in order to discriminate between experimentally introduced errors and true variants [87,100,101]. Data analysis issues are discussed in Section 2.9.

Several studies that employed 454 pyrosequencing for deep analysis of mutations in HIV protease and reverse transcriptase genes demonstrated the accuracy of this technique in detecting all drug-resistance mutations identified by population sequencing, and the ability to detect low-frequency mutations undetectable by population sequencing [100,102,103]. In addition, several studies demonstrated that drug-resistance mutations detected by 454 had a significant impact on virological failure [103–107] while others did not find a strong association of low-frequency mutations with clinical responses [108,109]. Deep sequencing using the 454 platform has been also applied to investigate drug-resistance mutations against the more recently approved integrase inhibitors and CCR5 antagonists.

Drug-resistance mutations to integrase inhibitors occur in the integrase gene. These mutations were detected by deep sequencing at very low levels if at all prior to initiating therapy [110] and could be selected by previous drug pressure [111]. Resistance to CCR5 antagonists, like maraviroc, occur by outgrowth of CXCR4-tropic HIV variants, i.e., viruses that use the CXCR4 coreceptor [112] or via mutations in the viral envelope protein [113–116]. Coreceptor usage can be screened using phenotypic coreceptor tropism assays, based on recombinant virus technology, or genotypic tests, based on sequencing of the V3 loop of HIV *env* gene [117]. Phenotypic assays have good sensitivity and specificity, but they are time consuming, expensive, and require special laboratory facilities; thus they are not convenient as diagnostic tests in clinical practice. Genotyping methods based on population sequencing represent a more feasible alternative, but their sensitivity for the detection of minority variants is lower than phenotypic assay (about 10–20%) and this represents a problem, since the proportion of CXCR4-tropic HIV variants before initiation of therapy is generally very low. In addition, the algorithms used for interpretation of sequencing results may underestimate the impact of some mutations in viral tropism [118]. Deep sequencing by using 454 has been used in several studies [119–123], including large clinical trials, to determine viral tropism and has been demonstrated to be comparable in sensitivity and specificity with phenotypic assays in detecting CXCR4-using variants. According to data reported to date, the clinical threshold for detection of CXCR4-tropic variants might range between 2–10% [118]. With this threshold, 454 pyrosequencing at 1% sensitivity for minority variants can represent a valuable diagnostic tool for viral tropism testing. In addition, deep sequencing of relatively long reads allows defining the contribution of multiple mutations in a single viral genome. This information could improve the performance of interpretation algorithms as compared with population sequencing.

Deep sequencing based on the 454 technology has been also applied for the detection of nucleoside and nucleotide reverse-transcriptase inhibitor resistance in HBV. The NGS method was more sensitive for the detection of rare HBV drug resistance mutations than conventional methods based on population sequencing or reverse hybridization [124,125]. In addition, deep sequencing allowed to identify G-to-A hypermutation mediated by the apolipoprotein B mRNA editing enzyme, which was estimated to be present in 0.6% of reverse-transcriptase genes [124].

Finally, with the availability of new drugs targeting HCV protease and polymerase, the experience of drug-resistance mutation and quasispecies analysis achieved with HIV is being translated to HCV. Also for HCV, deep sequencing technologies seem a promising tool for the study of minority variants present in the HCV quasispecies population at baseline and during antiviral drug pressure, giving new insights into the dynamics of resistance acquisition by HCV [126,127].

13.2.7 EPIDEMIOLOGY OF VIRAL INFECTIONS AND VIRAL EVOLUTION

High throughput sequencing is being used to investigate the epidemiology of viral infections and viral evolution, addressing issues such as viral superinfection (e.g., HIV superinfection, which occurs when a previously infected individual acquires a new distinct HIV strain) [128], tracing the evolution and spread of viral strains, such as the emergence, evolution and worldwide spread of HIV [88], tracing the transmission of viruses among individuals [129], or modeling the evolution of viruses within the host and the mechanism of immune escape, balanced with replication fitness, such as in the case of HIV and HCV infection [127,130,131].

13.2.8 QUALITY CONTROL OF LIVE-ATTENUATED VIRAL VACCINES

Intrinsic genetic instability of RNA viruses may lead to the accumulation of virulent revertants during manufacture of live viral vaccines, requiring rigorous quality control to ensure vaccine safety. High throughput deep sequencing methods have been proposed as tools for monitoring genetic consistency of live viral vaccines. Deep sequencing was used to analyze lots of oral poliovirus vaccine and the detected neurovirulence mutations were identical to the mutation detected with the standard method based on PCR and restriction enzyme cleavage [132]. Patterns of mutations present at a low level in vaccine preparations were characteristic of seed viruses used for their manufacture and could be used for identification of individual batches [132]. Deep sequencing was also used to examine eight live-attenuated viral vaccines, i.e., trivalent oral poliovirus, rubella, measles, yellow fever, varicella-zoster, multivalent measles/mumps/rubella, and two rotavirus live vaccines [133]. The method allowed identification of, not only mutations and minority variants relative to vaccine strains, but also sequences of adventitious viruses from the producer avian and primate cells. The results were in agreement with those obtained by using a panmicrobial microarray [133].

13.2.9 DATA ANALYSIS ISSUES

An aspect that should not be neglected when dealing with NGS data, is the bioinformatics analysis and issues concerning sequencing output. There are inherent strengths and weaknesses in the different platforms as reported in Table 1. For example, 454 technology is well suited for small de novo sequencing projects and amplicon studies, given its read length output that presently reaches the average length of sequences produced with Sanger method. The main issue to be aware of concerns the homopolymer length, due to signal thresholding of the incorporated nucleotides. SOLiD platform is not presently suitable for amplicon studies due to the short read length, but exhibits an extremely high throughput capacity. Illumina has a superior read length and is not affected by homopolymers but, as SOLiD, shows low coverage of AT rich regions [134]. Other platforms present in Table 1 (with the exclusion of the GS Junior, which shares the same features of 454 FLX but has a lower throughput) are still in development and not yet evaluated in their diagnostic potential.

Besides the specific limits of the different platforms, other common issues should be taken into account and carefully considered. The first sources of problems are certainly chimerical sequences, point mutations and insertions/deletions which occur during reverse transcription, PCR amplification or sequencing itself. In addition, PCR amplification bias might impact the relative frequencies of viral variants. The process of "data cleaning" consists of three main steps: sequence filtering, alignment and error correction, for which a panel of methods has been proposed [84,135,136]. Briefly, the filtering phase removes the low-quality sequences from the dataset, while the error correction separates true variants from those due to experimental noise. This step is based on the idea that errors are randomly distributed with low frequency, while sequences with real mutations can be clustered and their abundance quantified. A cluster of reads presenting the same mutations represents a haplotype and the size of the cluster is the haplotype frequency. Global haplotypes are more difficult to be identified, since the reads must be assembled in larger contigs and a unique solution in aligning overlapping reads is not guaranteed. To this respect, the advantage of 454 platform for haplotype reconstruction studies is evident, thanks to its longer reads output.

As concerns the data analysis step, a multitude of software has been developed for very different applications of NGS. Nevertheless, if on the one hand this availability of methods greatly eases the task, on the other hand available algorithms for both genome assembly and amplicon analysis present limitations or drawbacks [137] which require custom made scripting and in-house resolution of bioinformatics problems caused by specific needs [46]. The direct consequence is that data analysis can be no more sustained by the wet-lab researcher alone, but requires the acquisition of computer skills and bioinformatics expertise.

13.3 CONCLUSIONS

Next-generation high throughput sequencing technologies have become available in the last few years and are in continuous development and improvement. They have been widely used in many projects, e.g., whole genome sequencing, metagenomics, small RNA discovery and RNA sequencing. Their common feature is the extremely high throughput data generation. As a result, new issues have to be addressed in order to exploit the full potential of these new instruments: firstly, the data analysis step has become very time consuming and requires a competent amount of manpower and expertise in bioinformatics; secondly, adequate computing resources are necessary to handle the data produced.

Diagnostic virology is one of the most successful applications for NGS and exciting results have been achieved in the discovery and characterization of new viruses, detection of unexpected viral pathogens in clinical specimen, ultrasensitive monitoring of antiviral drug resistance, investigation of viral diversity, evolution and spread, and evaluation of the human virome. With the decrease of costs and improvement of turnaround time, these techniques will probably become essential diagnostic tools in clinical routines.

REFERENCES

1. Schadt, E.E.; Turner, S.; Kasarskis, A. A window into third-generation sequencing. Hum. Mol. Genet. 2010, 19, R227–R240.
2. Ansorge, W.J. Next-generation DNA sequencing techniques. N. Biotechnol. 2009, 25, 195–203.
3. Metzker, M.L. Sequencing technologies - the next generation. Nat. Rev. Genet. 2010, 11, 31–46.
4. Medini, D.; Serruto, D.; Parkhill, J.; Relman, D.A.; Donati, C.; Moxon, R.; Falkow, S.; Rappuoli, R. Microbiology in the post-genomic era. Nat. Rev. Microbiol. 2008, 6, 419–430.
5. Shendure, J.; Ji, H. Next-generation DNA sequencing. Nat. Biotechnol. 2008, 26, 1135–1145.
6. MacLean, D.; Jones, J.D.; Studholme, D.J. Application of "next-generation" sequencing technologies to microbial genetics. Nat. Rev. Microbiol. 2009, 7, 287–296.
7. Morens, D.M.; Folkers, G.K.; Fauci, A.S. The challenge of emerging and re-emerging infectious diseases. Nature 2004, 430, 242–249.
8. Chang, Y.; Cesarman, E.; Pessin, M.S.; Lee, F.; Culpepper, J.; Knowles, D.M.; Moore, P.S. Identification of herpesvirus-like DNA sequences in AIDS-associated Kaposi's sarcoma. Science 1994, 266, 1865–1869.
9. Simons, J.N.; Pilot-Matias, T.J.; Leary, T.P.; Dawson, G.J.; Desai, S.M.; Schlauder, G.G.; Muerhoff, A.S.; Erker, J.C.; Buijk, S.L.; Chalmers, M.L.; et al. Identification of two flaviviruslike genomes in the GB hepatitis agent. Proc. Natl. Acad. Sci. USA 1995, 92, 3401–3405.

10. Nishizawa, T.; Okamoto, H.; Konishi, K.; Yoshizawa, H.; Miyakawa, Y.; Mayumi, M. A novel DNA virus (TTV) associated with elevated transaminase levels in post-transfusion hepatitis of unknown etiology. Biochem. Biophys. Res. Commun. 1997, 241, 92–97.

11. Allander, T.; Tammi, M.T.; Eriksson, M.; Bjerkner, A.; Tiveljung-Lindell, A.; Andersson, B. Cloning of a human parvovirus by molecular screening of respiratory tract samples. Proc. Natl. Acad. Sci. USA 2005, 102, 12891–12896.

12. Jones, M.S.; Kapoor, A.; Lukashov, V.V.; Simmonds, P.; Hecht, F.; Delwart, E. New DNA viruses identified in patients with acute viral infection syndrome. J. Virol. 2005, 79, 8230–8236.

13. Gaynor, A.M.; Nissen, M.D.; Whiley, D.M.; Mackay, I.M.; Lambert, S.B.; Wu, G.; Brennan, D.C.; Storch, G.A.; Sloots, T.P.; Wang, D. Identification of a novel polyomavirus from patients with acute respiratory tract infections. PLoS Pathog. 2007, 3, doi:10.1371/journal.ppat.0030064.

14. Allander, T.; Andreasson, K.; Gupta, S.; Bjerkner, A.; Bogdanovic, G.; Persson, M.A.; Dalianis, T.; Ramqvist, T.; Andersson, B. Identification of a third human polyomavirus. J. Virol. 2007, 81, 4130–4136.

15. Pyrc, K.; Jebbink, M.F.; Berkhout, B.; van der Hoek, L. Detection of new viruses by VIDISCA. Virus discovery based on cDNA-amplified fragment length polymorphism. Methods Mol. Biol. 2008, 454, 73–89.

16. van der Hoek, L.; Pyrc, K.; Jebbink, M.F.; Vermeulen-Oost, W.; Berkhout, R.J.; Wolthers, K.C.; Wertheim-van Dillen, P.M.; Kaandorp, J.; Spaargaren, J.; Berkhout, B. Identification of a new human coronavirus. Nat. Med. 2004, 10, 368–373.

17. Wang, D.; Urisman, A.; Liu, Y.T.; Springer, M.; Ksiazek, T.G.; Erdman, D.D.; Mardis, E.R.; Hickenbotham, M.; Magrini, V.; Eldred, J.; et al. Viral discovery and sequence recovery using DNA microarrays. PLoS Biol. 2003, 1, 257–260.

18. Palacios, G.; Quan, P.L.; Jabado, O.J.; Conlan, S.; Hirschberg, D.L.; Liu, Y.; Zhai, J.; Renwick, N.; Hui, J.; Hegyi, H.; et al. Panmicrobial oligonucleotide array for diagnosis of infectious diseases. Emerg. Infect. Dis. 2007, 13, 73–81.

19. MacConaill, L.; Meyerson, M. Adding pathogens by genomic subtraction. Nat. Genet. 2008, 40, 380–382.

20. Tang, P.; Chiu, C. Metagenomics for the discovery of novel human viruses. Future Microbiol. 2010, 5, 177–189.

21. Palacios, G.; Druce, J.; Du, L.; Tran, T.; Birch, C.; Briese, T.; Conlan, S.; Quan, P. L.; Hui, J.; Marshall, J.; et al. A new arenavirus in a cluster of fatal transplant-associated diseases. N. Engl. J. Med. 2008, 358, 991–998.

22. Nakamura, S.; Yang, C.S.; Sakon, N.; Ueda, M.; Tougan, T.; Yamashita, A.; Goto, N.; Takahashi, K.; Yasunaga, T.; Ikuta, K.; et al. Direct metagenomic detection of viral pathogens in nasal and fecal specimens using an unbiased high-throughput sequencing approach. PLoS One 2009, 4, doi:10.1371/journal.pone.0004219.

23. Quan, P.L.; Wagner, T.A.; Briese, T.; Torgerson, T.R.; Hornig, M.; Tashmukhamedova, A.; Firth, C.; Palacios, G.; Baisre-De-Leon, A.; Paddock, C.D.; et al. Astrovirus encephalitis in boy with X-linked agammaglobulinemia. Emerg. Infect. Dis. 2010, 16, 918–925.

24. Briese, T.; Paweska, J.T.; McMullan, L.K.; Hutchison, S.K.; Street, C.; Palacios, G.; Khristova, M.L.; Weyer, J.; Swanepoel, R.; Egholm, M.; et al. Genetic detection

and characterization of Lujo Virus, a new Hemorrhagic Fever-associated Arenavirus from Southern Africa. PLoS Pathog. 2009, 5, doi:10.1371/journal.ppat.1000455.

25. de Vries, M.; Deijs, M.; Canuti, M.; van Schaik, B.D.C.; Faria, N.R.; van de Garde, M.D.B.; Jachimowski, L.C.M.; Jebbink, M.F.; Jakobs, M.; Luyf, A.C.M.; et al. A sensitive assay for virus discovery in respiratory clinical samples. PLoS One 2011, 6, doi:10.1371/journal.pone.0016118.

26. Dean, F.B.; Nelson, J.R.; Giesler, T.L.; Lasken, R.S. Rapid amplification of plasmid and phage DNA using phi29 DNA polymerase and multiply-primed rolling circle amplification. Genome Res. 2001, 11, 1095–1099.

27. Rector, A.; Tachezy, R.; Van Ranst, M. A sequence-independent strategy for detection and cloning of circular DNA virus genomes by using multiply primed rolling-circle amplification. J. Virol. 2004, 78, 4993–4998.

28. Schowalter, R.M.; Pastrana, D.V.; Pumphrey, K.A.; Moyer, A.L.; Buck, C.B. Merkel cell polyomavirus and two previously unknown polyomaviruses are chronically shed from human skin. Cell Host Microbe 2010, 7, 509–515.

29. van der Meijden, E.; Janssens, R.W.; Lauber, C.; Bouwes Bavinck, J.N.; Gorbalenya, A.E.; Feltkamp, M.C. Discovery of a new human polyomavirus associated with trichodysplasia spinulosa in an immunocompromized patient. PLoS Pathog. 2010, 6, doi:10.1371/journal.ppat.1001024.

30. Yongfeng, H.; Fan, Y.; Jie, D.; Jian, Y.; Ting, Z.; Lilian, S.; Jin, Q. Direct pathogen detection from swab samples using a new high-throughput sequencing technology. Clin. Microbiol. Infect. 2011, 17, 241–244.

31. Kuroda, M.; Katano, H.; Nakajima, N.; Tobiume, M.; Ainai, A.; Sekizuka, T.; Hasegawa, H.; Tashiro, M.; Sasaki, Y.; Arakawa, Y.; et al. Characterization of quasi-species of pandemic 2009 influenza A virus (A/H1N1/2009) by de novo sequencing using a next-generation DNA sequencer. PLoS One 2010, 5, doi:10.1371/journal.pone.0010256.

32. Greninger, A.L.; Chen, E.C.; Sittler, T.; Scheinerman, A.; Roubinian, N.; Yu, G.; Kim, E.; Pillai, D.R.; Guyard, C.; Mazzulli, T.; et al. A metagenomic analysis of pandemic influenza A (2009 H1N1) infection in patients from North America. PLoS One 2010, 5, doi:10.1371/journal.pone.0013381.

33. Yang, J.; Yang, F.; Ren, L.; Xiong, Z.; Wu, Z.; Dong, J.; Sun, L.; Zhang, T.; Hu, Y.; Du, J.; et al. Unbiased parallel detection of viral pathogens in clinical samples using a metagenomic approach. J. Clin. Microbiol. 2011, 49, 3463–3469.

34. Yozwiak, N.L.; Skewes-Cox, P.; Gordon, A.; Saborio, S.; Kuan, G.; Balmaseda, A.; Ganem, D.; Harris, E.; DeRisi, J.L. Human enterovirus 109: A novel interspecies recombinant enterovirus isolated from a case of acute pediatric respiratory illness in Nicaragua. J. Virol. 2010, 84, 9047–9058.

35. Cheval, J.; Sauvage, V.; Frangeul, L.; Dacheux, L.; Guigon, G.; Dumey, N.; Pariente, K.; Rousseaux, C.; Dorange, F.; Berthet, N.; et al. Evaluation of high throughput sequencing for identifying known and unknown viruses in biological samples. J. Clin. Microbiol. 2011, 49, 3268–3275.

36. Bishop-Lilly, K.A.; Turell, M.J.; Willner, K.M.; Butani, A.; Nolan, N.M.; Lentz, S.M.; Akmal, A.; Mateczun, A.; Brahmbhatt, T.N.; Sozhamannan, S.; et al. Arbovirus detection in insect vectors by rapid, high-throughput pyrosequencing. PLoS Negl. Trop. Dis. 2010, 4, e878:1– e878:12.

37. Wu, Q.; Luo, Y.; Lu, R.; Lau, N.; Lai, E.C.; Li, W.X.; Ding, S.W. Virus discovery by deep sequencing and assembly of virus-derived small silencing RNAs. Proc. Natl. Acad. Sci. USA 2010, 107, 1606–1611.

38. Kreuze, J.F.; Perez, A.; Untiveros, M.; Quispe, D.; Fuentes, S.; Barker, I.; Simon, R. Complete viral genome sequence and discovery of novel viruses by deep sequencing of small RNAs: A generic method for diagnosis, discovery and sequencing of viruses. Virology 2009, 388, 1–7.

39. Quan, P.L.; Firth, C.; Street, C.; Henriquez, J.A.; Petrosov, A.; Tashmukhamedova, A.; Hutchison, S.K.; Egholm, M.; Osinubi, M.O.; Niezgoda, M.; et al. Identification of a severe acute respiratory syndrome coronavirus-like virus in a leaf-nosed bat in Nigeria. mBio 2010, 1, e00208–e00210.

40. Weber, G.; Shendure, J.; Tanenbaum, D.M.; Church, G.M.; Meyerson, M. Identification of foreign gene sequences by transcript filtering against the human genome. Nat. Genet. 2002, 30, 141–142.

41. Xu, Y.; Stange-Thomann, N.; Weber, G.; Bo, R.; Dodge, S.; David, R.G.; Foley, K.; Beheshti, J.; Harris, N.L.; Birren, B.; et al. Pathogen discovery from human tissue by sequence-based computational subtraction. Genomics 2003, 81, 329–335.

42. Feng, H.; Shuda, M.; Chang, Y.; Moore, P.S. Clonal integration of a polyomavirus in human Merkel cell carcinoma. Science 2008, 319, 1096–1100.

43. Sauvage, V.; Foulongne, V.; Cheval, J.; Ar Gouilh, M.; Pariente, K.; Dereure, O.; Manuguerra, J.C.; Richardson, J.; Lecuit, M.; Burguiere, A.; Caro, V.; Eloit, M. Human polyomavirus related to african green monkey lymphotropic polyomavirus. Emerg. Infect. Dis. 2011, 17, 1364–1370.

44. Scuda, N.; Hofmann, J.; Calvignac-Spencer, S.; Ruprecht, K.; Liman, P.; Kuhn, J.; Hengel, H.; Ehlers, B. A novel human polyomavirus closely related to the african green monkey-derived lymphotropic polyomavirus. J. Virol. 2011, 85, 4586–4590.

45. Bernard, H.U.; Burk, R.D.; Chen, Z.; van Doorslaer, K.; Hausen, H.; de Villiers, E.M. Classification of papillomaviruses (PVs) based on 189 PV types and proposal of taxonomic amendments. Virology 2010, 401, 70–79.

46. Barzon, L.; Militello, V.; Lavezzo, E.; Franchin, E.; Peta, E.; Squarzon, L.; Trevisan, M.; Pagni, S.; Dal Bello, F.; Toppo, S.; et al. Human papillomavirus genotyping by 454 next generation sequencing technology. J. Clin. Virol. 2011, 52, 93–97.

47. Ekstrom, J.; Bzhalava, D.; Svenback, D.; Forslund, O.; Dillner, J. High throughput sequencing reveals diversity of Human Papillomaviruses in cutaneous lesions. Int. J. Cancer 2011, 129, 2643–2650.

48. Hacein-Bey-Abina, S.; Garrigue, A.; Wang, G.P.; Soulier, J.; Lim, A.; Morillon, E.; Clappier, E.; Caccavelli, L.; Delabesse, E.; Beldjord, K.; et al. Insertional oncogenesis in 4 patients after retrovirus-mediated gene therapy of SCID-X1. J. Clin. Invest. 2008, 118, 3132–3142.

49. Howe, S.J.; Mansour, M.R.; Schwarzwaelder, K.; Bartholomae, C.; Hubank, M.; Kempski, H.; Brugman, M.H.; Pike-Overzet, K.; Chatters, S.J.; de Ridder, D.; et al. Insertional mutagenesis combined with acquired somatic mutations causes leukemogenesis following gene therapy of SCID-X1 patients. J. Clin. Invest. 2008, 118, 3143–3150.

50. Bushman, F.; Lewinski, M.; Ciuffi, A.; Barr, S.; Leipzig, J.; Hannenhalli, S.; Hoffmann, C. Genome-wide analysis of retroviral DNA integration. Nat. Rev. Microbiol. 2005, 3, 848–858.

51. Wang, G.P.; Ciuffi, A.; Leipzig, J.; Berry, C.C.; Bushman, F.D. HIV integration site selection: analysis by massively parallel pyrosequencing reveals association with epigenetic modifications. Genome Res. 2007, 17, 1186–1194.

52. Varas, F.; Stadtfeld, M.; De Andres-Aguayo, L.; Maherali, N.; di Tullio, A.; Pantano, L.; Notredame, C.; Hochedlinger, K.; Graf, T. Fibroblast-derived induced pluripotent stem cells show no common retroviral vector insertions. Stem Cells 2009, 27, 300–306.

53. Winkler, T.; Cantilena, A.; Metais, J.Y.; Xu, X.L.; Nguyen, A.D.; Borate, B.; Antosiewicz-Bourget, J.E.; Wolfsberg, T.G.; Thomson, J.A.; Dunbar, C.E. No evidence for clonal selection due to lentiviral integration sites in human induced pluripotent stem cells. Stem Cells 2010, 28, 687–694.

54. Kane, N.M.; Nowrouzi, A.; Mukherjee, S.; Blundell, M.P.; Greig, J.A.; Lee, W.K.; Houslay, M.D.; Milligan, G.; Mountford, J.C.; von Kalle, C.; et al. Lentivirus-mediated reprogramming of somatic cells in the absence of transgenic transcription factors. Mol. Ther. 2010, 18, 2139–2145.

55. Cattoglio, C.; Pellin, D.; Rizzi, E.; Maruggi, G.; Corti, G.; Miselli, F.; Sartori, D.; Guffanti, A.; Di Serio, C.; Ambrosi, A.; et al. High-definition mapping of retroviral integration sites identifies active regulatory elements in human multipotent hematopoietic progenitors. Blood 2010, 116, 5507–5517.

56. Ciuffi, A.; Barr, S.D. Identification of HIV integration sites in infected host genomic DNA. Methods 2011, 53, 39–46.

57. Schmidt, M.; Schwarzwaelder, K.; Bartholomae, C.; Zaoui, K.; Ball, C.; Pilz, I.; Braun, S.; Glimm, H.; von Kalle, C. High-resolution insertion-site analysis by linear amplification-mediated PCR (LAM-PCR). Nat. Methods 2007, 4, 1051–1057.

58. Gabriel, R.; Eckenberg, R.; Paruzynski, A.; Bartholomae, C.C.; Nowrouzi, A.; Arens, A.; Howe, S.J.; Recchia, A.; Cattoglio, C.; Wang, W.; et al. Comprehensive genomic access to vector integration in clinical gene therapy. Nat. Med. 2009, 15, 1431–1436.

59. Brady, T.; Roth, S.L.; Malani, N.; Wang, G.P.; Berry, C.C.; Leboulch, P.; Hacein-Bey-Abina, S.; Cavazzana-Calvo, M.; Papapetrou, E.P.; Sadelain, M.; et al. A method to sequence and quantify DNA integration for monitoring outcome in gene therapy. Nucleic Acids Res. 2011, 39, doi: 10.1093/nar/gkr140.

60. Petrosino, J.F.; Highlander, S.; Luna, R.A.; Gibbs, R.A.; Versalovic, J. Metagenomic pyrosequencing and microbial identification. Clin. Chem. 2009, 55, 856–866.

61. Hamady, M.; Knight, R. Microbial community profiling for human microbiome projects: Tools, techniques, and challenges. Genome Res. 2009, 19, 1141–1152.

62. Delwart, E.L. Viral metagenomics. Rev. Med. Virol. 2007, 17, 115–131.

63. Turnbaugh, P.J.; Ley, R.E.; Hamady, M.; Fraser-Liggett, C.M.; Knight, R.; Gordon, J.I. The human microbiome project. Nature 2007, 449, 804–810.

64. Djikeng, A.; Kuzmickas, R.; Anderson, N.G.; Spiro, D.J. Metagenomic analysis of RNA viruses in a fresh water lake. PLoS One 2009, 4, doi:10.1371/journal.pone.0007264.

65. Lopez-Bueno, A.; Tamames, J.; Velazquez, D.; Moya, A.; Quesada, A.; Alcami, A. High diversity of the viral community from an Antarctic lake. Science 2009, 326, 858–861.

66. Angly, F.E.; Felts, B.; Breitbart, M.; Salamon, P.; Edwards, R.A.; Carlson, C.; Chan, A.M.; Haynes, M.; Kelley, S.; Liu, H.; et al. The marine viromes of four oceanic regions. PLoS Biol. 2006, 4, doi:10.1371/journal.pbio.0040368.

67. Rosario, K.; Nilsson, C.; Lim, Y.W.; Ruan, Y.; Breitbart, M. Metagenomic analysis of viruses in reclaimed water. Environ. Microbiol. 2009, 11, 2806–2820.

68. Kristensen, D.M.; Mushegian, A.R.; Dolja, V.V.; Koonin, E.V. New dimensions of the virus world discovered through metagenomics. Trends Microbiol. 2010, 18, 11–19.

69. Bibby, K.; Viau, E.; Peccia, J. Viral metagenome analysis to guide human pathogen monitoring in environmental samples. Lett. Appl. Microbiol. 2011, 52, 386–392.

70. Finkbeiner, S.R.; Allred, A.F.; Tarr, P.I.; Klein, E.J.; Kirkwood, C.D.; Wang, D. Metagenomic analysis of human diarrhea: viral detection and discovery. PLoS Pathog. 2008, 4, doi:10.1371/journal.ppat.1000011.

71. Svraka, S.; Rosario, K.; Duizer, E.; van der Avoort, H.; Breitbart, M.; Koopmans, M. Metagenomic sequencing for virus identification in a public-health setting. J. Gen. Virol. 2010, 91, 2846–2856.

72. Reyes, A.; Haynes, M.; Hanson, N.; Angly, F.E.; Heath, A.C.; Rohwer, F.; Gordon, J.I. Viruses in the faecal microbiota of monozygotic twins and their mothers. Nature 2010, 466, 334–338.

73. Greninger, A.L.; Runckel, C.; Chiu, C.Y.; Haggerty, T.; Parsonnet, J.; Ganem, D.; DeRisi, J.L. The complete genome of klassevirus - a novel picornavirus in pediatric stool. Virol. J. 2009, 6, doi:10.1186/1743-422X-6-82.

74. Legendre, M.; Santini, S.; Rico, A.; Abergel, C.; Claverie, J.M. Breaking the 1000-gene barrier for Mimivirus using ultra-deep genome and transcriptome sequencing. Virol. J. 2011, 8, 99.

75. Huang, Y.; Huang, X.; Liu, H.; Gong, J.; Ouyang, Z.; Cui, H.; Cao, J.; Zhao, Y.; Wang, X.; Jiang, Y.; Qin, Q. Complete sequence determination of a novel reptile iridovirus isolated from softshelled turtle and evolutionary analysis of Iridoviridae. BMC Genomics 2009, 10, doi:10.1186/1471-2164-10-224.

76. Hoper, D.; Hoffmann, B.; Beer, M. A comprehensive deep sequencing strategy for full-length genomes of influenza A. PLoS One 2011, 6, doi:10.1371/journal.pone.0019075.

77. Willerth, S.M.; Pedro, H.A.; Pachter, L.; Humeau, L.M.; Arkin, A.P.; Schaffer, D.V. Development of a low bias method for characterizing viral populations using next generation sequencing technology. PLoS One 2010, 5, doi:10.1371/journal.pone.0013564.

78. Monger, W.A.; Adams, I.P.; Glover, R.H.; Barrett, B. The complete genome sequence of Canna yellow streak virus. Arch. Virol. 2010, 155, 1515–1518.

79. Potgieter, A.C.; Page, N.A.; Liebenberg, J.; Wright, I.M.; Landt, O.; van Dijk, A.A. Improved strategies for sequence-independent amplification and sequencing of viral double-stranded RNA genomes. J. Gen. Virol. 2009, 90, 1423–1432.

80. Lauring, A.S.; Andino, R. Quasispecies theory and the behavior of RNA viruses. PLoS Pathog. 2010, 6, doi:10.1371/journal.ppat.1001005.

81. Novella, I.S.; Domingo, E.; Holland, J.J. Rapid viral quasispecies evolution: implications for vaccine and drug strategies. Mol. Med. Today 1995, 1, 248–253.

82. Ruiz-Jarabo, C.M.; Arias, A.; Baranowski, E.; Escarmis, C.; Domingo, E. Memory in viral quasispecies. J. Virol. 2000, 74, 3543–3547.

83. Eriksson, N.; Pachter, L.; Mitsuya, Y.; Rhee, S.Y.; Wang, C.; Gharizadeh, B.; Ronaghi, M.; Shafer, R.W.; Beerenwinkel, N. Viral population estimation using pyrosequencing. PLoS Comput. Biol. 2008, 4, e1000074.

84. Zagordi, O.; Geyrhofer, L.; Roth, V.; Beerenwinkel, N. Deep sequencing of a genetically heterogeneous sample: local haplotype reconstruction and read error correction. J. Comput. Biol. 2010, 17, 417–428.

85. Prosperi, M.C.; Prosperi, L.; Bruselles, A.; Abbate, I.; Rozera, G.; Vincenti, D.; Solmone, M.C.; Capobianchi, M.R.; Ulivi, G. Combinatorial analysis and algorithms for quasispecies reconstruction using next-generation sequencing. BMC Bioinform. 2011, 12, 5.

86. Beerenwinkel, N.; Zagordi, O. Ultra-deep sequencing for the analysis of viral populations. Curr. Opin. Virol. 2011, 1, 1–6.

87. Vrancken, B.; Lequime, S.; Theys, K.; Lemey, P. Covering all bases in HIV research: Unveiling a hidden world of viral evolution. AIDS Rev. 2010, 12, 89–102.

88. Tebit, D.M.; Arts, E.J. Tracking a century of global expansion and evolution of HIV to drive understanding and to combat disease. Lancet Infect. Dis. 2011, 11, 45–56.

89. Bruselles, A.; Rozera, G.; Bartolini, B.; Prosperi, M.; Del Nonno, F.; Narciso, P.; Capobianchi, M.R.; Abbate, I. Use of massive parallel pyrosequencing for near full-length characterization of a unique HIV Type 1 BF recombinant associated with a fatal primary infection. AIDS Res. Hum. Retrovir. 2009, 25, 937–942.

90. Wu, X.; Zhou, T.; Zhu, J.; Zhang, B.; Georgiev, I.; Wang, C.; Chen, X.; Longo, N.S.; Louder, M.; McKee, K.; et al. Focused evolution of HIV-1 neutralizing antibodies revealed by structures and deep sequencing. Science 2011, 333, 1593–1602.

91. Tapparel, C.; Cordey, S.; Junier, T.; Farinelli, L.; Van Belle, S.; Soccal, P.M.; Aubert, J.D.; Zdobnov, E.; Kaiser, L. Rhinovirus genome variation during chronic upper and lower respiratory tract infections. PLoS One 2011, 6, doi:10.1371/journal.pone.0021163.

92. Gorzer, I.; Guelly, C.; Trajanoski, S.; Puchhammer-Stockl, E. Deep sequencing reveals highly complex dynamics of human cytomegalovirus genotypes in transplant patients over time. J. Virol. 2010, 84, 7195–7203.

93. Renzette, N.; Bhattacharjee, B.; Jensen, J.D.; Gibson, L.; Kowalik, T.F. Extensive genome-wide variability of human cytomegalovirus in congenitally infected infants. PLoS Pathog. 2011, 7, doi:10.1371/journal.ppat.1001344.

94. Szpara, M.L.; Parsons, L.; Enquist, L.W. Sequence variability in clinical and laboratory isolates of herpes simplex virus 1 reveals new mutations. J. Virol. 2010, 84, 5303–5313.

95. Zemanick, M.C.; Strick, P.L.; Dix, R.D. Direction of transneuronal transport of herpes simplex virus 1 in the primate motor system is strain-dependent. Proc. Natl. Acad. Sci. USA 1991, 88, 8048–8051.

96. Palmer, S.; Kearney, M.; Maldarelli, F.; Halvas, E.K.; Bixby, C.J.; Bazmi, H.; Rock, D.; Falloon, J.; Davey, R.T., Jr.; Dewar, R.L.; et al. Multiple, linked human immunodeficiency virus type 1 drug resistance mutations in treatment-experienced patients are missed by standard genotype analysis. J. Clin. Microbiol. 2005, 43, 406–413.

97. Jourdain, G.; Ngo-Giang-Huong, N.; Le Coeur, S.; Bowonwatanuwong, C.; Kantipong, P.; Leechanachai, P.; Ariyadej, S.; Leenasirimakul, P.; Hammer, S.; Lallemant, M. Intrapartum exposure to nevirapine and subsequent maternal responses to nevirapine-based antiretroviral therapy. N. Engl. J. Med. 2004, 351, 229–240.

98. Lecossier, D.; Shulman, N.S.; Morand-Joubert, L.; Shafer, R.W.; Joly, V.; Zolopa, A.R.; Clavel, F.; Hance, A.J. Detection of minority populations of HIV-1 expressing the K103N resistance mutation in patients failing nevirapine. J. Acquir. Immune Defic. Syndr. 2005, 38, 37–42.

99. Palmer, S.; Boltz, V.; Martinson, N.; Maldarelli, F.; Gray, G.; McIntyre, J.; Mellors, J.; Morris, L.; Coffin, J. Persistence of nevirapine-resistant HIV-1 in women after single-dose nevirapine therapy for prevention of maternal-to-fetal HIV-1 transmission. Proc. Natl. Acad. Sci. USA 2006, 103, 7094–7099.

100. Wang, C.; Mitsuya, Y.; Gharizadeh, B.; Ronaghi, M.; Shafer, R.W. Characterization of mutation spectra with ultra-deep pyrosequencing: application to HIV-1 drug resistance. Genome Res. 2007, 17, 1195–1201.

101. Mild, M.; Hedskog, C.; Jernberg, J.; Albert, J. Performance of Ultra-Deep Pyrosequencing in Analysis of HIV-1 pol Gene Variation. PLoS One 2011, 6, doi:10.1371/journal.pone.0022741.

102. Hoffmann, C.; Minkah, N.; Leipzig, J.; Wang, G.; Arens, M.Q.; Tebas, P.; Bushman, F.D. DNA bar coding and pyrosequencing to identify rare HIV drug resistance mutations. Nucleic Acids Res. 2007, 35, doi: 10.1093/nar/gkm435.

103. Simen, B.B.; Simons, J.F.; Hullsiek, K.H.; Novak, R.M.; MacArthur, R.D.; Baxter, J.D.; Huang, C.L.; Lubeski, C.; Turenchalk, G.S.; Braverman, M.S.; et al. Low-abundance drugresistant viral variants in chronically HIV-infected, antiretroviral treatment-naive patients significantly impact treatment outcomes. J. Infect. Dis. 2009, 199, 693–701.

104. Metzner, K.J.; Giulieri, S.G.; Knoepfel, S.A.; Rauch, P.; Burgisser, P.; Yerly, S.; Gunthard, H.F.; Cavassini, M. Minority quasispecies of drug-resistant HIV-1 that lead to early therapy failure in treatment-naive and -adherent patients. Clin. Infect. Dis. 2009, 48, 239–247.

105. Johnson, J.A.; Li, J.F.; Wei, X.; Lipscomb, J.; Irlbeck, D.; Craig, C.; Smith, A.; Bennett, D.E.; Monsour, M.; Sandstrom, P.; et al. Minority HIV-1 drug resistance mutations are present in antiretroviral treatment-naive populations and associate with reduced treatment efficacy. PLoS Med. 2008, 5, doi:10.1371/journal.pmed.0050158.

106. Le, T.; Chiarella, J.; Simen, B.B.; Hanczaruk, B.; Egholm, M.; Landry, M.L.; Dieckhaus, K.; Rosen, M.I.; Kozal, M.J. Low-abundance HIV drug-resistant viral variants in treatment-experienced persons correlate with historical antiretroviral use. PLoS One 2009, 4, doi:10.1371/journal.pone.0006079.

107. Codoner, F.M.; Pou, C.; Thielen, A.; Garcia, F.; Delgado, R.; Dalmau, D.; Alvarez-Tejado, M.; Ruiz, L.; Clotet, B.; Paredes, R. Added value of deep sequencing relative to population sequencing in heavily pre-treated HIV-1-infected subjects. PLoS One 2011, 6, doi:10.1371/journal.pone.0019461.

108. Peuchant, O.; Thiebaut, R.; Capdepont, S.; Lavignolle-Aurillac, V.; Neau, D.; Morlat, P.; Dabis, F.; Fleury, H.; Masquelier, B. Transmission of HIV-1 minority-resistant variants and response to first-line antiretroviral therapy. AIDS 2008, 22, 1417–1423.

109. Jakobsen, M.R.; Tolstrup, M.; Sogaard, O.S.; Jorgensen, L.B.; Gorry, P.R.; Laursen, A.; Ostergaard, L. Transmission of HIV-1 drug-resistant variants: prevalence and effect on treatment outcome. Clin. Infect. Dis. 2010, 50, 566–573.

110. Mukherjee, R.; Jensen, S.T.; Male, F.; Bittinger, K.; Hodinka, R.L.; Miller, M.D.; Bushman, F.D. Switching between raltegravir resistance pathways analyzed by deep sequencing. AIDS 2011, 25, 1951–1959.

111. Codoner, F.M.; Pou, C.; Thielen, A.; Garcia, F.; Delgado, R.; Dalmau, D.; Santos, J.R.; Buzon, M.J.; Martinez-Picado, J.; Alvarez-Tejado, M.; et al. Dynamic escape of pre-existing raltegravir-resistant HIV-1 from raltegravir selection pressure. Antiviral Res. 2010, 88, 281–286.

112. Gulick, R.M.; Lalezari, J.; Goodrich, J.; Clumeck, N.; DeJesus, E.; Horban, A.; Nadler, J.; Clotet, B.; Karlsson, A.; Wohlfeiler, M.; et al. Maraviroc for previously treated patients with R5 HIV-1 infection. N. Engl. J. Med. 2008, 359, 1429–1441.

113. Ogert, R.A.; Hou, Y.; Ba, L.; Wojcik, L.; Qiu, P.; Murgolo, N.; Duca, J.; Dunkle, L.M.; Ralston, R.; Howe, J.A. Clinical resistance to vicriviroc through adaptive V3 loop mutations in HIV-1 subtype D gp120 that alter interactions with the N-terminus and ECL2 of CCR5. Virology 2010, 400, 145–155.

114. Tilton, J.C.; Amrine-Madsen, H.; Miamidian, J.L.; Kitrinos, K.M.; Pfaff, J.; Demarest, J.F.; Ray, N.; Jeffrey, J.L.; Labranche, C.C.; Doms, R.W. HIV type 1 from a patient with baseline resistance to CCR5 antagonists uses drug-bound receptor for entry. AIDS Res. Hum. Retrovir. 2010, 26, 13–24.

115. Tsibris, A.M.; Sagar, M.; Gulick, R.M.; Su, Z.; Hughes, M.; Greaves, W.; Subramanian, M.; Flexner, C.; Giguel, F.; Leopold, K.E.; et al. In vivo emergence of vicriviroc resistance in a human immunodeficiency virus type 1 subtype C-infected subject. J. Virol. 2008, 82, 8210–8214.

116. Tilton, J.C.; Wilen, C.B.; Didigu, C.A.; Sinha, R.; Harrison, J.E.; Agrawal-Gamse, C.; Henning, E.A.; Bushman, F.D.; Martin, J.N.; Deeks, S.G.; et al. A maraviroc-resistant HIV-1 with narrow cross-resistance to other CCR5 antagonists depends on both N-terminal and extracellular loop domains of drug-bound CCR5. J. Virol. 2010, 84, 10863–10876.

117. Hwang, S.S.; Boyle, T.J.; Lyerly, H.K.; Cullen, B.R. Identification of the envelope V3 loop as the primary determinant of cell tropism in HIV-1. Science 1991, 253, 71–74.

118. Poveda, E.; Alcami, J.; Paredes, R.; Cordoba, J.; Gutierrez, F.; Llibre, J.M.; Delgado, R.; Pulido, F.; Iribarren, J.A.; Garcia Deltoro, M.; et al. Genotypic determination of HIV tropism - clinical and methodological recommendations to guide the therapeutic use of CCR5 antagonists. AIDS Rev. 2010, 12, 135–148.

119. Swenson, L.C.; Mo, T.; Dong, W.W.Y.; Zhong, X.Y.; Woods, C.K.; Jensen, M.A.; Thielen, A.; Chapman, D.; Lewis, M.; James, I.; et al. Deep sequencing to infer HIV-1 co-receptor usage: Application to three clinical trials of Maraviroc in treatment-experienced patients. J. Infect. Dis. 2011, 203, 237–245.

120. Bunnik, E.M.; Swenson, L.C.; Edo-Matas, D.; Huang, W.; Dong, W.; Frantzell, A.; Petropoulos, C.J.; Coakley, E.; Schuitemaker, H.; Harrigan, P.R.; et al. Detection of inferred CCR5- and CXCR4-using HIV-1 variants and evolutionary intermediates using ultra-deep pyrosequencing. PLoS Pathog. 2011, 7, doi:10.1371/journal.ppat.1002106.

121. Abbate, I.; Vlassi, C.; Rozera, G.; Bruselles, A.; Bartolini, B.; Giombini, E.; Corpolongo, A.; D'Offizi, G.; Narciso, P.; Desideri, A.; et al. Detection of quasispecies variants predicted to use CXCR4 by ultra-deep pyrosequencing during early HIV infection. AIDS 2011, 25, 611–617.

122. Archer, J.; Braverman, M.S.; Taillon, B.E.; Desany, B.; James, I.; Harrigan, P.R.; Lewis, M.; Robertson, D.L. Detection of low-frequency pretherapy chemokine (CXC motif) receptor 4 (CXCR4)-using HIV-1 with ultra-deep pyrosequencing. AIDS 2009, 23, 1209–1218.

123. Tsibris, A.M.; Korber, B.; Arnaout, R.; Russ, C.; Lo, C.C.; Leitner, T.; Gaschen, B.; Theiler, J.; Paredes, R.; Su, Z.; et al. Quantitative deep sequencing reveals dynamic HIV-1 escape and large population shifts during CCR5 antagonist therapy in vivo. PLoS One 2009, 4, doi:10.1371/journal.pone.0005683.

124. Margeridon-Thermet, S.; Shulman, N.S.; Ahmed, A.; Shahriar, R.; Liu, T.; Wang, C.; Holmes, S.P.; Babrzadeh, F.; Gharizadeh, B.; Hanczaruk, B.; et al. Ultra-deep pyrosequencing of hepatitis B virus quasispecies from nucleoside and nucleotide reverse-transcriptase inhibitor (NRTI)-treated patients and NRTI-naive patients. J. Infect. Dis. 2009, 199, 1275–1285.

125. Solmone, M.; Vincenti, D.; Prosperi, M.C.F.; Bruselles, A.; Ippolito, G.; Capobianchi, M.R. Use of massively parallel ultradeep pyrosequencing to characterize the genetic diversity of Hepatitis B Virus in drug-resistant and drug-naive patients and to detect minor variants in reverse transcriptase and Hepatitis B S antigen. J. Virol. 2009, 83, 1718–1726.

126. Verbinnen, T.; Van Marck, H.; Vandenbroucke, I.; Vijgen, L.; Claes, M.; Lin, T.I.; Simmen, K.; Neyts, J.; Fanning, G.; Lenz, O. Tracking the evolution of multiple in vitro hepatitis C virus replicon variants under protease inhibitor selection pressure by 454 deep sequencing. J. Virol. 2010, 84, 11124–11133.

127. Delang, L.; Vliegen, I.; Froeyen, M.; Neyts, J. Comparative study of the genetic barriers and pathways towards resistance of selective inhibitors of Hepatitis C Virus replication. Antimicrob. Agents Chemother. 2011, 55, 4103–4113.

128. Redd, A.D.; Collinson-Streng, A.; Martens, C.; Ricklefs, S.; Mullis, C.E.; Manucci, J.; Tobian, A.A.; Selig, E.J.; Laeyendecker, O.; Sewankambo, N.; et al. Identification of HIV superinfection in seroconcordant couples in Rakai, Uganda, by use of next-generation deep sequencing. J. Clin. Microbiol. 2011, 49, 2859–2867.

129. Campbell, M.S.; Mullins, J.I.; Hughes, J.P.; Celum, C.; Wong, K.G.; Raugi, D.N.; Sorensen, S.; Stoddard, J.N.; Zhao, H.; Deng, W.J.; et al. Viral linkage in HIV-1 seroconverters and their partners in an HIV-1 prevention clinical trial. PLoS One 2011, 6, doi:10.1371/journal.pone.0016986.

130. Fischer, W.; Ganusov, V.V.; Giorgi, E.E.; Hraber, P.T.; Keele, B.F.; Leitner, T.; Han, C.S.; Gleasner, C.D.; Green, L.; Lo, C.C.; et al. Transmission of single HIV-1 genomes and dynamics of early immune escape revealed by ultra-deep sequencing. PLoS One 2010, 5, doi:10.1371/journal.pone.0012303.

131. Bull, R.A.; Luciani, F.; McElroy, K.; Gaudieri, S.; Pham, S.T.; Chopra, A.; Cameron, B.; Maher, L.; Dore, G.J.; White, P.A.; et al. Sequential bottlenecks drive viral evolution in early acute hepatitis C virus infection. PLoS Pathog. 2011, 7, doi:10.1371/journal.ppat.1002243.

132. Neverov, A.; Chumakov, K. Massively parallel sequencing for monitoring genetic consistency and quality control of live viral vaccines. Proc. Natl. Acad. Sci. USA 2010, 107, 20063–20068.

133. Victoria, J.G.; Wang, C.; Jones, M.S.; Jaing, C.; McLoughlin, K.; Gardner, S.; Delwart, E.L. Viral nucleic acids in live-attenuated vaccines: detection of minority variants and an adventitious virus. J. Virol. 2010, 84, 6033–6040.

134. Harismendy, O.; Ng, P.C.; Strausberg, R.L.; Wang, X.; Stockwell, T.B.; Beeson, K.Y.; Schork, N.J.; Murray, S.S.; Topol, E.J.; Levy, S.; et al. Evaluation of next generation sequencing platforms for population targeted sequencing studies. Genome Biol. 2009, 10, R32.

135. Zagordi, O.; Klein, R.; Daumer, M.; Beerenwinkel, N. Error correction of next-generation sequencing data and reliable estimation of HIV quasispecies. Nucleic Acids Res. 2010, 38, 7400–7409.

136. Quince, C.; Lanzen, A.; Davenport, R.J.; Turnbaugh, P.J. Removing noise from pyrosequenced amplicons. BMC Bioinform. 2011, 12, 38.

137. Finotello, F.; Lavezzo, E.; Fontana, P.; Peruzzo, D.; Albiero, A.; Barzon, L.; Di Camillo, B.; Toppo, S. Comparative analysis of algorithms for whole genome assembly of pyrosequencing data. Brief. Bioinform. 2011, doi:10.1093/bib/bbr063.

CHAPTER 14

MASS SPECTROMETRY-BASED PROTEOMICS IN MOLECULAR DIAGNOSTICS: DISCOVERY OF CANCER BIOMARKERS USING TISSUE CULTURE

DEBASISH PAUL, AVINASH KUMAR, AKSHADA GAJBHIYE, MANAS K. SANTRA, AND RAPOLE SRIKANTH

14.1 INTRODUCTION

Cancer is a genetically and clinically diverse disease. The concept of early detection has attracted the attention of both physicians and researchers for decades and thus evolved the concept of "Biomarker" [1]. According to the definition of National Cancer Institute (USA), "biomarker is a biological molecule found in blood, other body fluids, or tissues that is a sign of a normal or abnormal process, or of a condition or disease." The ideal biomarker should be easily detectable, highly sensitive and specific for its target phenotype as well as economically feasible [2]. A biomarker may be used to monitor the body responses to a treatment for a disease or condition. It is also referred to as a molecular marker or biosignature. It can be

This chapter was originally published under the Creative Commons Attribution License. Paul D, Kumar A, Gajbhiye A, Santra MK, and Srikanth R. Current Status and Prospects of Biodiesel Production from Microalgae. BioMed Research International *2013* (2013). http://dx.doi.org/10.1155/2013/783131.

any molecule like DNA, RNA, proteins, or metabolites [3]. Although the survival rate of cancer patients has increased in the last 20 years, newer diagnostic methods with improved sensitivity and specificity are essential for the proper detection and prognosis of this fatal disease.

Discovery of biomarkers through the analysis of patient serum or tissue is a conventional approach being used since the beginning of diagnosis of cancer, but the broad range of serum proteome and availability of patient tissue samples are the major hurdles. Thus, the use of tumor cell lines becomes an attractive option for the study and discovery of candidate biomarkers since the cells possess a rich source of secreted as well as cellular proteins. Secretome comprising the secretory proteins in the culture media, also referred to as conditioned media (CM), serve as a potent source for biomarkers due to ease and effectiveness of detection; however, nowadays even cellular proteins are also providing important information about disease conditions. Thus, this model system can serve as an early provider of potential biomarkers. An overview of tissue culture-based model system for candidate cancer biomarker discovery is represented in Figure 1. A number of studies have used the cell culture-based system to identify the potential biomarkers [4–6]. The clinical significance of using cell lines to understand biological functions lies in the fact that they can be examined through various techniques and that they display the same heterogeneity as the primary tumors as well as different grades [7, 8].

We have witnessed a tremendous improvement in the past decade in the field of high-throughput research that heralds the initiation of a new era in the area of biological science research. Almost all proteomic biomarker discovery platforms use mass spectrometry (MS) as the central technique in association with other proteomic approaches. MS has certain advantages like prediction of molecular mass with the highest specificity and sensitivity with the use of smallest amount of sample [9–11]. Different mass spectrometry-based proteomic approaches have been used to identify biomarkers from various sources and are broadly classified into two categories: gel-based (2-DE and 2D-DIGE) and gel-free (SILAC, iTRAQ) techniques [12–14]. Detection of biomarkers through two-dimensional gel electrophoresis (2-DE) is the most widely used gel-based approach [15]. Improvements over the years have provided us with a more sensitive and high-throughput gel-based technique termed as two-dimensional dif-

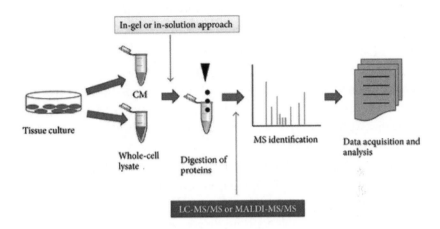

FIGURE 1: An overview of biomarker discovery using tissue culture. Cancer cells are cultured in plates. The CM as well as cells is collected separately. Extracted proteins from each fraction are processed for either in-gel or in-solution digestion followed by the detection of peptides by mass spectrometric approach. Data analysis leads to detection of candidate biomarkers.

ference gel electrophoresis (2D-DIGE). This is based on the differential excitation-emission properties of fluorescent dyes such as Cy2, Cy3, and Cy5 [16]. Apart from the gel-based techniques, gel-free techniques have been dominating the field of biomarker discovery in the last decade. Stable isotope labelling by amino acids in cell culture (SILAC), which relies on the incorporation of amino acids with substituted stable isotopic nuclei such as H^2, C^{13}, and N^{15}, is highly suitable for tissue culture-based model system [17]. Another very sensitive gel free technique known as isobaric tags for relative and absolute quantitation (iTRAQ) is also a method of choice [18].

Moreover, these MS-based proteomic tools have advanced satisfyingly since the last decade and hence have become capable of simultaneously identifying thousands of proteins even from very small amounts. MS advancement has helped enormously in the identification and delivery of candidate biomarkers for cancer diagnosis, prognosis and monitoring of treatment regimen.

14.2 MASS SPECTROMETRY-BASED PROTEOMICS

MS has increasingly become the method of choice for all the proteomic approaches available to date. As the name indicates, "mass spectrometry" determines the molecular mass of a charged particle by measuring its mass-to-charge (m/z) ratio. Basically, a mass spectrum is a plot of ion abundance versus m/z. A mass spectrometer consists of an ion source that converts molecules to ionized analytes, a mass analyser that resolves ions according to m/z ratio, and a detector that registers the number of ions at respective m/z value. The mass analyser depends on three key parameters: sensitivity, resolution, and mass accuracy. The sensitivity, resolution, and accuracy of advanced mass spectrometers allow the detection of femtogram levels of individual proteins in complex mixtures. As recognized by the 2002 Nobel Prize in Chemistry, innovation of electrospray ionization (ESI) and matrix-assisted laser desorption/ionization (MALDI) techniques has made it possible to ionize big molecules such as proteins, peptides, and nucleotides for mass spectrometric analysis. ESI generates ions at atmospheric pressure by injecting a solution-based sample through a small capillary (Figure 2(a)). MALDI produces ions by pulsed-laser irradiation of a sample which is cocrystallized with a solid matrix that can absorb the wavelength of light emitted by the laser (Figure 2(b)). Protonation or deprotonation is the main source of charging for the ions generated in ESI/MALDI. MS-based proteomics is a widely used approach to find protein sequence from unknown samples by correlating the sequence ions generated from tandem mass spectral data with sequence information available in protein databases. MS-based proteomics analyses of complex protein mixture usually require a starting amount in the range of 0.1–10 µg, depending on the experimental setup and the type of mass spectrometer used. ESI is playing an increasingly conspicuous role in the study of the protein structure, folding, and noncovalent interactions [19]. Recently, MALDI imaging has allowed biomolecular profiling of tissue sections and single cells [20]. In combination with chromatographic separation techniques, MS is playing an important role in discovering the biomarkers for various diseases. Many research groups have been using MS-based techniques in order to identify potential cancer biomarkers for diagnostic as well as therapeutic purposes [21–25].

14.3 MASS SPECTROMETRY-BASED QUANTITATIVE PROTEOMIC STRATEGIES TOWARDS BIOMARKER DISCOVERY

Cancer remains a major cause of mortality worldwide despite the progress in detection, diagnosis, and therapy. Early diagnosis of cancer improves the likelihood of successful treatment and can save many lives. Thus, early diagnostic biomarkers are highly important for detection and diagnosis in cancer, but due to the lower sensitivity and lack of specific biomarkers, there is an urgent need to discover new and better biomarkers that would be helpful in improving cancer diagnosis, prognosis and treatment. Proteomics is the most powerful technique which can help to discover novel candidate biomarkers for cancer. Current progress in proteomics has been largely due to recent advancements in MS-based technologies. This powerful MS-based quantitative proteomic technologies can aid in the identification of all differentially expressed proteins and their posttranslational modifications during cancer progression which can be used as biomarkers for early diagnosis and monitoring disease treatment in cancers. Moreover, the candidate biomarkers for other diseases, like diabetes, cardiovascular,

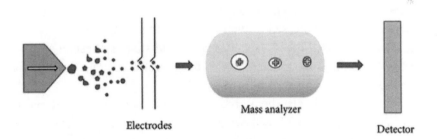

Electrodes   Mass analyzer   Detector

FIGURE 2: (a) A schematic representation of ESI-MS—solvent along with sample flows from the needle with electrical potential generating charged droplets. The droplets carry the sample, and they are desolvated by applying heat and nebuliser gas to produce ions. These ions are now separated according to m/z ratio in the mass analyzer and registered by detector. (b) A schematic representation of MALDI-MS—the sample is mixed with the matrix and allowed to crystalize on the MALDI plate, when the laser hits the sample-matrix mixture on the plate, matrix absorbs the energy of the laser to get vaporized along with samples. Next, the charge exchange takes place from matrix and sample ions are generated.

and so forth, are also discovered with the help of these techniques [26, 27]. This section focuses on different mass spectrometry-based proteomic strategies and explores their applications in potential biomarker discovery.

14.4 2D GEL ELECTROPHORESIS (2-DE)

The 2-DE method is a primary technique regularly used in proteomic investigations [15]. In this method, extracted proteins are resolved in the first-dimension based on their isoelectric point (pI) followed by molecular weight in the second-dimension (Figure 3). The gels are then stained by either Coomassie Brilliant Blue or silver stain to visualize the protein spots. Using 2-DE software, differentially expressed protein spots are excised and identified by mass spectrometry [28]. This approach could lead to separation and identification of about 2000 unique spots. Using the 2-DE, Braun et al. successfully identified 64 differentially regulated proteins in cancer by mass spectrometry and showed that microfilamental network-associated proteins are frequently downregulated in leukocytes of breast cancer patients [29]. These are functionally important for all central processes and highly relevant for all stages of tumorigenesis-like metastasis [29]. Similarly, Cancemi et al. identified S100 group of proteins that are preferentially expressed in tumor samples than their normal counterpart. They have used breast cancer as subject of study and established for the first time the importance of the S100 group of proteins as potential biomarkers [30].

This technique is also being routinely used for the proteomic profiling of cancer cells treated with drugs (in vitro). Strong et al. studied the differential regulation of mitochondrial proteome of Adriamycin-resistant MCF-7 breast cancer cells. They have identified 156 unique proteins and established coproporphyrinogen III oxidase and ATP synthase alpha chain to be responsible for the chemotherapeutic resistance [31]. Similar kind of study has been carried out by Lee and coworkers to show hnRNPA2 and GDI2 proteins to be associated with paclitaxel resistance in ovarian cancer cell lines [32]. They have established a paclitaxel resistance subline SKpac from the sensitive counterpart SKOV3 followed by quantitative proteomic analysis and further validated their findings by western blotting. These

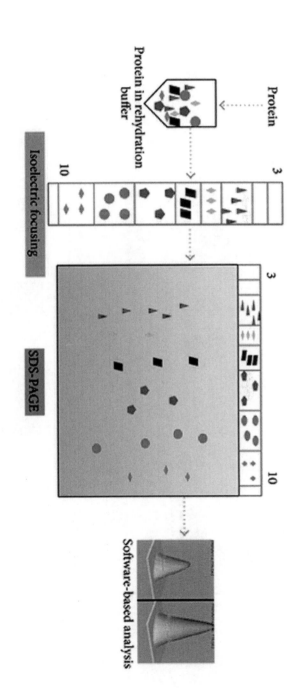

FIGURE 3: An outline of 2-dimensional gel electrophoresis (2-DE). The extracted proteins are solubilized in rehydration buffer. The proteins are immobilized on IPG strips of different pH ranges depending on the requirement of the experiment. In the first-dimension, the proteins are separated on the basis of their isoelectric points (pI) and are further resolved according to their molecular weight in the second-dimension. Finally, protein spots of interest are excised and subjected to tryptic digestion followed by MS.

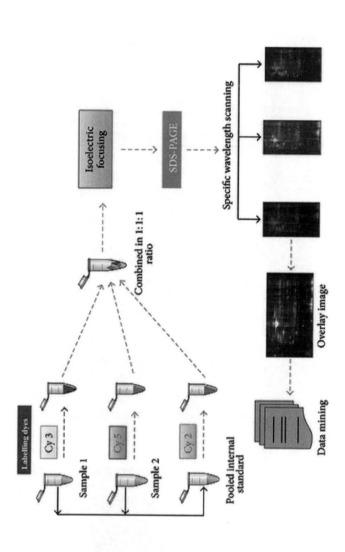

FIGURE 4: An outline figure of 2D-DIGE. Proteins are extracted from the samples and are labelled with different fluorophores as Cy 3 for sample 1, Cy 5 for sample 2, and Cy 2 for the pooled internal standard. All the samples are resolved in the same 2D gel followed by protein spot pattern detection by scanning the gel in respective wavelength for the Cy dyes; the merging of all of them yields an overlay image consisting of all three Cy dyes. The images are analyzed to get potential candidates of interest.

examples demonstrate the potency of 2-DE approach in the discovery of novel proteins involved in tumorigenesis and chemotherapeutic drug response. However, 2-DE still has its limitations like the inability to resolve too basic, too acidic, and hydrophobic proteins. The ampholytes used for the generation of pH gradient are not stable at extreme acidic and basic pH and are therefore unsuitable for use. In addition, the membrane proteins due to their highly hydrophobic nature pose problems in solubilisation, making them difficult to resolve. Reproducibility and low relative quantification accuracy are other major obstacles which arise due to factors such as run to run variation and limitation of the detection methods available [33]. Requirement of huge amount of sample and inability to detect low abundance proteins is also a major drawback. Though 2-DE has its limitations, still it will be a method of choice for proteomic study because of its robustness and simplicity.

14.5 2D DIFFERENCE GEL ELECTROPHORESIS (2D-DIGE)

The 2D-DIGE method is an improved version of 2-DE technique. In this technique, two different protein samples (control and diseased) and one internal control (mixture of control and diseased sample in equal proportion) are labelled with any of the three fluorophores: Cy2, Cy3, or Cy5. These fluorophores have the identical charge and molecular mass but unique fluorescent properties. This allows us to discriminate them during scanning using appropriate optical filters [16, 34]. The labelled samples are then mixed together and separated on a single gel. The best part of this technique is the use of the same internal pool for all the gels that serves as an internal control for normalization (Figure 4) [16, 34]. The gel is scanned by an advanced scanner which can resolve the three different wavelengths: 488 nm (Cy2), 532 nm (Cy3), and 633 nm (Cy5). Each of the samples generates its unique image. This technique eliminates gel-to-gel variation, enhances sensitivity (order of 4 magnitudes), and is less laborious [35, 36]. However, the sample source variation of 2D-DIGE is as vivid as 2-DE. This technique is routinely used for the discovery of candidate biomarkers as well as any quantitative proteomic data generation and therapeutic drug development. Zhang et al. used this technique for the identification of differentially expressed proteins between early submucosal noninvasive and invasive colorectal cancer [37]. They have established a Fischer-344 rat model for the invasive and noninvasive colorectal cancer and found two candidates, transgelin (upregulated) and carbonic anhydrase 2 (CSII) to play significant role in CRC. They have also validated these candidates through fluorescence-based quantitative polymerase chain reaction, western blotting and immunohistochemistry assays [37]. In a similar kind of study, isocitrate dehydrogenase 1 (IDH1) was detected and validated as a potential biomarker for nonsmall cell lung carcinoma [38]. They have identified IDH1 as a potential biomarker in different NSCLC cell lines and further validated it using patient tissue samples via different techniques like western blotting, immunohistochemistry, knockdown assay, and xenograft model. Although the relevance of IDH1 via different genomic and molecular biology techniques is well established now, the basis of its potential was established by this kind of proteomic studies [38]. In another study, Banerjee et al. used 2D-DIGE in combination with MS for the identification of prognostic biomarkers in glioblastoma multiforme using human astrocyte cells and HTB12 human astrocytoma cells [39]. Similarly, Sinclair et al. used this technique to identify the candidate tumor suppressor biomarker in ovarian cancer cell lines: TOV-112D and TOV-21G. They have employed 2D-DIGE and 2D-LC-MS/MS with tandem mass tagging (TMT) to identify potential tumor suppressors in cell lysate [40]. In a separate study, Wilmes et al. compared the proteomic profile of paclitaxel and peloruside-A-treated HL-60 promyelocytic leukemic cells [41]. This technique is a widely used and accepted one in the field of quantitative proteomics. Although the major limitations of 2-DE still apply to 2D-DIGE, but the introduction of more sensitive 2-D DIGE technique has overcome most of the limitations such as requirement of huge amount of sample and inability to detect low abundance proteins. Although in the past decade gel-free techniques have developed immensely, 2D-DIGE has kept its position in proteomic research and will be there for years to come.

14.6 STABLE ISOTOPE LABELLING BY AMINO ACIDS IN CELL CULTURE (SILAC)

The use of quantitative proteomic techniques for the identification of potential biomarkers is a fast gaining ground. For cell culture-based comparative proteomic studies, SILAC is a method of choice [17]. A number of amino acids such as arginine, leucine, and lysine with stable isotope are suitable for use in SILAC, but lysine and arginine are the two most commonly used labelled amino acids. This method solely relies on metabolic incorporation of labelled (heavy) amino acids during cell proliferation. Two different populations of cells (tumor cells and normal cells) are cultured in vitro under similar conditions except that tumor cells are grown in media containing heavy isotope of an amino acid (e.g., C^{13} labelled arginine) and the normal cell line is grown in usual media. The cells are allowed to grow as usual for over five to seven passages to ensure >95% labelling [42]. Once the cell lysates are prepared, the samples are combined in a 1 : 1 stoichiometric ratio. Prepared samples are then separated on a SDS-PAGE and further subjected to in-gel trypsin digestion followed by MS analysis. The samples may also be digested in-solution before analysis. During the analysis by mass spectrometer, different isotope composition can be differentiated as the labelled amino acids will induce a shift in the m/z ratio in comparison to the unlabelled amino acids. This process ensures that a particular peptide fragment of diseased sample differs from its normal counterpart in m/z ratio and hence enabling them to be detected by mass spectrometry (Figure 5). Geiger et al. identified prognostic biomarkers such as IDH2, CRABP2, and SEC14L2 for overall breast cancer survival [43]. They have done the stage-specific analysis of proteome using tissue culture-based model system and further validated them using the patient tissue samples. They validated the candidates via immunohistochemistry and tissue array of human tumor samples. These kinds of holistic studies have helped us to find the potential biomarker for monitoring disease progression and prognosis (CRABP2 and IDH2 are markers of poor prognosis and SEC14L2 is a marker of good prognosis) [43]. Kashyap et al. used SILAC-based proteomic investigation for the discovery of new candidate biomarkers in oral squamous cancer using tissue culture-based system [44]. In a similar type of study, Wang et al. established the regulatory network of karyopherin subunit alpha-2 (KPNA2) as a novel cargo protein in nonsmall cell lung carcinoma (NSCLC) to further establish KPNA2 as a candidate biomarker for NSCLC [45]. In a different kind of approach, Cuomo et al. used this versatile technique for the identification of histone signatures in breast cancer cell lines. They specifically focused on histone H3 and H4 and came up with "breast cancer-specific epigenetic signature," with implications for the characterization of histone-related biomarkers [46]. Moreover, the use of this technique is no longer confined to in vitro cell culture. Recently, the founder of this technique, Matthias Mann, has come up with a variation for the use of SILAC in vivo [47]. Here, the authors labelled the mice by continuous feeding of either natural or heavy isotope lysine-containing food for four generations. They isolated blood samples and organs to evaluate the incorporation of heavy isotope and found that all the proteins were labelled in the second generation. Further, they validated their result by comparing the proteomes from platelets, heart and erythrocytes from β1-integrin, β-parvin, and kindlin-3 deficient mice, respectively [47]. They proposed that it is a novel technique, which can be used to monitor the function(s) of a gene at a proteomic level in vivo by generating knockout mouse of that gene. Although in vivo SILAC mouse model is a great advancement, the same technique cannot be applied to human subjects. SILAC's advantage lies in the nonrequirement of targeted analysis of specific proteins or peptides, as every peptide is labelled and can be quantified depending on the degree of resolution and instrument sensitivity. It is also more robust and accurate than other quantitative techniques. However, SILAC also has few drawbacks like it cannot be used directly to human tissue samples as well as autotrophic cells (plant cells). Moreover, costly reagents are also an obstacle [48]. Although SILAC has its own set of disadvantages, it has immense potential and is yet to be exploited fully. It is gaining popularity quickly and will continue to be used as a significant tool in quantitative proteomic studies.

Super-SILAC is an improved version of SILAC. As a single-cell line cannot represent the heterogeneity of tumor tissue, super-SILAC helps to enhance the sensitivity and robustness of tissue culture-based model sys-

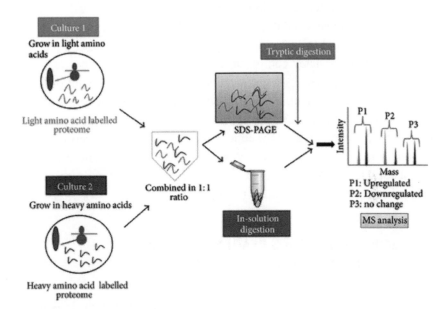

FIGURE 5: A schematic overview of SILAC. Cells are grown in normal and heavy amino acids containing media for 6 generations to achieve maximal incorporation of heavy amino acids. The proteins are extracted from both populations of cells and mixed in equal proportion and then subjected to either in-gel or in-solution digestion. Relative abundance of the digested peptides is determined from the ratio of heavy-to-light peptide signals as obtained from MS.

tem for quantitative proteomic approach [49]. This method relies on the use of a mix of several SILAC-labelled cell lines as an internal standard for more comprehensive representation of the tumor proteome. Geiger et al. have used this method to show that it represents the tumor heterogeneity better than SILAC. They have used a panel of breast cancer, glioblastoma, and astrocytoma cell lines that represent the internal standards for these tumor types [50]. Boersema et al. used super-SILAC with LC-MS/MS to identify N-glycosylated proteins in cell secretome and patient blood samples. They have used 11 breast cancer cell lines that represent different stages of breast cancer and took few cell lines representing the super-SILAC mix as internal standard for more accurate quantification. Enriched

N-glycosylated proteome mainly comprised the membrane and secretory proteins. They have validated the identified candidates in human blood samples [51]. Lund et al. have used it for the study of metastatic markers in primary tumors. They compared the proteome of tumors derived from inoculation of a panel of isogenic human cancer cell lines with different metastatic capabilities into the mammary fat pad of immunodeficient mice [52]. As of now, it does show a great potential to serve as a relative proteomic quantitation method for understanding molecular aspects of cancer biology and perhaps as a convenient approach for candidate biomarker discovery. Due to its high accuracy and low error rate, it is becoming the method of choice in quantitative proteomics.

14.7 ISOBARIC TAGGING REAGENT FOR ABSOLUTE QUANTITATION (ITRAQ)

Another popular and comprehensive quantitative technology is Isobaric Tagging Reagent for Absolute Quantitation (iTRAQ) introduced by Ross et al. [18]. iTRAQ label consists of a reporter group (variable mass of 114–117 Da), a balance group, and an amine-reactive group that reacts at lysine side chains and NH_2-terminal. In iTRAQ, samples are labelled after trypsin digestion with four independent iTRAQ reagents. The labelled samples are pooled and the tagged peptides are fractionated by strong cation exchange (SCX) chromatography, and each desalted fraction is subjected to tandem mass spectrometry [53]. The reporter groups of the iTRAQ reagent generate reporter ions for each sample with mass/charge (m/z) of 114, 115, 116, and 117 during MS/MS. These reporter ions allow the differentiation of the different samples in MS and furnish the necessary quantitative information (Figure 6). Recently, electrostatic repulsion-hydrophilic interaction chromatography (ERLIC) and off-gel fractionation have evolved as an alternative to the cumbersome process of SCX chromatography [54, 55]. ERLIC method separates peptides on the basis of electrostatic repulsion and hydrophilic interaction and is found to be increasing the proteome coverage [54]. In off-gel fractionation, the samples are rehydrated on a gel strip and further separated up to 24 fractions according to pI [55]. iTRAQ method can also be improved to perform absolute quantification by adding internal standard peptide. Recently, Eight-plex iTRAQ reagents have also become commercially available that allows the quantification of eight different samples in a single run. The advantage of iTRAQ labelling is that signal obtained from combined peptides enhances the sensitivity of detection in MS and MS/MS. However, the variability in labelling efficiencies and the costly reagents are major limitations of this high fidelity technique [56]. The use of this powerful technique is gradually becoming the method of choice in the field of biomarker discovery. In a study by Rehman and coworkers, this technique was used for candidate biomarker discovery associated with metastasis using both patient sample as well as prostate cancer cell lines [57]. They have pooled serum samples from three different stages of prostate cancer patient group as nonprogressing samples, progressing samples, and metastatic samples followed by identification of a set of potential prognostic biomarkers that may be involved in disease progression and metastasis. They have identified eEF1A1 as a novel candidate biomarker significantly showing increase in all three groups of samples when compared to benign prostatic hyperplasia (BPH) samples. They have also validated their results in 11 frequently used prostate cancer cell lines to show eEF1A1 expression validation both at the translational and transcriptional levels. Further, they have also identified C-reactive protein (CRP) that is already established as a potential marker for bone metastatic prostate cancer [57]. In a similar type of study, metastasis-related candidate biomarkers have been identified in colorectal cancer cell lines. They labelled the whole-cell lysates of SW480 (primary cell line) and SW620 (lymph node metastatic variant of SW480) with 4-plex iTRAQ followed by 2D-LC and MALDI-TOF/TOF to identify β-catenin and calcyclin binding protein (CacyBP) as differentially expressed. CacyBP degrades β-catenin. Thus, these two proteins show a very nice inverse correlation in the progression of metastasis and hence are potential candidate biomarkers [58]. In another study, it has been used for the profiling of tyrosine phosphorylation level in breast cancer progression using MCF10AT breast cancer cell line [59]. Using complementary MALDI- and ESI-based mass spectrometry, they have identified 57 unique proteins comprising tyrosine kinases, phosphatases, and other signaling network proteins that might play significant role during disease progression. For the first time, they have identified SLC4A7

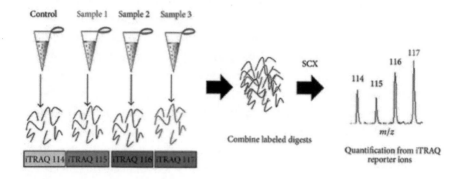

FIGURE 6: Experimental outline of iTRAQ. Proteins are extracted from either tissue samples or cultured cells and subjected to proteolytic digestion. The digested peptides are then labelled with isobaric tags followed by the pooling of the samples. The samples are then fractioned through SCX followed by tandem mass spectrometry analysis.

(sodium bicarbonate cotransporter) and TOLLIP (Toll interacting protein) as novel tyrosine kinase substrates associated with cancer development providing valuable insights into the disease progression [59].

14.8 LABEL-FREE TECHNIQUES

To overcome the difficulties in labelling techniques such as high cost of the reagents, higher concentration of sample requirement, and incomplete labelling, researchers are turning to mass spectrometry-based label-free shotgun proteomic technology. It is a very high throughput technique that opens up a new era in the discovery of potential biomarkers. Label-free technology is based on the assumption that the peak area of a peptide in the chromatogram is directly proportional to its concentration [60, 61]. This strategy is generally based on two classes of measurements; the first is based on the measurements of ion intensity changes like peptide peak areas or peak heights in chromatogram, and the second is the spectral counting in the MS/MS analysis. Recently, label-free approaches have been used for absolute quantification in addition to the relative quantification of peptides/proteins. Initially, protein abundance was estimated using protein abundance index (PAI), but later on it was converted to exponentially modified PAI (emPAI) which is routinely used for determining absolute protein abundance. Recently, a modified way of spectral counting termed absolute protein expression (APEX) profiling has been used to measure the absolute protein concentration. Decyder MS from GE Healthcare, Protein Lynx from Waters, and SIEVE from Thermo Electron are some of the commercially available softwares for label-free analysis. This technology is applied for candidate biomarker discovery mostly using clinical samples. Ishihara et al. have used it to identify N-glycoproteins as potential biomarkers in hepatocellular carcinoma [62]. Similarly, it was used by Old and coworkers to identify differentially expressed proteins in K562 human erythroleukemia cells [63]. They used peptide spectral counts and LC-MS/MS to study the simulation effect under different conditions that promote cell differentiation by mitogen-activated protein kinase pathway activation [63].

Like other techniques, this technique also has its own advantages and limitations. It seems to be a promising technique for shotgun quantitation and cheap, simplistic, and less complicated in terms of analysis. The limitation of this technique is redundancy in detection which may arise from the peptides which are shared between more than one protein, leading to the suppression effect. In addition, label-free quantification methods suffer from less accurate, semiquantitative, and are not suitable for low abundance and short proteins. Another drawback is in the normalization of the data while exploring multiple samples in multiple reactions [23, 60, 61]. While considering this technique for the quantitation, one should consider that the correlation of MS/MS spectra with a protein is only an approximation owing to the errors arising due to false identification. Proteins of low abundance could still be present in the sample in spite of the spectral count being zero and also larger proteins can give rise to more tryptic digests, hence more spectral counts. The fact that the signal for a given peptide is governed by many factors like efficiency of fragmentation and ionization in electrospray should also be taken into consideration. Thus, the spectra in MS/MS accounting for the identification of a protein can only be used as an indication of its abundance in the sample [64]. These limitations have left us with the scope for more improvement.

14.9 STABLE ISOTOPE DILUTION MASS SPECTROMETRY (SID-MS)

In contrast to the relative quantification proteomic approaches, tandem mass spectrometry-based selected reaction monitoring (SRM), and multiple-reaction monitoring (MRM) techniques have been used for absolute quantification of proteins in combination with stable isotope dilution. This MS-based absolute quantification method relies on the incorporation of known quantities of isotope-labelled standards, which display very similar chromatographic properties to the target compounds but can be distinguished by their difference in m/z [65]. This isotope dilution method is generally a targeted approach which is focused on a limited set of proteins. In this method, first initial analysis requires identification of signature peptides for targeted proteins followed by an internal standardization performed by spiking stable isotope-labelled peptides into the samples in defined amount before analysis. Quantification is performed by comparing the peak height or peak area in the extracted ion chromatograph of the isotope-labelled and the native forms of a signature peptide [65]. The major advantage of this method is good linearity and excellent precision, but the accuracy and ability to determine the true abundance of target protein strongly depend on the choice of signature peptide selected and the purity of internal standard. The disadvantage of this method is that it is limited to small number of proteins because suitable internal standards need to be purchased/synthesized. The second disadvantage is that this kind of experiments can be preferably done in triple quadrupole mass spectrometers but not all available tandem mass spectrometers. First time, Kippen and coworkers used this method for precise determination of insulin, C-peptide, and proinsulin levels in blood of nondiabetic and type II individuals [66]. Gerber et al. successfully used this method for absolute quantification of proteins and phosphoproteins from cell lysates [67]. Kuzyk and coworkers used this technique to develop a method for the quantitation of 45 serum proteins in human plasma [68]. Recently, Jiang et al. quantified endogenous cystic fibrosis transmembrane conductance regulator (CFTR) in HT29 and BHK cells using MRM-MS and oxygen stable isotope dilution [69]. Apart from these notable studies, SID-MS has been used for the quantification and verification of potential biomarkers in pancreatic [70], prostate cancer [71], and cardiovascular diseases [72]. SID-MS-based quantification is filling the gap between the discovery and validation phases, which may promote potential biomarkers towards clinical trials and thereby their development as diagnostic tools.

14.10 ROLE OF TISSUE CULTURE AND PROTEOMICS IN CANDIDATE BIOMARKER DISCOVERY

Despite the development of the omic technologies, the search for candidate biomarkers that would provide detailed information on diagnosis, prognosis, and disease monitoring has remained largely elusive. The serum proteome of patient samples is largely (>95%) covered by the most abundant 20 proteins and the potential biomarkers are from the remaining 5% of the proteins, thereby yielding very few cancer biomarkers which are in current clinical use [73]. The availability and number of patient tissue samples are also a limitation. Therefore, researchers thought of an alternative approach comprising tissue culture-based candidate biomarker discovery systems to gain insight into different cancers.

The cancer proteome can be classified into two broad groups: secretome and cellular proteome. Secretome, comprising the secretory proteins, plays important roles in vital cellular functions and they can act locally as well as systemically. The secretome reflects the functionality of a cell in a given environment [74]. The proteins or their fragments are secreted from cancer cells into the media termed as conditioned media (CM). Therefore, secretory proteins can function as novel candidate tumor markers for different cancers and can be extracted from tissue culture media of human cancer cell lines. CM, as a source for potential biomarkers, is increasingly becoming popular as revealed by the surge in the number of recent publications [75–77]. On the other hand, analysis of cellular proteome has also given insight into the pathogenesis and has helped us to come up with candidate biomarkers such as Hv1 (voltage-gated proton channel) [78]. The authors reported that Hv1 is specifically expressed in highly metastatic human breast tumor tissues and cell lines and that its level significantly correlated with tumor size, classification, and disease-free survival [78, 79].

Breast cancer cell lines, specifically MCF7, have been widely used as a model to explore potential breast cancer biomarkers [80]. Jung et al. identified potential biomarkers in lung cancer using tissue culture-based approaches [81]. It is becoming increasingly clearer that cell lines are as heterogeneous as primary tumors [8].

Although the in vitro cell culture model provides us with great advantages, it also has its own set of disadvantages as reported by Kulasingam and Diamandis, like a single-cell line has multiple variants that makes this system complex. Moreover, it is yet to reach the stage where it can mimic the tumor microenvironment as well as its real characteristics, and others [6]. The three-dimensional (3D) cell culture techniques have made the things more reliable and versatile because of mimicking the in vivo conditions [82]. Therefore, their usage for candidate biomarker discovery is more relevant. Moreover, the field of drug discovery and disease prevention largely depends on the tissue culture-based model system, majorly relying on high-throughput proteomics techniques, because there is no scope for direct human trials of newly developed lead molecules [83, 84]. Recently, there have been a lot of reports where researchers have tried to find out the working mechanism of a drug through proteomic, genomic, and many other techniques [85, 86]. Currently, there is a major concern regarding the drug resistance, which implies the nonresponsiveness of a disease for a certain drug at its working concentration. Researchers are relying heavily on robust proteomic approaches for finding the probable "culprits" for this drug resistance [87–91]. The studies using tissue culture-based candidate biomarker discovery platform are shown in Table 1.

14.11 ADVANTAGES OF TISSUE CULTURE IN BIOMARKER DISCOVERY AND DIAGNOSIS

Cancer cell lines are the most widely used models to study the deregulation in cancer as well as the identification of potential biomarkers for the early detection and prognosis of cancer [92, 100]. Both the cellular milieu and conditioned medium (CM) serve as a rich source of potential biomarkers. The clinical relevance of using cell lines is already well established [101, 102]. As PSA (prostate-specific antigen), the existing biomarker for

Table 1: Tissue culture-based candidate biomarkers discovery in different cancers.

Cancer types	Cell lines used	Biomarker identified	Clinical relevance	Quantitative techniques used	Reference
Breast cancer	21T series of Breast cancer cell lines HMT-3522-S1, MFM223, HCC202 and HCC2218, HCC1599, HCC1143, HCC1937, MCF7, MCF10A, MDA-MB-453	TIMM17A	Disease prognosis	SILAC and LC-MS/MS	[5]
		IDH2, CRABP2, SEC14L2	Disease progression and monitoring	SILAC and MALDI-MS/MS	[44]
Prostate cancer	PC3, LnCAP, 22Rv1	Follistatin, chemokine (C-X-C motif) ligand 16, Pentraxin 3, and spondin 2	Disease progression and monitoring	Two-dimensional chromatography and tandem mass spectrometry 2D-DIGE, MALDI-MS/MS	[76] [92]
Lung cancer	1198 and 1170-I, BEAS-2B and 1799 CL1-0 and CL1-5	PGP9.5, TCTP, TIMP-2, and TPI KPNA2	Disease monitoring Disease detection and progression monitoring	2DE and MS SILAC, LC-MS/MS	[81] [46]
Gastric cancer	AGS and MKN7	GRN	Disease detection and monitoring	2D-LC-MS/MS and iTRAQ	[93]
Pancreatic cancer	PANC1, BxPc3, MIA-PaCa2, SU.86.86	Anterior gradient homolog 2, syn-collin, olfactomedin-4, polymeric immunoglobulin receptor, and collagen alpha-1(VI) chain	Early disease detection and monitoring	LC-MS/MS, ELISA	[94]
Colorectal cancer	Tumor samples were cultured in vitro	EFEMP2	Detection and monitoring	1D-LC-MS/MS	[95]

Table 1: *Cont.*

Cancer types	Cell lines used	Biomarker identified	Clinical relevance	Quantitative techniques used	Reference
Head and Neck cancer	SCC4, HSC2, SCC38, and AMOSIII	alpha-enolase, peptidyl prolyl isomerase A/cyclophilin A, 14-3-3 z, heterogeneous ribonucleoprotein K, and 14-3-3 s	Disease detection and progression monitoring	LC-MS/MS, western blot	[96]
Oral cancer	OEC-M1 and SCC4 OEC-M1 and SCC4	Mac-2 BP Guanylate-binding protein 1 (GBP1)	Early detection of disease Disease detection and progression	MALDI-TOF MS 1D and LC-MS/MS	[97] [98]
Renal cell carcinoma	786-O, Caki-1, A498, ACHN OS-RC-2, HK-2, HUVEC	FoxM1	Detection and potential drug target	IHC, western blot, ELISA	[99]

detection of prostate cancer poses problems; there is a need for a more accurate biomarker. Qian et al. identified Spondin 2 (Spon-2) as a candidate biomarker for prostate cancer [92]. They first identified the extra-cellular proteins by 2-DE coupled with LC-MS/MS. Further, they concentrated on Spon-2 as it was consistently overexpressing in prostate cancer cell secretome, and then they validated their findings in human prostate cancer tissue samples. Moreover, they have checked the sensitivity and specificity of Spon-2 by receiver operator characteristic (ROC) curve analysis. Spon-2 also out rated PSA in the patient samples in terms of sensitivity and specificity [92]. Similarly, Lee et al. showed high-mobility group protein B1 (HMGB1) as a better prognostic marker over carcinoembryonic antigen (CEA) for colorectal carcinoma [102]. They have used 10 colon cancer cell lines along with a normal colon cell line CCD18Co and detected the presence of HMGB1 in the secreted medium. Further, they validated their findings in patient sera also. They have proven the diagnostic value of HMGB1 in a cohort of 219 colorectal patient samples along with 75 control samples. HMGB1 showed more stage-specific diagnosis value than CEA. When HMGB1 and CEA are combined, the overall diagnostic sensitivity is improved when compared to CEA alone (42% versus 25.6%) and the stage 1 cancer diagnosis (47% versus 5.9%) [102]. This kind of study sets the platform for the identification of potential new prognostic biomarkers that might be a tedious job using patient samples directly. The cell culture-based model system possesses its own uniqueness and benefits. The availability and the number of patient tissue samples always present a challenge for the researchers at least in countries with poor public awareness. This is where the easy availability of cell lines (cancerous and normal) comes in. They can also be easily propagated compared to the patient sample. The other advantage is that the cell culture-based model system is cost effective compared to the patient sample system. This system also has the versatility that patient sample system does not have. The cell culture-based system can be used to check the potential efficacy of a novel lead molecule which can be a prospective drug over various types of cancers. This kind of studies also allows us to get an insight into a drug's mechanism of action.

CM of cancer cells allows us to search for potential biomarkers at the level of secretome. This approach offers various advantages like removal of the potential infectious sources. Few of the currently available biomarkers also pose problems, as for pancreatic cancer, the best available marker is CA19.9; however, the false positive rates of this marker are high as they are also elevated in nonneoplastic conditions like acute and chronic pancreatitis, hepatitis, and biliary obstruction [103]. The cell secretome possesses a great advantage for the dissection of potential biomarkers; as for the clinical use, the best biomarkers are those that can be detected in body fluids. The cell secretome indirectly represents the proteins that can be found in the body fluids of a patient, so the identified secretory proteins can be a good biomarker. Moreover, the dynamic range of the secretome is very low compared to cell lysate, so it is a better source for the profiling of biomarkers for diseases like cancer. It is a noninvasive method for the detection of biomarkers rather than directly encountering the patient samples, and the availability of so many cell lines that represent the different stages of the disease helps to provide relevant information [104]. It also effectively bypasses the large amount of serum proteins present in the body fluids of patients. Importin alpha subunit-2 (also called KPNA-2) was identified as a candidate biomarker by Wang and colleagues using CL1-0 and CL1-5 lung cancer cell lines. They have integrated the data of cancer cell secretome and transcriptome of adenocarcinoma tissues. Further, they have validated their results by immunohistochemistry, and, moreover, they have shown that KPNA-2 and CEA in combination produce more efficient diagnostic capacity in the patients [105]. A similar approach was taken by Kulasingam and Diamandis to identify the candidate biomarkers in breast cancer cell lines using a panel of three breast cancer cell lines: MCF-10A, BT474, and MDA-MB-468. They have identified low abundant proteins like elafin and kallikrein family of proteins along with highly abundant proteins by using "bottom-up" proteomic technique via 2D-LC-MS/MS on a linear ion trap (LTQ) as a potential drug target as well as candidate biomarker [106]. Using this technique, Ahmed et al. have identified a candidate biomarker, immunoreactive integrin-linked kinase (ILK) for ovarian cancer [107]. Similarly, the cell line established from human prostate cancer was confirmed to release PSA when cultured in serum-free CM [108]. This system can be easily modified to allow us to study the prognostic and diagnostic markers under different conditions. If we wish to study the differential regulation of a candidate prognostic biomarker in different disease conditions, it is only possible by the use of tissue culture model system. Another very important advantage of this system over the patient tissue sample is the relatively easy detection of the less abundant proteins, which are the source of potential biomarkers. In patient sample, the high abundant proteins like albumin and immunoglobulin create problem for the detection of less abundant proteins through high throughput techniques like mass spectrometry.

Nowadays there are ways to remove high abundant proteins. In most cases, it seems to affect the protein concentration in a big way and people are still trying to find a way to improve this technique. It is often cumbersome to reproduce the data using patient samples because of the heterogeneity. The physical as well as physiological status of the patient plays important role in the tumor biology, but cell culture-based system offers a better way to solve this problem as we have a way to propagate the cells for passages and the results can be more easily reproduced in this system as we can use the same lineage of cells for the study. This system allows us to detect the alterations at proteome level which is also possible for patient sample study but again it is more labour intensive, time consuming, and expensive. In well-defined experimental conditions, the proteome of a cell line should reflect the genetic changes of a cell. To get an in vivo insight into the disease, researchers use cancer cell xenograft model system. More recently, 3D cell culture system has become a model of choice. Mikesh et al. have used this system to successfully identify molecular markers associated with melanoma [109]. CD151 was identified as a potential prognostic marker for breast cancer. The researchers have used MDA-MB-231 as a model system. In tumor xenograft model, CD-151 knockdown cells showed reduced tumorigenecity compared to normal tumor cells. CD-151 also affects the tumor vasculature. Moreover, the overall survival rate of CD-151 positive patients was 45.8% compared to CD-151 negative patients. Further, they have deciphered its molecular modulator network to establish it as a novel drug target [110]. In a similar kind of study, Yao et al. have used a lectin affinity-based approach to enrich as well as increase the detectable number of secreted proteins in the CM of cultured tissues followed by LC-MS/MS and identified EFEMP2 as a potential marker for early detection of colorectal carcinoma (CRC). They have also proven it to be superior to CRC biomarker and CEA and validated their results by immunohistochemistry [95]. Lee and coworkers have established H^+-myo-inositol transporter SLC2A13 as a potential biomarker for cancer stem cell (CSC) in oral squamous cell carcinoma (OSCC) [111]. Head and neck carcinoma is one of the poorly understood cancer and there is a need of biomarkers for its diagnosis and prognosis at early stages. Ralhan et al. have used proteomic-based approaches to identify new potential biomarkers for head and neck carcinoma [96]. They have analyzed the secretory medium of different head and neck cancer cell lines via LC-MS/MS and identified a panel of potential biomarkers. Further, they have validated their results via immunoblotting in patient sera also [96]. Once identified, few of these potential biomarkers can be undertaken for clinical trials to further investigate their potential as biomarkers. Similarly, tissue culture-based model system has been used to mine for potential biomarkers in other cancers as well (Table 1) [93, 94, 97–99].

As stated, there are various advantages of using tissue culture-based candidate biomarker discovery but ultimately the studies have to be carried out in patient sample to validate a potential candidate as a biomarker for diagnosis, prognosis, or disease monitoring. This in no way undermines the potential of tissue culture-based model in potential biomarker discovery as the validation can be achieved by alternative means, but the identification is less cumbersome using this system. The initial studies which include the study of differential expression of a candidate in normal versus malignant cells, their mode of action, or whether they can be used as a potential drug target, have to be done using tissue culture-based model system. It creates the foundation based upon which we can carry forward the hunt for novel biomarkers not only in the field of oncology but also for other prevalent diseases.

14.12 FUTURE PERSPECTIVE

The inherent capability of mass spectrometry along with its sensitivity, speed, and specificity when combined with tissue culture-based model provides a promising tool for the discovery of candidate potential biomarkers (Table 2). In this paper, we have tried to emphasize the use of tissue culture as model for biomarker discovery along with brief outline of different mass spectrometry-based quantitative proteomic techniques that are routinely used in such studies. With the advancement of mass spectrometry-based proteomic techniques and bioinformatics tools, tissue culture-based model system becomes the most beneficial choice for the identification of potential biomarkers. The CM of these cell lines also serves as a potent source of biomarkers. The contemporary biomarkers generally used in clinics such as carbohydrate antigen CA 125, CA 19.9, and PSA were discovered using cancerous cell lines or tumor extracts [112]. It is likely that the tumor microenvironment or the tumor itself can be a source of biomarkers allowing for better sensitivity and specificity as well as proper diagnosis of the disease. However, in tissue culture-based system, the role of tumor microenvironment in biomarker discovery is yet to reach its peak. The 3D culture methods are currently being used that can be considered as an alternative to 2D culture system which receives criticism for its inability to mimic tumor microenvironment. The CM enriched with secretory proteins is largely used for the identification of potential cancer biomarkers. It acts as a perfect source for the potential biomarkers, and to date the majority of the biomarkers being used clinically are secretory proteins. Proteome profiles of many cancers are influenced by hormones, and tissue culture-based model system serves as a promising approach to study this process. Hormonal stimulation of the cells followed by different gel-based or gel-free proteomic approaches to identify differentially expressed proteins serves as an approach to search for the "cause-effect" candidates. Tissue culture-based model system can also be used in the field of pharmacokinetics and drug discovery. The potential effect of a drug can be assessed by using tissue culture-based system. The differential expression of proteins upon drug treatment also provides the insight into the mechanism of action as well as potential drug targets. Moreover, these differentially expressed proteins can serve as potential biomarkers for drug response in clinics.

There has been a rapid fruitful development of MS-based proteomic techniques in the last decade that has immensely helped researchers in candidate biomarker discovery. First, there was 2-DE and then its 2D-DIGE that enhances the accuracy of quantitation utilizing very littel amount of sample. Now, there are techniques like SILAC and iTRAQ which are more advanced versions of labelling techniques in combina-

TABLE 2: Different mass spectrometry-based proteomic approaches with its merits, demerits, and compatibility towards tissue culture.

Proteomic approach	Merits	Demerits	Compatibility with tissue culturea	References
2-DE	(i) Robust	(i) Involves large amount of sample	***	[15, 33]
	(ii) Simplistic	(ii) Low throughput		
	(iii) Highly suitable for MS analysis	(iii) Poor recovery of hydrophobic proteins		
2D-DIGE	(i) Multiplexed	(i) Not suitable for MS analysis	****	[16, 34]
	(ii) Better quantitation	(ii) Expensive Cy dyes		
	(iii) Minimized gel to gel variation	(iii) Poor recovery of hydrophobic proteins		
SILAC	(i) High-throughput	(i) Only suitable for tissue culture model	*****	[25, 48]
	(ii) Robust and accurate	(ii) Costly reagents		
	(iii) Sensitivity and simplicity	(iii) Not applicable to tissue samples		
Super-SILAC	(i) Better representation of tumor heterogeneity	(i) Only suitable to tissue culture model	*****	[50]
	(ii) Accurate quantitation	(ii) Costly reagents		
	(iii) Less error rate	(iii) Internal standard library required		
iTRAQ	(i) Multiplexed	(i) Incomplete labelling	****	[18, 56]
	(ii) Applicable to versatile samples	(ii) Involves high amount of sample		
	(iii) Better quantitation	(iii) Expensive reagents		
Label free	(i) Involves less amount of sample	(i) High-throughput instrumentation	****	[61, 64]
	(ii) Broader applicability	(ii) Redundancy in detection		
	(iii) Avoid labelling	(iii) Not suitable for low abundant proteins		

Proteomic approach	Merits	Demerits	Compatibility with tissue culture[a]	References
SID-MS	(i) Absolute quantitation	(i) Applicable to limited number of proteins	***	[65, 68]
	(ii) Targeted approach	(ii) Internal standards are required		
	(iii) Applicable to versatile samples	(iii) Generally used for validation		

[a]Number of "" indicates extent of compatibility.*

tion with improved chromatographic, and mass spectrometric techniques provide better resolution. Recently, people have started moving towards label-free quantitation, which is the most advanced form of relative quantitation-based proteomic technique. With this advancement, the number of potential biomarkers will certainly increase, but we have to be very careful and critical in choosing the biomarkers that can be used clinically. It is not tough to anticipate more development in the near future that will make tissue culture-based systems for potential biomarker discovery more robust, sensitive, and reliable. This will lead to the discovery of useful biomarkers for patient diagnosis, prognosis, treatment, and monitoring not only for cancer but also for other diseases.

REFERENCES

1. R. Etzioni, N. Urban, S. Ramsey et al., "The case for early detection," Nature Reviews Cancer, vol. 3, no. 4, pp. 243–252, 2003.
2. V. Kulasingam and E. P. Diamandis, "Strategies for discovering novel cancer biomarkers through utilization of emerging technologies," Nature Clinical Practice Oncology, vol. 5, no. 10, pp. 588–599, 2008.
3. D. F. Hayes, R. C. Bast, C. E. Desch et al., "Tumor marker utility grading system: a framework to evaluate clinical utility of tumor markers," Journal of the National Cancer Institute, vol. 88, no. 20, pp. 1456–1466, 1996.
4. S. Minamida, M. Iwamura, Y. Kodera et al., "Profilin 1 overexpression in renal cell carcinoma," International Journal of Urology, vol. 18, no. 1, pp. 63–71, 2011.

5. X. Xu, M. Qiao, Y. Zhang et al., "Quantitative proteomics study of breast cancer cell lines isolated from a single patient: discovery of TIMM17A as a marker for breast cancer," Proteomics, vol. 10, no. 7, pp. 1374–1390, 2010.

6. V. Kulasingam and E. P. Diamandis, "Tissue culture-based breast cancer biomarker discovery platform," International Journal of Cancer, vol. 123, no. 9, pp. 2007–2012, 2008.

7. E. Charafe-Jauffret, C. Ginestier, F. Monville et al., "Gene expression profiling of breast cell lines identifies potential new basal markers," Oncogene, vol. 25, no. 15, pp. 2273–2284, 2006.

8. R. M. Neve, K. Chin, J. Fridlyand et al., "A collection of breast cancer cell lines for the study of functionally distinct cancer subtypes," Cancer Cell, vol. 10, no. 6, pp. 515–527, 2006.

9. E. S. Boja and H. Rodriguez, "Mass spectrometry-based targeted quantitative pro-teomics: achieving sensitive and reproducible detection of proteins," Proteomics, vol. 12, no. 8, pp. 1093–1110, 2012.

10. G. L. Glish and R. W. Vachet, "The basics of mass spectrometry in the twenty-first century," Nature Reviews Drug Discovery, vol. 2, no. 2, pp. 140–150, 2003.

11. B. F. Cravatt, G. M. Simon, and J. R. Yates III, "The biological impact of mass spectrometry-based proteomics," Nature, vol. 450, no. 7172, pp. 991–1000, 2007.

12. C. Y. Chen, L. M. Chi, H. C. Chi et al., "Stable isotope labeling with amino acids in cell culture (SILAC)-based quantitative proteomics study of a thyroid hormone-regulated secretome in human hepatoma cells," Molecular and Cellular Proteomics, vol. 11, no. 4, Article ID M111.011270, 2012.

13. S. Leong, M. J. McKay, R. I. Christopherson, and R. C. Baxter, "Biomarkers of breast cancer apoptosis induced by chemotherapy and TRAIL," Journal of Proteome Research, vol. 11, no. 2, pp. 1240–1250, 2012.

14. G. Q. Zeng, P. F. Zhang, C. Li et al., "Comparative proteome analysis of human lung squamous carcinoma using two different methods: two-dimensional gel electropho-resis and iTRAQ analysis," Technology in Cancer Research and Treatment, vol. 11, no. 4, pp. 395–408, 2012.

15. P. Meleady, "2D gel electrophoresis and mass spectrometry identification and analy-sis of proteins," Methods in Molecular Biology, vol. 784, pp. 123–137, 2011.

16. J. F. Timms and R. Cramer, "Difference gel electrophoresis," Proteomics, vol. 8, no. 23-24, pp. 4886–4897, 2008.

17. M. Mann, "Functional and quantitative proteomics using SILAC," Nature Reviews Molecular Cell Biology, vol. 7, no. 12, pp. 952–958, 2006.

18. P. L. Ross, Y. N. Huang, J. N. Marchese et al., "Multiplexed protein quantitation in Saccharomyces cerevisiae using amine-reactive isobaric tagging reagents," Molecu-lar and Cellular Proteomics, vol. 3, no. 12, pp. 1154–1169, 2004.

19. J. L. P. Benesch, B. T. Ruotolo, D. A. Simmons, and C. V. Robinsons, "Protein complexes in the gas phase: technology for structural genomics and proteomics," Chemical Reviews, vol. 107, no. 8, pp. 3544–3567, 2007.

20. A. F. M. Altelaar, S. L. Luxembourg, L. A. McDonnell, S. R. Piersma, and R. M. A. Heeren, "Imaging mass spectrometry at cellular length scales," Nature Protocols, vol. 2, no. 5, pp. 1185–1196, 2007.

21. M. Schirle, M. Bantscheff, and B. Kuster, "Mass spectrometry-based proteomics in preclinical drug discovery," Chemistry and Biology, vol. 19, no. 1, pp. 72–84, 2012.

22. M. Zhu, H. Zhang, and W. G. Humphreys, "Drug metabolite profiling and identification by high-resolution mass spectrometry," The Journal of Biological Chemistry, vol. 286, no. 29, pp. 25419–25425, 2011.

23. H. Zhang, S. Chen, and L. Huang, "Proteomics-based identification of proapoptotic caspase adapter protein as a novel serum marker of non-small cell lung cancer," Chinese Journal of Lung Cancer, vol. 15, no. 5, pp. 287–293, 2012.

24. X. Lou, T. Xiao, K. Zhao et al., "Cathepsin D is secreted from M-BE cells: Its potential role as a biomarker of lung cancer," Journal of Proteome Research, vol. 6, no. 3, pp. 1083–1092, 2007.

25. M. H. Elliott, D. S. Smith, C. E. Parker, and C. Borchers, "Current trends in quantitative proteomics," Journal of Mass Spectrometry, vol. 44, no. 12, pp. 1637–1660, 2009.

26. F. Di Girolamo, F. Del Chierico, G. Caenaro, I. Lante, M. Muraca, and L. Putignani, "Human serum proteome analysis: new source of markers in metabolic disorders," Biomarkers in Medicine, vol. 6, no. 6, pp. 759–773, 2012.

27. E. P. Rhee and R. E. Gerszten, "Metabolomics and cardiovascular biomarker discovery," Clinical Chemistry, vol. 58, no. 1, pp. 139–147, 2012.

28. F. Bertucci, D. Birnbaum, and A. Goncalves, "Proteomics of breast cancer: principles and potential clinical applications," Molecular and Cellular Proteomics, vol. 5, no. 10, pp. 1772–1786, 2006.

29. M. Braun, M. Fountoulakis, A. Papadopoulou et al., "Down-regulation of microfilamental network-associated proteins in leukocytes of breast cancer patients: potential application to predictive diagnosis," Cancer Genomics and Proteomics, vol. 6, no. 1, pp. 31–40, 2009.

30. P. Cancemi, G. Di Cara, N. N. Albanese et al., "Large-scale proteomic identification of S100 proteins in breast cancer tissues," BMC Cancer, vol. 10, article 476, 2010.

31. R. Strong, T. Nakanishi, D. Ross, and C. Fenselau, "Alterations in the mitochondrial proteome of adriamycin resistant MCF-7 breast cancer cells," Journal of Proteome Research, vol. 5, no. 9, pp. 2389–2395, 2006.

32. D. H. Lee, K. Chung, J. A. Song et al., "Proteomic identification of paclitaxel-resistance associated hnRNP A2 and GDI 2 proteins in human ovarian cancer cells," Journal of Proteome Research, vol. 9, no. 11, pp. 5668–5676, 2010.

33. T. Rabilloud and C. Lelong, "Two-dimensional gel electrophoresis in proteomics: a tutorial," Journal of Proteomics, vol. 74, no. 10, pp. 1829–1841, 2011.

34. J. X. Yan, A. T. Devenish, R. Wait, T. Stone, S. Lewis, and S. Fowler, "Fluorescence two-dimensional difference gel electrophoresis and mass spectrometry based proteomic analysis of Escherichia coli," Proteomics, vol. 2, no. 12, pp. 1682–1698, 2002.

35. S. C. Wong, C. M. L. Chan, B. B. Y. Ma et al., "Advanced proteomic technologies for cancer biomarker discovery," Expert Review of Proteomics, vol. 6, no. 2, pp. 123–134, 2009.

36. J. Blonder, H. J. Issaq, and T. D. Veenstra, "Proteomic biomarker discovery: it's more than just mass spectrometry," Electrophoresis, vol. 32, no. 13, pp. 1541–1548, 2011.

37. J. Zhang, M. Q. Song, J. S. Zhu et al., "Identification of differentially-expressed proteins between early submucosal non-invasive and invasive colorectal cancer using 2D-dige and mass spectrometry," International Journal of Immunopathology and Pharmacology, vol. 24, no. 4, pp. 849–859, 2011.

38. F. Tan, Y. Jiang, N. Sun et al., "Identification of isocitrate dehydrogenase 1 as a potential diagnostic and prognostic biomarker for non-small cell lung cancer by proteomic analysis.," Molecular & cellular proteomics : MCP, vol. 11, no. 2, Article ID M111.008821, 2012.

39. H. N. Banerjee, K. Mahaffey, E. Riddick, A. Banerjee, N. Bhowmik, and M. Patra, "Search for a diagnostic/prognostic biomarker for the brain cancer glioblastoma multiforme by 2D-DIGE-MS technique," Molecular and Cellular Biochemistry, vol. 367, no. 1-2, pp. 59–63, 2012.

40. J. Sinclair, G. Metodieva, D. Dafou, S. A. Gayther, and J. F. Timms, "Profiling signatures of ovarian cancer tumour suppression using 2D-DIGE and 2D-LC-MS/MS with tandem mass tagging," Journal of Proteomics, vol. 74, no. 4, pp. 451–465, 2011.

41. A. Wilmes, A. Chan, P. Rawson, T. W. Jordan, and J. H. Miller, "Paclitaxel effects on the proteome of HL-60 promyelocytic leukemic cells: comparison to peloruside A," Investigational New Drugs, vol. 30, no. 1, pp. 121–129, 2012.

42. S. E. Ong, B. Blagoev, I. Kratchmarova et al., "Stable isotope labeling by amino acids in cell culture, SILAC, as a simple and accurate approach to expression proteomics," Molecular and Cellular Proteomics, vol. 1, no. 5, pp. 376–386, 2002.

43. T. Geiger, S. F. Madden, W. M. Gallagher, J. Cox, and M. Mann, "Proteomic portrait of human breast cancer progression identifies novel prognostic markers," Cancer Research, vol. 72, no. 9, pp. 2428–2439, 2012.

44. M. K. Kashyap, H. C. Harsha, S. Renuse et al., "SILAC-based quantitative proteomic approach to identify potential biomarkers from the esophageal squamous cell carcinoma secretome," Cancer Biology and Therapy, vol. 10, no. 8, pp. 796–810, 2010.

45. C. I. Wang, K. Y. Chien, C. L. Wang et al., "Quantitative proteomics reveals regulation of KPNA2 and its potential novel cargo proteins in non-small cell lung cancer," Molecular and Cellular Proteomics, vol. 11, no. 11, pp. 1105–1122, 2012.

46. A. Cuomo, S. Moretti, S. Minucci, and T. Bonaldi, "SILAC-based proteomic analysis to dissect the "histone modification signature" of human breast cancer cells," Amino Acids, vol. 41, no. 2, pp. 387–399, 2011.

47. M. Krüger, M. Moser, S. Ussar et al., "SILAC mouse for quantitative proteomics uncovers kindlin-3 as an essential factor for red blood cell function," Cell, vol. 134, no. 2, pp. 353–364, 2008.

48. S. E. Ong and M. Mann, "Mass spectrometry-based proteomics turns quantitative," Nature chemical biology, vol. 1, no. 5, pp. 252–262, 2005.

49. Y. Ishihama, T. Sato, T. Tabata et al., "Quantitative mouse brain proteomics using culture-derived isotope tags as internal standards," Nature Biotechnology, vol. 23, no. 5, pp. 617–621, 2005.

50. T. Geiger, J. Cox, P. Ostasiewicz, J. R. Wisniewski, and M. Mann, "Super-SILAC mix for quantitative proteomics of human tumor tissue," Nature Methods, vol. 7, no. 5, pp. 383–385, 2010.

51. P. J. Boersema, T. Geiger, J. R. Wiśniewski, and M. Mann, "Quantification of the N-glycosylated secretome by super-SILAC during breast cancer progression and in human blood samples," Molecular and Cellular Proteomics, vol. 12, no. 1, pp. 158–171, 2013.

52. R. R. Lund, M. G. Terp, A.-V. Lænkholm, O. N. Jensen, R. Leth-Larsen, and H. J. Ditzel, "Quantitative proteomics of primary tumors with varying metastatic capabilities using stable isotope-labeled proteins of multiple histogenic origins," Proteomics, vol. 12, no. 13, pp. 2139–2148, 2012.

53. Y. T. Chen, C. L. Chen, H. W. Chen et al., "Discovery of novel bladder cancer biomarkers by comparative urine proteomics using iTRAQ technology," Journal of Proteome Research, vol. 9, no. 11, pp. 5803–5815, 2010.

54. P. Hao, T. Guo, X. Li et al., "Novel application of electrostatic repulsion-hydrophilic interaction chromatography (ERLIC) in shotgun proteomics: comprehensive profiling of rat kidney proteome," Journal of Proteome Research, vol. 9, no. 7, pp. 3520–3526, 2010.

55. S. Elschenbroich, V. Ignatchenko, P. Sharma, G. Schmitt-Ulms, A. O. Gramolini, and T. Kislinger, "Peptide separations by on-line MudPIT compared to isoelectric focusing in an off-gel format: application to a membrane-enriched fraction from C2C12 mouse skeletal muscle cells," Journal of Proteome Research, vol. 8, no. 10, pp. 4860–4869, 2009.

56. L. V. DeSouza, A. D. Romaschin, T. J. Colgan, and K. W. M. Siu, "Absolute quantification of potential cancer markers in clinical tissue homogenates using multiple reaction monitoring on a hybrid triple quadrupole/linear ion trap tandem mass spectrometer," Analytical Chemistry, vol. 81, no. 9, pp. 3462–3470, 2009.

57. I. Rehman, C. A. Evans, A. Glen et al., "iTRAQ identification of candidate serum biomarkers associated with metastatic progression of human prostate cancer," PLoS ONE, vol. 7, no. 2, Article ID e30885, 2012.

58. D. Ghosh, H. Yu, X. F. Tan et al., "Identification of key players for colorectal cancer metastasis by iTRAQ quantitative proteomics profiling of isogenic SW480 and SW620 cell lines," Journal of Proteome Research, vol. 10, no. 10, pp. 4373–4387, 2011.

59. Y. Chen, L. Y. Choong, Q. Lin et al., "Differential expression of novel tyrosine kinase substrates during breast cancer development," Molecular and Cellular Proteomics, vol. 6, no. 12, pp. 2072–2087, 2007.

60. D. Chelius and P. V. Bondarenko, "Quantitative profiling of proteins in complex mixtures using liquid chromatography and mass spectrometry," Journal of Proteome Research, vol. 1, no. 4, pp. 317–323, 2002.

61. W. Zhu, J. W. Smith, and C. M. Huang, "Mass spectrometry-based label-free quantitative proteomics," Journal of Biomedicine and Biotechnology, vol. 2010, Article ID 840518, 6 pages, 2010.

62. T. Ishihara, I. Fukuda, A. Morita et al., "Development of quantitative plasma N-glycoproteomics using label-free 2-D LC-MALDI MS and its applicability for biomarker discovery in hepatocellular carcinoma," Journal of Proteomics, vol. 74, no. 10, pp. 2159–2168, 2011.

63. W. M. Old, K. Meyer-Arendt, L. Aveline-Wolf et al., "Comparison of label-free methods for quantifying human proteins by shotgun proteomics," Molecular and Cellular Proteomics, vol. 4, no. 10, pp. 1487–1502, 2005.

64. S. P. Mirza and M. Olivier, "Methods and approaches for the comprehensive charac-
    terization and quantification of cellular proteomes using mass spectrometry," Physi-
    ological Genomics, vol. 33, no. 1, pp. 3–11, 2008.

65. V. Brun, C. Masselon, J. Garin, and A. Dupuis, "Isotope dilution strategies for ab-
    solute quantitative proteomics," Journal of Proteomics, vol. 72, no. 5, pp. 740–749,
    2009.

66. A. D. Kippen, F. Cerini, L. Vadas et al., "Development of an isotope dilution assay
    for precise determination of insulin, C-peptide, and proinsulin levels in non-diabetic
    and type II diabetic individuals with comparison to immunoassay," The Journal of
    Biological Chemistry, vol. 272, no. 19, pp. 12513–12522, 1997.

67. S. A. Gerber, J. Rush, O. Stemman, M. W. Kirschner, and S. P. Gygi, "Absolute
    quantification of proteins and phosphoproteins from cell lysates by tandem MS,"
    Proceedings of the National Academy of Sciences of the United States of America,
    vol. 100, no. 12, pp. 6940–6945, 2003.

68. M. A. Kuzyk, D. Smith, J. Yang et al., "Multiple reaction monitoring-based, multi-
    plexed, absolute quantitation of 45 proteins in human plasma," Molecular and Cel-
    lular Proteomics, vol. 8, no. 8, pp. 1860–1877, 2009.

69. H. Jiang, A. A. Ramos, and X. Yao, "Targeted quantitation of overexpressed and
    endogenous cystic fibrosis transmembrane conductance regulator using multiple re-
    action monitoring tandem mass spectrometry and oxygen stable isotope dilution,"
    Analytical Chemistry, vol. 82, no. 1, pp. 336–342, 2010.

70. K. H. Yu, C. G. Barry, D. Austin et al., "Stable isotope dilution multidimensional
    liquid chromatography-tandem mass spectrometry for pancreatic cancer serum bio-
    marker discovery," Journal of Proteome Research, vol. 8, no. 3, pp. 1565–1576,
    2009.

71. D. R. Barnidge, M. K. Goodmanson, G. G. Klee, and D. C. Muddiman, "Absolute
    quantification of the model biomarker prostate-specific antigen in serum by LC-MS/
    MS using protein cleavage and isotope dilution mass spectrometry," Journal of Pro-
    teome Research, vol. 3, no. 3, pp. 644–652, 2004.

72. H. Keshishian, T. Addona, M. Burgess et al., "Quantification of cardiovascular bio-
    markers in patient plasma by targeted mass spectrometry and stable isotope dilu-
    tion," Molecular and Cellular Proteomics, vol. 8, no. 10, pp. 2339–2349, 2009.

73. N. L. Anderson and N. G. Anderson, "The human plasma proteome: history, char-
    acter, and diagnostic prospects," Molecular and Cellular Proteomics, vol. 1, no. 11,
    pp. 845–867, 2002.

74. Y. Hathout, "Approaches to the study of the cell secretome," Expert Review of Pro-
    teomics, vol. 4, no. 2, pp. 239–248, 2007.

75. P. Dowling and M. Clynes, "Conditioned media from cell lines: a complementary
    model to clinical specimens for the discovery of disease-specific biomarkers," Pro-
    teomics, vol. 11, no. 4, pp. 794–804, 2011.

76. G. Sardana, K. Jung, C. Stephan, and E. P. Diamandis, "Proteomic analysis of condi-
    tioned media from the PC3, LNCaP, and 22Rv1 prostate cancer cell lines: discovery
    and validation of candidate prostate cancer biomarkers," Journal of Proteome Re-
    search, vol. 7, no. 8, pp. 3329–3338, 2008.

77. R. L. Shreeve, R. E. Banks, P. J. Selby, and N. S. Vasudev, "Proteomic study of conditioned media: cancer biomarker discovery," International Journal of Genomics and Proteomics, vol. 3, no. 1, pp. 50–56, 2012.

78. Y. Wang, S. J. Li, X. Wu, Y. Che, and Q. Li, "Clinicopathological and biological significance of human voltage-gated proton channel Hv1 protein overexpression in breast cancer," The Journal of Biological Chemistry, vol. 287, no. 17, pp. 13877–13888, 2012.

79. Y. Wang, S. J. Li, J. Pan, Y. Che, J. Yin, and Q. Zhao, "Specific expression of the human voltage-gated proton channel Hv1 in highly metastatic breast cancer cells, promotes tumor progression and metastasis," Biochemical and Biophysical Research Communications, vol. 412, no. 2, pp. 353–359, 2011.

80. M. C. Hinestrosa, K. Dickersin, P. Klein et al., "Shaping the future of biomarker research in breast cancer to ensure clinical relevance," Nature Reviews Cancer, vol. 7, no. 4, pp. 309–315, 2007.

81. E. K. Jung, H. K. Kyung, H. K. Yeul, J. Sohn, and G. P. Yun, "Identification of potential lung cancer biomarkers using an in vitro carcinogenesis model," Experimental and Molecular Medicine, vol. 40, no. 6, pp. 709–720, 2008.

82. K. M. Yamada and E. Cukierman, "Modeling tissue morphogenesis and cancer in 3D," Cell, vol. 130, no. 4, pp. 601–610, 2007.

83. U. Kruse, M. Bantscheff, G. Drewes, and C. Hopf, "Chemical and pathway proteomics: powerful tools for oncology drug discovery and personalized health care," Molecular and Cellular Proteomics, vol. 7, no. 10, pp. 1887–1901, 2008.

84. J. M. Lee, J. J. Han, G. Altwerger, and E. C. Kohn, "Proteomics and biomarkers in clinical trials for drug development," Journal of Proteomics, vol. 74, no. 12, pp. 2632–2641, 2011.

85. E. E. Balashova, M. I. Dashtiev, and P. G. Lokhov, "Proteomic footprinting of drug-treated cancer cells as a measure of cellular vaccine efficacy for the prevention of cancer recurrence," Molecular and Cellular Proteomics, vol. 11, no. 2, Article ID M111.014480, 2012.

86. E. K. Yim, J. S. Bae, S. B. Lee et al., "Proteome analysis of differential protein expression in cervical cancer cells after paclitaxel treatment," Cancer Research and Treatment, vol. 36, no. 6, pp. 395–399, 2004.

87. K. O'Connell, M. Prencipe, A. O'Neill et al., "The use of LC-MS to identify differentially expressed proteins in docetaxel-resistant prostate cancer cell lines," Proteomics, vol. 12, no. 13, pp. 2115–2126, 2012.

88. Y. Liu, H. Liu, B. Han, and J. T. Zhang, "Identification of 14-3-3σ as a contributor to drug resistance in human breast cancer cells using functional proteomic analysis," Cancer Research, vol. 66, no. 6, pp. 3248–3255, 2006.

89. L. Murphy, M. Henry, P. Meleady, M. Clynes, and J. Keenan, "Proteomic investigation of taxol and taxotere resistance and invasiveness in a squamous lung carcinoma cell line," Biochimica et Biophysica Acta, vol. 1784, no. 9, pp. 1184–1191, 2008.

90. S. L. Li, F. Ye, W. J. Cai et al., "Quantitative proteome analysis of multidrug resistance in human ovarian cancer cell line," Journal of Cellular Biochemistry, vol. 109, no. 4, pp. 625–633, 2010.

91. N. P. Chappell, P. N. Teng, B. L. Hood et al., "Mitochondrial proteomic analysis of cisplatin resistance in ovarian cancer," Journal of Proteome Research, vol. 11, no. 9, pp. 4605–4614, 2012.

92. X. Qian, C. Li, B. Pang, M. Xue, J. Wang, and J. Zhou, "Spondin-2 (SPON2), a more prostate-cancer-specific diagnostic biomarker," PLoS ONE, vol. 7, no. 5, Article ID e37225, 2012.

93. H. Loei, H. T. Tan, T. K. Lim et al., "Mining the gastric cancer secretome: identification of GRN as a potential diagnostic marker for early gastric cancer," Journal of Proteome Research, vol. 11, no. 3, pp. 1759–1772, 2012.

94. S. Makawita, C. Smith, I. Batruch et al., "Integrated proteomic profiling of cell line conditioned media and pancreatic juice for the identification of pancreatic cancer biomarkers," Molecular and Cellular Proteomics, vol. 10, no. 10, Article ID M111.008599, 2011.

95. L. Yao, W. Lao, Y. Zhang et al., "Identification of EFEMP2 as a serum biomarker for the early detection of colorectal cancer with lectin affinity capture assisted secretome analysis of cultured fresh tissues," Journal of Proteomic Research, vol. 11, no. 6, pp. 3281–3294, 2012.

96. R. Ralhan, O. Masui, L. V. Desouza, A. Matta, M. Macha, and K. W. M. Siu, "Identification of proteins secreted by head and neck cancer cell lines using LC-MS/MS: strategy for discovery of candidate serological biomarkers," Proteomics, vol. 11, no. 12, pp. 2363–2376, 2011.

97. L. P. Weng, C. C. Wu, B. L. Hsu et al., "Secretome-based identification of Mac-2 binding protein as a potential oral cancer marker involved in cell growth and motility," Journal of Proteome Research, vol. 7, no. 9, pp. 3765–3775, 2008.

98. C. J. Yu, K. P. Chang, Y. J. Chang et al., "Identification of guanylate-binding protein 1 as a potential oral cancer marker involved in cell invasion using omics-based analysis," Journal of Proteome Research, vol. 10, no. 8, pp. 3778–3788, 2011.

99. Y. J. Xue, R. H. Xiao, D. Z. Long et al., "Overexpression of FoxM1 is associated with tumor progression in patients with clear cell renal cell carcinoma," Journal of Translational Medicine, vol. 10, article 200, 2012.

100. Y. X. Zheng, M. Yang, T. T. Rong et al., "CD74 and macrophage migration inhibitory factor as therapeutic targets in gastric cancer," World Journal of Gastroenterology, vol. 18, no. 18, pp. 2253–2261, 2012.

101. S. Chakraborty, S. Kaur, S. Guha, and S. K. Batra, "The multifaceted roles of neutrophil gelatinase associated lipocalin (NGAL) in inflammation and cancer," Biochimical and Biophysical Acta, no. 1, pp. 129–169, 1826.

102. H. Lee, M. Song, N. Shin et al., "Diagnostic significance of serum HMGB1 in colorectal carcinomas," PLoS ONE, vol. 7, no. 4, Article ID e34318, 2012.

103. M. Akdogan, N. Sasmaz, B. Kayhan, I. Biyikoglu, S. Disibeyaz, and B. Sahin, "Extraordinarily elevated CA19-9 in benign conditions: a case report and review of the literature," Tumori, vol. 87, no. 5, pp. 337–339, 2001.

104. P. G. Righetti, A. Castagna, F. Antonucci et al., "The proteome: anno domini 2002," Clinical Chemistry and Laboratory Medicine, vol. 41, no. 4, pp. 425–438, 2003.

105. C. I. Wang, C. L. Wang, C. W. Wang et al., "Importin subunit alpha-2 is identified as a potential biomarker for non-small cell lung cancer by integration of the cancer

cell secretome and tissue transcriptome," International Journal of Cancer, vol. 128, no. 10, pp. 2364–2372, 2011.

106. V. Kulasingam and E. P. Diamandis, "Proteomics analysis of conditioned media from three breast cancer cell lines: a mine for biomarkers and therapeutic targets," Molecular and Cellular Proteomics, vol. 6, no. 11, pp. 1997–2011, 2007.

107. N. Ahmed, K. Oliva, G. E. Rice, and M. A. Quinn, "Cell-free 59 kDa immunoreactive integrin-linked kinase: a novel marker for ovarian carcinoma," Clinical Cancer Research, vol. 10, no. 7, pp. 2415–2420, 2004.

108. J. T. Hsieh, H. C. Wu, M. E. Gleave, A. C. von Eschenbach, and L. W. K. Chung, "Autocrine regulation of prostate-specific antigen gene expression in a human prostatic cancer (LNCaP) subline," Cancer Research, vol. 53, no. 12, pp. 2852–2857, 1993.

109. L. M. Mikesh, M. Kumar, G. Erdag et al., "Evaluation of molecular markers of mesenchymal phenotype in melanoma," Melanoma Research, vol. 20, no. 6, pp. 485–495, 2010.

110. R. Sadej, H. Romanska, G. Baldwin et al., "CD151 regulates tumorigenesis by modulating the communication between tumor cells and endothelium," Molecular Cancer Research, vol. 7, no. 6, pp. 787–798, 2009.

111. D. G. Lee, J. H. Lee, B. K. Choi et al., "H+-myo-inositol transporter SLC2A13 as a potential marker for cancer stem cells in an oral squamous cell carcinoma," Current Cancer Drug Targets, vol. 11, no. 8, pp. 966–975, 2011.

112. R. C. Bast Jr., T. L. Klug, E. St John et al., "A radioimmunoassay using a monoclonal antibody to monitor the course of epithelial ovarian cancer," The New England Journal of Medicine, vol. 309, no. 15, pp. 883–887, 1983.

AUTHOR NOTES

CHAPTER 1

Competing Interests
The authors declare that they have no competing interests.

Author Contributions
Both authors participated in drafting and editing the manuscript. Both authors read and apporoved the final manuscript.

Acknowledgments
This study was supported by: Natural Science Youth Foundation of Shandong Province, China (No. ZR2011HQ009); Natural Science Foundation of Shandong Province, China (No. 2007 C053); Project of Scientific and Technological Development of Shandong Province, China (N2007GG10).

CHAPTER 2

Acknowledgments
This work was supported by Grant MIUR-COFIN 20089SRS2X and by Grant Regione Campania 2008.

CHAPTER 3

Competing Interests
The authors declare that they have no competing interests.

Author Contributions
QL led the project. DS drafted the manuscript and QL revised the manuscript. All authors read and approved the final manuscript.

Acknowledgments

This work was supported by National Cancer Institute grants U01 CA163056, P30 CA068485, P50 CA098131, and P50 CA090949 and QL's work was partially supported by the State Key Program of National Natural Science of China (no. 31230058) and the National Natural Science Foundation of China (no. 31070746).

CHAPTER 4

Conflict of Interests

The author does not have a direct financial conflict of interests with any of the commercial identities mentioned in this paper.

CHAPTER 5

Competing Interests

The authors declare that they have no competing interests.

Author Contributions

Authors EL, LM, JLP and ML carried out Clinical Proteomics and Translational Medicine studies for this short-review, in order to develop future Oncohematology Proteomic-OMICS research studies and publish this article. All authors read and approved the final manuscript.

Acknowledgments

EL is Scientist (Dr.) and a recipient of a Post-doctoral fellowship of Ministerio de Ciencia e Innovación de España (MICINN). LM is Prof. at Universidad Autónoma de Medicina UAM (Madrid) and PhD MD, Department of Oncohematology-Pediatric, Hospital Universitario Niño Jesús (Madrid). JLP is PhD MD, Department of Hematology, Hospital Universitario 12 de Octubre (Madrid). ML is Professor at the Proteogenomics Research Institute for Systems Medicine, San Diego, USA.

Special thanks to Prof. Xiangdong Wang (Honorary Prof. of Medicine, Fudan University, Prof. of Mol. Bioscience, NCSU, USA and Director of Biomedical Research Center).

CHAPTER 6

Funding

This work was funded by the Royal College of Surgeons in Ireland SYNERGY 2008 award and the Irish Government under its Programme for Research in Third Level Institutions Centre for Synthesis and Chemical Biology. The funders had no role in study design, data collection and analysis, decision to publish, or preparation of the manuscript.

Competing Interests

The authors have declared that no competing interests exist.

Author Contributions

Conceived and designed the experiments: DT AJC. Performed the experiments: DT. Analyzed the data: HH MPB. Contributed reagents/materials/analysis tools: KBN. Wrote the paper: AJC.

CHAPTER 7

Competing Interests

The authors declared that they have no competing interests.

Author Contributions

ED helped carry out the economic studies and drafted the manuscript. KK initiated the studies. Both authors read and approved the final manuscript.

Author Information

ED is a student at the University of Applied Science in Vienna and will graduate with a master's degree in Molecular Biotechnology soon. She specialises in biopharmaceutical technologies for biopharmaceutical production, drug delivery and bioanalytical chemistry. She is a freelance writer with experience in writing for the pharmaceutical industry. KK has been involved in pharmaceutical research for over 12 years as a scientist and project leader within different leading pharmaceutical companies.

CHAPTER 8

Competing Interests
The authors declare no competing interests.

Author Contributions
The concept of this article was conceived by BE and FE. Whereas all authors contributed to some extent to all aspects of this work, some authors provided the bulk contribution to specific sections. Section When will genome sequences, expression profiles and computer vision for bioimage interpretation be routinely used in clinical medicine? was written by BE and FE with contributions by MJM and SP. FE, VK and SP provided the main contributions to section Bioinformatics moving towards clinical oncology: biomarkers for cancer classification, early diagnostics, prognosis and personalized therapy. Section Sequence-structure-function relationships for pathogenic viruses and bacteria and their role in combating infections is mainly the result of SMS's and FE's work with some contribution from SP. Section Impact of Bioimage Informatics on Healthcare was written mainly by HKL and edited by BE and FE. All authors read and approved the final manuscript.

Author Information
VK, HKL, SMS and BE are principal investigators at the Bioinformatics Institute (BII) of the Agency for Science, Technology and Research in Singapore; FE is the director of this institute. MJM is the director of the Institute of Genomic Medicine and Rare Disorders at Semmelweis University in Budapest. SP is a senior scientist and group leader of the Protein Structure and Bioinformatics Laboratory at the ICGEB in Trieste and professor at the Pázmány Péter Catholic University in Budapest.

Acknowledgments
This work was partially supported by grants A*STAR-NKTH 10/1/06/24635, IAF311010, A*STAR IMAGIN and IAF311011 at the Singaporean side and by the grants TÉT 10-1-2011-0058 (bilateral cooperation between Singapore and Hungary) as well as TÁMOP-4.2.1/B-11/2/KMR-2011-0002 and TÁMOP-4.2.2/B-10/1-2010-0014 at the Hungarian side.

CHAPTER 10

Funding
This work was supported by the Modernising Medical Microbiology Consortium funded under the UK Clinical Research Collaboration Translational Infection Research Initiative supported by the Medical Research Council; Biotechnology and Biological Sciences Research Council; National Institute for Health Research on behalf of the UK Department of Health Grant G0800778; and Wellcome Trust Grant 087646/Z/08/Z. The funders had no role in study design, data collection and analysis, decision to publish, or preparation of the manuscript.

Competing Interests
The author has declared that no competing interests exist.

Acknowledgments
I would like to thank Adam Auton, Bethany Dearlove, Sarah Walker, Bernadette Young, and two anonymous reviewers for helpful comments on the manuscript, and Roy Kishony, Johannes Krause, and Nature Publishing Group for permission to reproduce Figures 2 and 3.

CHAPTER 11

Funding
This project was mainly funded by a grant from the Swiss National Science Foundation no. 3200BO-116445. This project was also partially supported by a grant from the Infectigen Foundation (In010) and by the CNRS (UPRESA 6020). Gilbert Greub is supported by the Leenards Foundation through a career award entitled "Bourse Leenards pour la relève académique en médecine clinique à Lausanne". The funders had no role in study design, data collection and analysis, decision to publish, or preparation of the manuscript.

Competing Interests
The antigenic proteins and corresponding polypeptides identified as immunogenic thanks to this work have been patented for the detection of

antibodies directed against Parachlamydia acanthamoebae as well as for their use in related diagnostic tests (i.e. immunohistochemistry) and vaccines (European patent n° 08172133.4-1223, 12th December 2008). This patent does not alter the authors' adherence to all the PLoS ONE policies on sharing data and materials, and the authors of the present work will make freely available any materials and information associated with their publication that are reasonably requested by others for the purpose of academic, non-commercial research.

Acknowledgments

We thank C. Robert (Unité des Rickettsies, Marseille) for the technical realization of the GS20 sequences as well as M. Quadroni and the proteomic facility of the University of Lausanne for assisting with mass spectrometry analyses. We also thank I. Riederer (Proteomics Unit, Prilly-Lausanne) and F. Auderset (Institute of Microbiology, Lausanne) for technical help, as well as L. Farinelli (Fasteris, Plan-les-Ouates), J. Schrenzel and D. Hernandez (University of Geneva) for helpful discussions. Gilbert Greub is supported by the Leenards Foundation through a career award entitled "Bourse Leenards pour la relève académique en médecine clinique à Lausanne".

Author Contributions

Conceived and designed the experiments: GG BMR DR. Performed the experiments: CKB CB FC CY AC. Analyzed the data: GG CKB CB FC BMR CY AC DR. Contributed reagents/materials/analysis tools: GG BMR DR. Wrote the paper: GG CKB CB. Improved the manuscript and approved the final version: FC BMR CY AC DR.

CHAPTER 12

Acknowledgments

This work was partly supported by a Research Grants Council Grant, The University of Hong Kong and Consultancy Service for Enhancing Laboratory Surveillance of Emerging Infectious Disease for the HKSAR Department of Health.

CHAPTER 13

Acknowledgments

This study was supported by the European Commission under FP7, Project 261426 (WINGS West Nile Integrated Shield Project).

CHAPTER 14

Acknowledgments

The authors acknowledge Ms. Parul Dutta for her valuable suggestions. This work is supported by research grants from NCCS and Department of Biotechnology, India. M. K. Santra is a Ramalingaswamy Fellow. D. Paul and A. Gajbhiye are UGC Junior Research Fellows. A. Kumar is DBT Postdoctoral Fellow.

INDEX

α –enolase